Process Development Division 2021

Held at the 2021 AIChE Annual Meeting

Boston, Massachusetts, USA and Online
7 - 11 November and 15 - 19 November 2021

ISBN: 978-1-7138-5707-5

Printed from e-media with permission by:

Curran Associates, Inc.
57 Morehouse Lane
Red Hook, NY 12571

Some format issues inherent in the e-media version may also appear in this print version.

Printed with permission by Curran Associates, Inc. (2022)

For permission requests, please contact AIChE
at the address below.

AIChE
120 Wall Street, FL 23
New York, NY 10005-4020

Phone: (800) 242-4363
Fax: (203) 775-5177

www.aiche.org

Additional copies of this publication are available from:

Curran Associates, Inc.
57 Morehouse Lane
Red Hook, NY 12571 USA
Phone: 845-758-0400
Fax: 845-758-2633
Email: curran@proceedings.com
Web: www.proceedings.com

TABLE OF CONTENTS

624940 Effect of Pyrolysis Conditions on Producing Mesophase Pitch from Varying Ranks of Coal and Correlating Their Product Properties with Machine Learning Models

Abstract:

Coal tar pitch (CTP) is the heavy fraction of the coal tar by-products of coal coking processes and is often used as a feedstock for high-value products, such as high-modulus carbon fiber or needle coke. During the production of these high-value products from CTP, the CTP must undergo conversion to an intermediate product called mesophase. Mesophase formation in carbon materials has been widely studied over the years, and it is well known that mesophase conversion and quality is highly dependent on the chemical properties of the feedstock. Some of the important chemical properties that impact mesophase conversion are aromaticity, heteroatom functional groups, viscosity, and molecular weight. Because these chemical properties critically affect mesophase conversion, some feedstocks are intrinsically more suitable for mesophase conversion than others.

Previous work in our group has shown that feedstocks with chemistry originally not suitable for mesophase conversion (e.g. coal tar from non-coking coals) could be modified to be more suitable for mesophase conversion through the use of controlled secondary gas-phase reactions during pyrolysis. In this work, this approach was investigated more broadly on multiple coals of varying ranks (Wyoming PRB Wyodak, Utah Sufco, Illinois #6, and West Virginia Flying Eagle). The intermediate coal tar products were analyzed for their general chemical functionality through Fourier-transform infrared spectroscopy, and their respective molecular weight distributions were measured through laser desorption ionization mass spectrometry. After converting the coal tar samples into mesophase CTP, the samples were then viewed under a polarized microscope to analyze the mesophase optical textures to determine their mesophase qualities. In addition to investigating this pyrolysis approach on multiple coals, the data collected was used in a machine learning approach to predict mesophase qualities based on the precursor coal tar chemical properties.

Our results show that for all coals investigated, improvements were seen in the mesophase CTP qualities with an increase in the secondary gas-phase reaction temperature and residence time. The most marked mesophase improvements were observed for Utah Sufco, PRB Wyodak and Flying Eagle CTPs, but the Illinois #6 CTP had little improvement. A preliminary model based upon machine learning was constructed using the well-known Elastic-Net regression algorithm from the Scikit-learn Python module. Model training and testing was performed on randomized training/testing data splits at least 50 times and the average training and testing accuracies were 0.819 +/- 0.048 and 0.556 +/- 0.216, respectively.

The results of this work show that the use of controlled secondary gas-phase reactions in pyrolysis may be more broadly applied to various carbon feedstocks to improve their ability to produce desirable mesophase carbon intermediates. In addition, the Elastic-Net model supports that there is a measurable correlation between coal tar feedstock properties and their resulting mesophase CTP qualities. However, the current model displays instability and high variance on the test score, due to the limited dataset collected thus far. Model accuracy and robustness will be improved on the Elastic-Net model once the targeted dataset is complete and smoothing techniques are applied. Other non-linear models, such as random forests and neural networks, will also be explored for comparison on the completed dataset.

622293 Mild Solvolysis Liquefaction of Low-Rank Coal into a Feedstock of Value-Added Carbon Materials

Abstract:

Coal tar is a liquid product derived from the thermochemical conversion (metallurgical coking, pyrolysis, and gasification) of coal. Relative to direct use as fuel, conversion of coal tar into fine chemicals and high-performance carbon materials is considered a value-adding utilization pathway for coal. However, producing coal tar via coking requires a high-rank coal as the feedstock, while pyrolysis and gasification methods need high temperatures (over 500 °C). In addition, these methods lead to a relatively low tar yield (around 10 wt%). In contrast, the direct coal liquefaction (DCL) process has no strict requirement of coal rank and provides a very high liquid yield (over 50 wt%). The main costs of DCL come from higher liquefaction temperature, extra hydrogen consumption, and the wastage of catalysts. These conditions are primarily used to crack aromatic rings in coal and to reform liquefaction fragments into gasoline-like liquid fuel. Notably, the abundant aromatic structures in coal make coal tar an excellent source to prepare value-added carbon materials. Therefore, liquefying coal at mild conditions can be a promising method to achieve high coal tar yield and improve the performance of preparing coal tar-derived carbon materials.

In the research presented here, low-rank coals from various American coal-producing areas were liquefied by solvolysis at mild temperature conditions. Macromolecules in coal were dissolved, fragmented, rearranged, partially hydrogenated, and converted into DCL tar with high yields (over 50%), while most of the aromatic or naphthenic structures in raw coals were preserved. The DCL tar contained more heavy fractions and hydrogen transferability attributes than pyrolysis tar and coking tar. Results of subsequent thermal treatment indicated that this kind of DCL tar can be effectively converted into mesophase pitch with excellent anisotropic characteristics. The obtained mesophase pitch is of promising potential to produce high-performance carbon materials such as carbon fiber, needle coke, carbon foam, and supercapacitors. Products and byproducts were characterized analytically (TGA, FTIR, NMR, BET, SEM, etc.), and a mass balance was evaluated for the process. The influence of operating parameters (liquefaction temperature, solvent types, initial gas atmosphere, and solvent/coal mass ratio) on the yield, composition, and carbonization performance of DCL tar was studied and will be discussed. The possibility of replacing hydrogen donor solvents and H_2 gas sources with cheap and renewable feedstocks was also investigated. Our research indicates that mild solvolysis liquefaction of low-rank coals can be used to generate value-added product precursors in coal-producing areas of the United States.

628900 Modular Processing of Flare Gas for Carbon Nanoproducts

Abstract:

In the United States, the "shale boom" and wide implementation of hydraulic fracturing has dramatically reduced the price of natural gas. As a result, oil producers increasingly choose to flare or vent associated natural gas rather than sell it, as investing in expensive transportation infrastructure, such as pipelines, is less economically prudent. Moreover, tight oil wells exhaust quickly as production from a single well depletes rapidly, thus imposing difficulties in predicting the optimal pipeline location[1]. A profitable alternative to harnessing this waste of natural resources is needed.

A one-step Catalytic-Chemical Vapor Deposition (CCVD) process to produce carbon nanofibers (CNFs), carbon nanoparticles (CNPs), and hydrogen is a financially promising repurposing of the flared natural gas. An economically viable CCVD process largely depends on the production of a cheap catalyst. Particle Atomic Layer Deposition (ALD) enables the formation of highly dispersed metal nanocatalysts on a given support. The highly dispersed low-cost catalyst is essential as this process contains a *sacrificial catalyst*; the catalyst, support, CNFs and CNPs combine as one product to improve the strength and durability of concrete. The sacrificial catalyst circumvents the well-known disadvantages of rapid catalyst deactivation in methane thermal cracking[2] and avoids the costly step of catalyst/product separation. To combat the challenges of rapid well depletion, this process will be scaled up to a modular unit that can be transferred between hydraulic fracturing sites. If each flare site in the United States transitioned to this modular CCVD process, over 30 million metric tons of CO_2 emissions per annum could be eliminated[3].

In this work, transition metals are deposited onto a silica fume support via ALD. CNFs and CNPs are then produced via CCVD of methane with the ALD nanocatalyst. The carbon product is evaluated for its improvements to concrete durability via various compression and deformation tests. The carbon product evaluation and empirical reaction data inform both the modular unit design and technoeconomic analysis.

[1] Energy, B. *The Williston Basin: Greasing the Gears for Growth in North Dakota*; BENTEK Energy: 2012, 2012

[2] Abanades, A.; Ruiz, E.; Ferruelo, E. M.; Hernandez, F.; Cabanillas, A.; Martinez-Val, J. M.; Rubio, J. A.; Lopez, C.; Gavela, R.; Barrera, G.; Rubbia, C.; Salmieri, D.; Rodilla, E.; Gutierrez, D., Experimental analysis of direct thermal methane cracking. *International Journal of Hydrogen Energy* 2011, *36* (20), 12877-12886.

[3] U.S. Energy Information Administration. Natural Gas Gross Withdrawals and Production.

http://www.eia.gov/dnav/ng/ng_prod_sum_a_epg0_vgv_mmcf_a.htm (accessed April, 2021)

629147 Innovative Conceptual Design for an Industrial Complex Coupling Allam Power Cycle, Air Separation Unit, Ammonia Production, and Fertilizer Manufacturing Processes

Abstract:

The human population is expected to reach about 9.2 billion by 2050, when the land availability for farming could only keep the current level or even decrease. Thus, the demand and deployment of fertilizers, especially nitrogen fertilizers, must increase steadily in the future to enable the increase of the average land productivity. In addition to this challenge, the sustainable development of global economy will also need a fast-paced energy demand with less carbon footprint and cost-effective manufacturing methods. Natural gas is an abundant, cheap, and clean energy/material source in this planet that could be immediately available to help human to meet these two major challenges, which should be extensively studied for its comprehensive utilization.

In this paper, a super industrial complex (A3U) based on based on comprehensive utilization of natural gas has been developed, modeled, and analyzed, which include four subsystems: (i) an Allam-cycle power plant (ACPP); (ii) an air separation unit (ASU); (iii) an ammonia plant (AP), and (iv) a urea (UFP) fertilizer plant. The A3U complex employs natural gas and air as the major feedstocks. By optimally sharing energy/material streams among four subsystems, the A3U complex can simultaneously produce multiple products such as net power, ammonia, fertilizer, pipeline-ready CO2, pure nitrogen, and steam efficiently. Equally important, the A3U complex itself has very low carbon emissions during its manufacturing process. The developed A3U complex has been virtually demonstrated by rigorous modeling and simulation with ASPEN PLUS. Its corresponding economic and environmental benefits have also been provided.

632373 Parametric Study on Physicochemical Properties of Ozone Microbubble - Applications to Fresh Produce Washing

Abstract:

To overcome the adverse effects of unhealthy and environmentally unfriendly by-products from conventional washing agents for fresh produce, ozone has been used as a sustainable technology. Although ozone is a powerful oxidant that has a strong biocidal effect and decomposes to harmless oxygen, its application is limited. This is because ozone is highly unstable, reactive with low solubility in water, which makes it difficult to control its concentration during the processing. Microbubble technology has the potential to enhance the washing efficacy because of its unique physicochemical properties such as extended stability, enhanced mass transfer rate of gas dissolution, and promotion of hydroxyl radical generation. Therefore, ozone microbubble technology has been suggested to be an effective method for fresh produce that not only enhances the safety and quality of the fresh produce but also provides a sustainable alternative for humans and environment.

The biocidal effect by combining both microbubble and ozone largely depends on its unique physicochemical properties. However, very little is known about the influence of washing processing conditions on physicochemical properties. Here, we provide a systematic study on the influence of various washing processing conditions including bubble generation methods, fluid to gas ratio, pH level, temperature, ozone concentration, and organic compound load. We focus on those physicochemical properties that are relevant to the washing efficacy including bubble size distribution, number density, mass transfer rate, stability, zeta-potential, oxidation reduction potential. The large sets of experiments allow us to develop a data-driven model that will provide a better prediction and optimization of the system. Furthermore, this model can be used as a critical tool for a broad spectrum of applications using ozone microbubbles.

624181 Predicting the Diameters of Droplets Produced in Turbulent Liquid-Liquid Dispersion

Abstract:

Liquid-liquid dispersion is a common chemical manufacturing batch process typically performed in agitated tanks. The physics informing the droplet size distribution is a complex convolution of impeller speed, impeller type, fluid properties, and flow conditions. In this work, we present three *a priori* mechanistic modeling approaches for predicting the droplet diameter distributions as a function of system operating conditions. In the first approach, called the two-fluid approach, we use high-resolution solutions to the Navier-Stokes equations to directly model the flow of each phase and the corresponding droplet breakup/coalescence events. In the second approach, based on an Eulerian-Lagrangian model, we describe the dispersed fluid as individual spheres undergoing ongoing breakup and coalescence events per user-defined interaction kernels. In the third approach, called the Eulerian-Parcel model, we model a sub-set of the droplets in the Eulerian-Lagrangian model to estimate the overall behavior of the entire droplet population. We discuss output from each model within the context of predictions from first principles turbulent theory and measured data. We then discuss the scalability of each method to various operating volumes/conditions.

624622 Computational Fluid Dynamics Study to Develop a Scale-up Strategy for Non-Newtonian Fluids in Pharmaceutical Industry

Abstract:

The generic pharmaceutical industry relies on the speed and rigor in the drug development process. Getting it right the first time requires detailed product and process understanding. Various unit operations like sterilization, milling, mixing, etc. govern the rapidity of the process. Mixing is one such operation, developing a proper technical understanding can yield highly productive results and can reduce experimental work thereby proving to be cost and time effective. This work provides a comprehensive approach for understanding mixing in different equipment for several parameters and scale-up strategy using computational fluid dynamics.

The first question to ask is what comprises good mixing. The judging parameters include the uniform distribution of energy with less dissipation and getting the right CQAs in the samples. The major challenges in mixing can be attributed to non-uniformity and unpredictability in scale-up operations. Computational fluid dynamics (CFD) is an approach that is coming into action in the pharmaceutical industry. CFD Software (ANSYS in this example) can be leveraged to understand mixing profiles and help in analyzing the technical aspects and provide a generalized methodology for the problem.

As the complexity of molecules has increased and with process intensification, the need to develop technical solutions faster and having a standardized approach will support the entire industry to reach its goal of serving people better. This study gives a comparison of equipment based on histograms, contour plots, and vector plots for shear stress, velocity, and molecular viscosity, along with the discrete phase model to draw inferences and conclusions for non-Newtonian fluid systems. Transient simulations are used to determine the particle distribution in the mixing vessel volume being dispersed from a specified location. This helps in finding time to achieve uniformity throughout the volume. Using CFD the problem at lab scale is resolved by determining the CPPs which in turn gives us the ability to scale-up the product in a plant to achieve desired targets.

The impact of this study can be huge, as it can contribute not only to reduce experimental work at lab scale but can also improve visibility in the scale-up plant operations. It further adds as a standardized method for such issues.

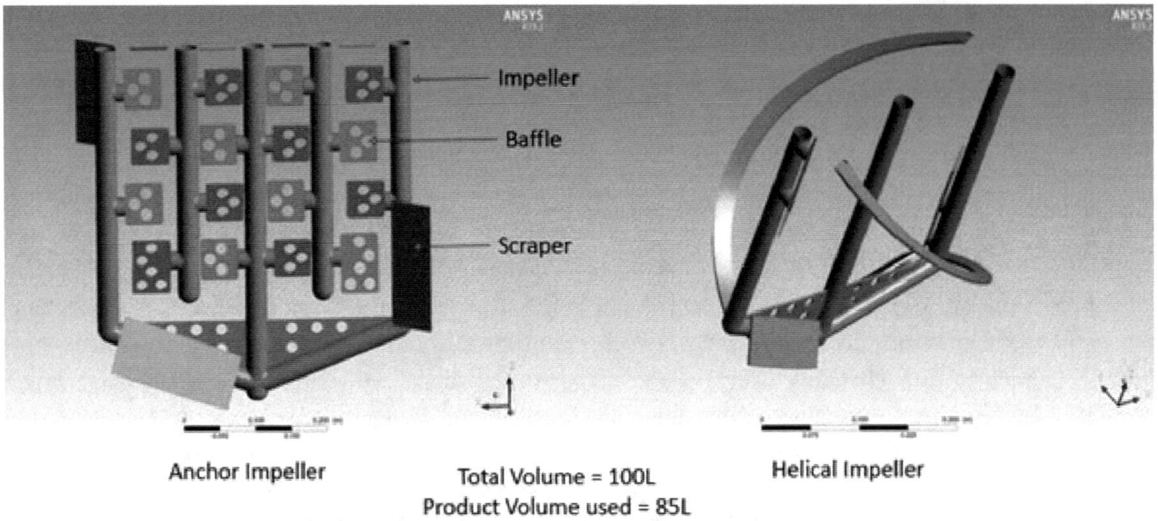

Figure 1: CAD models of Anchor and Helical impellers available at plant scale

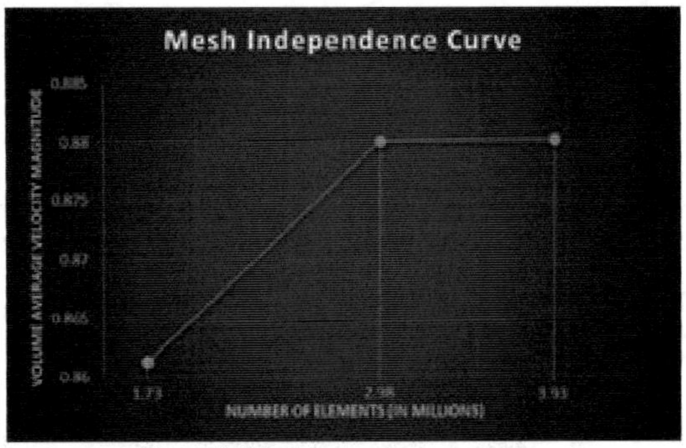

Figure 2: Mesh independence study

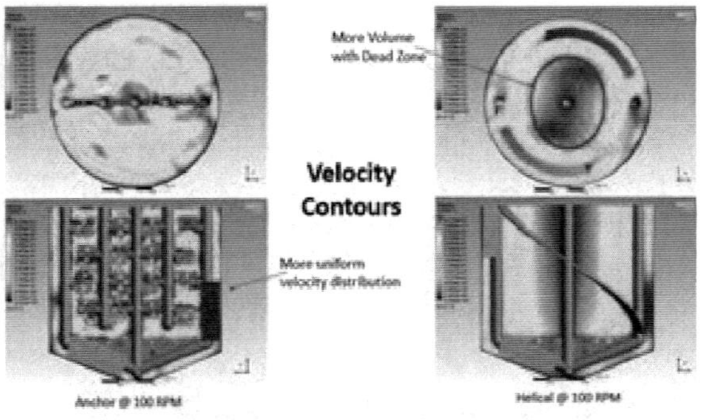

Figure 3: Velocity Contours of plant scale study

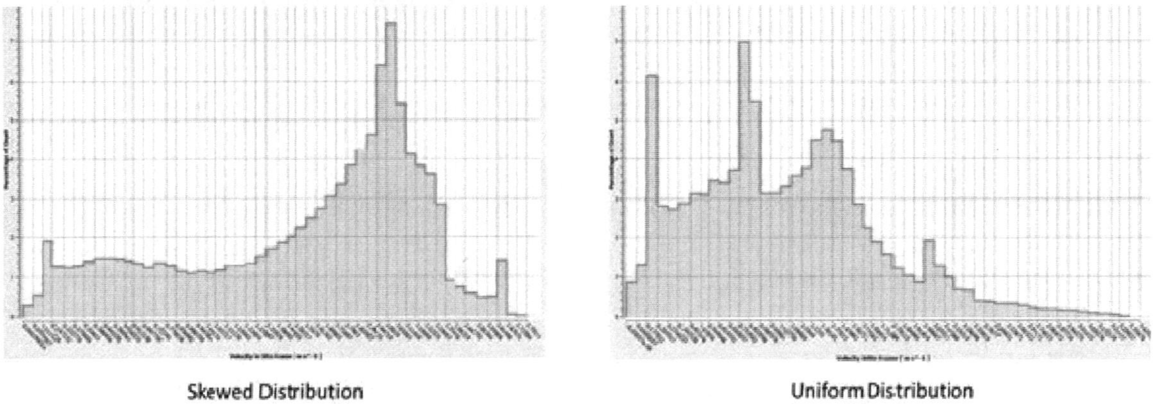

Figure 4: Velocity Histograms for Helical and Anchor impellers for plant scale study

627762 Modelling the Effect of Ice Nucleation Stochasticity on Batch Heterogeneity of Freezing Processes of Vials on a Shelf

Abstract:

Introduction

Freezing and freeze-drying are widely used processes to improve the stability and thus shelf life of pharmaceutical formulations. Ice nucleation is the first step of the freezing process; it is an activated process that requires cooling of the product beyond its equilibrium freezing temperature. It occurs at different times and temperatures for individual vials, even in case they experience identical cooling profiles. This variance was identified as a main source of batch heterogeneity across frozen vials on a shelf [1]. Such heterogeneity is a major risk for industrial processes involving freezing steps since several product attributes depend strongly on the freezing history of each vial: The ice crystal morphology in a vial depends both on the temperature at which ice nucleation occurs and on the time required for complete solidification after nucleation, i.e. the solidification time [2]. Finally, stochasticity of ice nucleation depends, to some extent, on the process conditions: Generally, one observes lower average nucleation temperatures and a broader distribution in dust-free GxP settings compared to the laboratory, thus rendering the transfer of a process design from laboratory to production scale challenging. To mitigate the effects of this stochasticity, process technologies have been developed that artificially induce ice nucleation at the same time in all vials, so called controlled nucleation techniques [1].

While there is experimental evidence that these techniques indeed reduce heterogeneity across the shelf, experimental studies to quantify heterogeneity are challenging due to the requirement of analyzing several product attributes in a statistically significant number of vials. Therefore, a shelf-scale model of the freezing process is required to deepen the understanding of batch heterogeneity and to optimize the process design. To the best of our knowledge, such model is not available in literature and the theoretical understanding of the relevant mechanisms leading to heterogeneity is limited. Besides the stochasticity of ice nucleation, variability in heat transfer was identified as major source of heterogeneity, in the form of spatial differences across the shelf and due to geometric imperfections of vials [4].

We thus propose a modelling framework for the freezing process of an arbitrary number of vials on a shelf to understand and identify all potential mechanisms of batch heterogeneity with a focus on the stochasticity of ice nucleation. With the help of this model, we investigate the effect of various process parameters on heterogeneity to aid process design and upscaling.

Methods

The model describes the freezing process of vials on a shelf in a mechanistic way taking into account the geometry of the system, heat transfer, and ice nucleation. Each vial is considered as thermally homogeneous while it evolves from liquid to a partially frozen, and to a completely frozen state. The governing equations account for the vial's physical state and ice nucleation is described as an inhomogeneous Poisson process [5].

While experimental studies on batch heterogeneity mostly focus on measuring product attributes, we characterize heterogeneity based on three characteristic quantities of the freezing process. These are the nucleation time, nucleation temperature and solidification time for each vial of the simulated system.

We have simulated sets of freezing experiments and generated the corresponding histograms of these quantities. To identify potential mechanisms for heterogeneity, we investigated the effect of ice nucleation parameters and their variability, heat transfer parameters and their variability and process control strategies on these distributions.

Results

We have performed simulation studies of a system comprising 7x7 vials on a shelf, a typical batch size for experimental studies in laboratory scale freeze-dryers. In addition to the direct effect of the stochasticity of ice nucleation on heterogeneity, we account also for an indirect role played by ice nucleation, namely heat transfer between vials triggered by the heat released upon nucleation. Since nucleation is stochastic, the occurrence of such thermal effects adds to the heterogeneity across the shelf. Figure 1 illustrates both these effects on the quantities characterizing the freezing process, expressed in terms of the intensity of the interaction. We use the heat transfer coefficient between vials, as parameter to quantify this interaction.

In case of thermally independent vials, i.e. $k_{int} = 0$, the distributions of the characteristic quantities correspond to the freezing process of single vials: The only source of variability between vials is the stochasticity of ice nucleation. For sufficiently strong interaction, i.e. $k_{int} > 10\ Wm^{-2}K^{-1}$, we observed a bimodal nucleation time distribution as shown in Fig. 1a. In this case, a first group of vials nucleated randomly across the shelf and delayed the nucleation events of the remaining vials. These early nucleating vials then act as "local hotspots" on the shelf that temporarily slow down the cooling process of neighboring vials. The formation of distinct populations corresponding to early and late nucleating vials can be seen in the bivariate distribution of nucleation times and solidification times in Fig. 1d. These populations differ both in nucleation and solidification times, thus creating a significant heterogeneity across the batch.

The effect of thermal interaction on the nucleation temperature distribution was smaller than for the other two quantities; we only identified a weak trend towards higher temperatures for increasing interaction as shown in Fig 1a-c. This observation, namely that in some situations the distributions of nucleation time and of temperature behave differently, is of importance for process monitoring. Commonly, the two distributions are used interchangeably; nucleation times in literature are detected with a camera and converted into temperatures based on reference temperature measurements [3]. Based on our model results, we hypothesize that such approach may be misleading and suggest the direct measurement of both quantities, e.g. via infrared thermography.

We can use these outcomes to improve the freezing process design. In case of freeze-drying processes, a homogeneous crystal morphology across the shelf is required. The crystal morphology correlates directly with the solidification time, so that a narrow distribution of the solidification times across the shelf is desired. Consequently, one should aim to operate the process in a regime with low thermal interaction. In practice, this may be achieved by arranging vials in a geometry with less direct contact and by arranging them directly on the shelf without tray.

In general, we can apply this modelling framework to study not only the impact of the stochasticity of ice nucleation, but all potential sources of heterogeneity. Our model can be extended with ease to simulate the additional heat transfer to edge vials, as well as a random variability of the vial-to-shelf heat transfer coefficient.

Conclusion

We present for the first time a mechanistic model for the freezing of an arbitrary number of vials on a shelf that is capable of capturing the stochasticity of ice nucleation and its effects on the freezing process. It proves to be a useful tool in improving the understanding of batch heterogeneity in freezing processes. The model enables us to assess the impact of process parameters, such as varying cooling profiles or techniques like controlled nucleation, on the freezing outcome. It also allows the simulation of the freezing process for both GxP and non-GxP settings based on the specific ice nucleation kinetics of each setting; thus aiding upscaling and technology transfer between freezing devices. The developed model thus serves as a powerful simulation tool in the design and optimization of freezing and freeze-drying.

Acknowledgement

The authors thank The Janssen Pharmaceutical Companies of Johnson & Johnson for the support in the course of this project.

References

[1] A. Arsiccio et al.: Vacuum Induced Surface Freezing as an effective method for improved inter- and intra-vial product homogeneity, *European Journal of Pharmaceutics and Biopharmaceutics* 128 (2018) 210–219.

[2] A. Arsiccio, A. A. Barresi, and R. Pisano: Prediction of Ice Crystal Size Distribution after Freezing of Pharmaceutical Solutions, *Crystal Growth & Design* (2017), 17, 4573−4581.

[3] L. C. Capozzi, R. Pisano: Looking inside the 'black box': Freezing engineering to ensure the quality of freeze-dried biopharmaceuticals, *European Journal of Pharmaceutics and Biopharmaceutics* 129 (2018) 58–65.

[4] B. Scutella et al.: How Vial Geometry Variability Influences Heat Transfer and Product Temperature During Freeze-Drying, *Journal of Pharmaceutical Sciences* 106 (2017) 770-778.

[5] G. M. Maggioni and M. Mazzotti: Modelling the stochastic behaviour of primary nucleation, *Faraday Discussions* (2015), 179, 359.

629683 API Crystallization Process Development for Improved Consistency, Bulk Properties, and to Enable Downstream Processing

Abstract:

A revised crystallization process for an Active Pharmaceutical Ingredient (API) was developed to improve the physiochemical and bulk powder properties of the isolated product. The first generation process consisted of an unseeded cooling crystallization that provided material with low bulk density, a propensity to agglomerate on handling and storage, and was not amenable to formulation. Additionally, a clarifying filtration was not possible due to solubility limitations of the API at the target operating volume. Development of the crystallization process explored the use of seed crystals generated by wet-milling and jet-milling, the use of in-situ wet milling during the crystallization, and thermalcycling to improve the bulk properties of the resultant powders. Optimization of these parameters led to a second generation process that is run at a higher operating volume to accommodate a clarifying filtration, distilled to a target concentration, seeded with jet-milled seed material, and subjected to a series of thermalcycles to ripen the product crystals. The resulting API is isolated with an increased bulk density, higher aspect ratio, and improved handling during formulation. The second generation process was studied as part of a fractional factorial Design of Experiments to assess and rank the impact of all of the processing parameters on the isolated product's quality attributes. It was determined that the physical properties of the seed crystals, the seeding point, and the temperature targets of the thermal cycling had the greatest impact on the product particle size distribution and bulk powder properties.

622378 Loss-in-Weight Control of a Low Dose Powder Feeding System

Abstract:

Accurate and robust powder feeding is the key for successful continuous pharmaceutical manufacturing. Feeding systems available on the market are typically suitable for feed rates above 200 g/h. We show the development of loss-in-weight (LIW) control for a novel feeding system capable of continuously feeding powder at feed rates in the range of few g/h. Our in-house developed prototype is equipped with a LIW scale, a powder pump and a scraping device. Additionally, a high precision gain-in-weight (GIW) scale is used as reference. First, a feed-forward (FF) control strategy was established to compensate for the material dependent displacement density along the powder container. This approach is similar to the feed factor correction based on the hopper fill level of commercially available large scale feeders. In a second step, an iterative learning (IL) control strategy was developed, which updates the model for FF control after regularly spaced time intervals based on the measured feed rate deviation from the set-point.

In the presentation, we compare operation with constant powder pump speed to feed-forward controlled experiments. It can be shown that the feed rate deviations, caused by the reproducible displacement density, can be mitigated. Furthermore, we show the principle of IL control in order to compensate model inaccuracies as a result of changing material behavior due to batch-to-batch variability or inconsistent powder conditioning by the operator.

A LIW control strategy for a novel low dose feeding system was developed, successfully transferred to the prototype and tested. Accurate and robust feeding for feed rates as low as 1.5 g/h could be achieved and by means of IL control, the effect of changing material behavior could be significantly reduced.

624791 A Generalized Framework for Reactor Network Synthesis: A Graph Theoretic Approach

Abstract:

In general, the design of the reactor network has arguably the greatest influence on the overall process flowsheet, determining the necessary features of subsequent separation units and utilities required to carry out the chemical process. However, most studies that focused on the synthesis of reactor networks either considered the reactor system in isolation from the synthesis of other systems, or addressed a narrow problem statement. A narrow problem statement can obstruct the full potential of simultaneous optimization strategies for process synthesis, such as superstructure-based approaches.

To address this limitation, we propose a generalized superstructure-based framework for reactor network synthesis that allows seamless coupling with approaches for separation network synthesis. We develop a method based on graph theory tools to systematically build the reactor network superstructure. The proposed framework addresses a generalized problem statement characterized by variable input data of the reactor system, both in terms of inlet streams (e.g., feed and recycle streams) and candidate reactions modeled through *tasks* assigned to reactors. A *task* is defined in terms of a subset of reactions, range of operating conditions, and catalyst required. A single task is assigned to each selected (active) reactor.

Our method consists of three steps. First, we use graph theoretic ideas to represent the reaction network and task information through multiple graphs. We define the *component–reaction* and *component–task* graphs as bipartite directed graphs. In the component–reaction graph, components and reactions constitute the set of nodes, and arcs identify the reactants and products of each reaction. Similarly, in the component–task graph, components and tasks constitute the set of nodes, and arcs indicate the components that are reactants or products of at least one reaction associated to a task. Furthermore, we identify competition among tasks. A group of competing tasks includes tasks that can play similar roles in the reactor network, and the simultaneous selection and assignment (to reactors) of these tasks is not beneficial. We represent the competition information through an undirected *task competition graph*, where tasks are represented as nodes, and the groups of competing tasks are represented as cliques of this graph.

Second, we identify and solve a graph cover problem based on the task competition graph to reduce the number of candidate reactors required in the superstructure. Each candidate reactor has a subset of tasks available to be assigned to that reactor. We solve the cover problem to select the minimum number of cliques, thus minimizing the number of candidate reactors. Additional constraints can be incorporated into the cover problem to simultaneously minimize the number of reactors as well as the reactor–reactor streams required to have an efficient and yet rich superstructure.

Finally, we formulate the optimization model for the integrated synthesis problem. The proposed framework is flexible in terms of task modeling. We can use multiple task types associated to different reactor models (e.g., detailed kinetic models, or simpler extent-based linear models). Additional constraints are proposed to eliminate symmetric solutions with respect to reactor activation and task-to-reactor assignments. The applicability and flexibility of the framework is illustrated through several examples.

623272 Modeling and Performance Assessment of Rotating Packed Beds for the Production of Precipitated Carbonate Nanoparticles

Abstract:

The global warming due to anthropogenic CO_2 emissions is a pressing issue that attracts significant research efforts for the development of efficient CO_2 capture and utilization (CCU) technologies. The exploitation of CO_2 as a raw material of added value will be key to the wide industrial adoption of CCU solutions. Among the numerous different utilization options that are being investigated, the production of carbonate nanoparticles such as precipitated calcium carbonate (PCC) is a very appealing option for carbon leakage industries such as cement, lime, pulp and paper and steel. PCC nanoparticles have a broad range of applications in building materials, cement, plastics, paper, pharmaceuticals, dyes etc.[1]. Their commercial value ranges on average within $1150 - 2350 / ton, depending on their size. The nanoparticle market is very dynamic and is expected to need about 40 million tons/yr by 2024, with a total value of $10.5 Billion[2]. The vast majority of PCC production worldwide takes place in conventional CSTR or bubble reactors[3]. In such systems the production requires chemical additives for size control, whereas the reactors are not appropriate for large capacities as they require large residence times in larger than 50 m^3 volumes[3]. Rotating Packed Beds (RPBs)[4] have emerged as a promising alternative for PCC production[5]. Centrifugal forces are exploited in a spinning packing material to intensify mixing and mass transfer, and to significantly reduce the residence time and reactor volume, among other advantages. Despite the high industrial potential, RPB-based PCC production is currently very limited[5]. Few details are available in published literature regarding operating parameters and conditions, while limited experimental information includes lab-scale results on precipitation and particle size distribution (PSD)[6].

The use of model-based approaches is very important for the scaling-up of RPBs, but there are currently no models for RPB-based PCC production, that can combine the mixing and precipitation phenomena with macroscopic, RPB operating parameters for performance assessment. In the production of nanoparticles, the prediction of the particle size distribution of the product is equally important to the assessment of the operating efficiency of the process. The majority of published works focus on barium sulfate production with computational fluid dynamics (CFD) simulations on various types of mixing equipment. For example, Gradl et al.[7] investigated the precipitation of barium sulfate in a T-mixer by coupling the particle population dynamics with CFD, using direct numerical simulations. Gavi et al.[8] performed CFD simulations of barium sulfate precipitation in confined impinging jet reactors, using the Reynolds averaged Navier–Stokes approach. Steyer et al.[9] studied the applicability of different activity coefficient models to barium sulfate precipitation. The work of Pan et al.[10] investigated the carbonation of calcium ions in an RPB, but focused on gas-liquid mass transfer without considering precipitation. The work of Xiang et al.[11] is the only one that investigated precipitation in combination with a mixing model in an RPB, but it refers to barium sulfate.

In the current work, a robust model that combines micro- and macro-mixing with population balances in an RPB used for PCC production is developed. The model captures nucleation, molecular growth and aggregation in the representation of the precipitation process, whereas the engulfment model for micro-mixing is coupled with the population balance equation to investigate mixing effects during scale-up. The proposed model essentially associates the PSD that is attained under different operating conditions in the RPB, with key process performance indicators such as the rotor speed, the energy consumption and so forth. The model predictions are validated against experimental data from published literature, considering various packing types and similar rotor speeds, and flowrates. The simulated results presented in Table 1 indicate a very good match with the experimental results. Using the proposed model, a parametric analysis of the process is performed. Among other parameters, the sensitivity of the nucleation rate to variations of the interfacial energy constant and the supersaturation are investigated. The results indicate that the variation of the interfacial energy constant causes a change of the nucleation rate by more than 3 orders of magnitude. Furthermore, a wide range of operating conditions are explored in order to determine the particle sizes and distribution that can be attained within a given range of energy dissipation. According to the present analysis, the mean particle size decreases with increase of the energy dissipation rate and rotor speed respectively, and is found to vary between 5.5 – 7.5 nm for cases of fast mixing. Overall, the proposed model is capable of fast and efficient prediction of the particle size and distribution.

Acknowledgements

This research has been co-financed by the European Union and Greek national funds through the Operational Program Competitiveness, Entrepreneurship and Innovation, under the call Research–Create–Innovate, Project T1EDK-02472.

References

1. Sargheini, J.; Ataie, A.; Salili, S. M.; Hoseinion, A. A. One-Step Facile Synthesis of $CaCO_3$ Nanoparticles via Mechano-Chemical Route. *Powder Technol.* **2012**, *219*, 72–77. https://doi.org/10.1016/j.powtec.2011.12.011.

2. GVR. Nano Calcium Carbonate Market to Reach $9.65 Billion By 2024.

3. Jimoh, O. A.; Ariffin, K. S.; Hussin, H. Bin; Temitope, A. E. Synthesis of Precipitated Calcium Carbonate: A Review. *Carbonates and Evaporites*. 2018, pp 331–346. https://doi.org/10.1007/s13146-017-0341-x.

4. Neumann, K.; Gladyszewski, K.; Groß, K.; Qammar, H.; Wenzel, D.; Górak, A.; Skiborowski, M. A Guide on the Industrial Application of Rotating Packed Beds. *Chemical Engineering Research and Design*. 2018, pp 443–462. https://doi.org/10.1016/j.cherd.2018.04.024.

5. Chen, J.; Shao, L. Mass Production of Nanoparticles by High Gravity Reactive Precipitation Technology with Low Cost. *China Particuology* **2003**, *1* (2), 64–69. https://doi.org/10.1016/s1672-2515(07)60110-9.

6. Chen, J. F.; Wang, Y. H.; Guo, F.; Wang, X. M.; Zheng, C. Synthesis of Nanoparticles with Novel Technology: High-Gravity Reactive Precipitation. *Ind. Eng. Chem. Res.* **2000**, *39* (4), 948–954. https://doi.org/10.1021/ie990549a.

7. Gradl, J.; Schwarzer, H. C.; Schwertfirm, F.; Manhart, M.; Peukert, W. Precipitation of Nanoparticles in a T-Mixer: Coupling the Particle Population Dynamics with Hydrodynamics through Direct Numerical Simulation. *Chem. Eng. Process. Process Intensif.* **2006**, *45* (10), 908–916. https://doi.org/10.1016/j.cep.2005.11.012.

8. Gavi, E.; Rivautella, L.; Marchisio, D. L.; Vanni, M.; Barresi, A. A.; Baldi, G. CFD Modelling of Nano-Particle Precipitation in Confined Impinging Jet Reactors. *Chem. Eng. Res. Des.* **2007**, *85* (5 A), 735–744. https://doi.org/10.1205/cherd06176.

9. Steyer, C.; Mangold, M.; Sundmacher, K. Modeling of Particle Size Distribution for Semibatch Precipitation of Barium Sulfate Using Different Activity Coefficient Models. *Ind. Eng. Chem. Res.* **2010**, *49* (5), 2456–2468. https://doi.org/10.1021/ie901306r.

10. Pan, S. Y.; Eleazar, E. G.; Chang, E. E.; Lin, Y. P.; Kim, H.; Chiang, P. C. Systematic Approach to Determination of Optimum Gas-Phase Mass Transfer Rate for High-Gravity Carbonation Process of Steelmaking Slags in a Rotating Packed Bed. *Appl. Energy* **2015**, *148*, 23–31. https://doi.org/10.1016/j.apenergy.2015.03.047.

11. Xiang, Y.; Wen, L.; Chu, G.; Shao, L.; Xiao, G.; Chen, J. Modeling of the Precipitation Process in a

Rotating Packed Bed and Its Experimental Validation. *Chinese J. Chem. Eng.* **2010**, *18* (2), 249–257. https://doi.org/10.1016/S1004-9541(08)60350-X.

12. Emami-Meibodi, M.; Soleimani, M.; Bani-Najarian, S. Toward Enhancement of Rotating Packed Bed (RPB) Reactor for CaCO3 Nanoparticle Synthesis. *Int. Nano Lett.* **2018**, *8* (3), 189–199. https://doi.org/10.1007/s40089-018-0244-4.

Table 1: Model predictions compared to Emami-Meibodi et al. [12] and predicted PSD

Test No.	Rotor speed [rpm]	Liquid flow rate [L/min]	Reaction time [s]	Experimental PSD in the liquid phase [nm] *[12]*	Average of PSD of experimental results [nm]	Average of predicted PSD in the liquid phase [nm]	ARD% between Exp & Pre PSD
Saddle Type	*packing*						
1	3000	6	260	46-60	53	**55.3**	**3.7 %**
2	4000	6	370	37-70	53.5	**48.2**	**9.8 %**
3	4000	9	389	64	64	**60.8**	**6.2 %**
Pall type	*packing*						
4	3000	6	430	78	78	**71.2**	**9.5 %**
5	3500	9	437	48-115	81.5	**78.3**	**3.9 %**
6	4000	9	380	36-138	87	**85**	**2.3 %**

626066 Process Synthesis and Intensification for Shale Resources Valorization

Abstract:

Shale gas is a promising energy resource and chemical feedstock for the transition period towards a sustainable economy and has the potential to be a carbon source for the long term. However, huge amount of shale gas at remote shale gas basins is directly flared[1] due to the lack of infrastructure to transport the gas from the well head to the central processing plant. To avoid long distance transportation, it is essential to covert shale gas to value-added products on site at a distributed scale. The value-added product could be gasoline, diesel, aromatics, alcohols, or any other liquid chemicals which can be easily shipped. Unlike current shale gas process where large scales are preferred, simple and intensified processes with least processing steps and least pieces of equipment are favored for remote shale plays.

Natural gas liquids contained in shale gas are especially of interest since they are comparably easier to activate and transform to liquid products. In the current work, novel processes are developed to achieve the upgrading and more importantly, several key process intensification principles are identified which can be potentially applied to other processes that require a modular and intensified design. While conventional shale gas processing usually follows a hierarchy of "Front end separation-NGL activation – NGL upgrading", our novel process introduces a new hierarchy of "NGL activation - NGL upgrading - Back end separation". This paradigm shift of hierarchy intensifies and simplifies the entire process by eliminating repeated or unnecessary unit operations and associated equipment. A small scale process at the gas gathering station (10 MMSCFD) and a medium scale process at the gas processing plant (100 MMSCFD) are developed in this work. These processes are demonstrated, through detailed simulation and economic analysis, to be economically attractive.

Reference

1. Natural centers for environmental information. Global gas flaring observed from space. https://ngdc.noaa.gov/eog/viirs/download_global_flare.html (accessed 04/06/20)
2. Ridha, T., Li, Y., Gençer, E., Siirola, J. J., Miller, J. T., Ribeiro, F. H., & Agrawal, R. (2018). Valorization of Shale Gas Condensate to Liquid Hydrocarbons through Catalytic Dehydrogenation and Oligomerization. Processes, 6(9), 139.

630079 A Short-Cut Method for Synthesis of Solvent-Based Separations

Abstract:

In the process industry, solvent-based separations, e.g., extraction, extractive distillation, and azeotropic distillation, are widely applied to separate nonideal mixtures by adding an external component to the system. The optimization of solvent-based separations includes two sequential activities, solvent selection and process design/control. First, the solvent is selected based on the physical properties of the solvent, such as solubility power, boiling point, selectivity, etc. [1]. Then, the process variables such as solvent flowrate, column stages, and feed inlet stages are determined by sensitivity analysis or simulation-based optimization for the selected solvent. For example, Luyben [2] designed the optimum acetone/chloroform azeotropic distillation system by varying the solvent flowrate and reflux ratio with a fixed number of stages. Similarly, You et al. [3] designed an acetone/methanol extractive distillation process by minimizing the process energy consumption. However, these process design methods are based on knowing the identity of the solvent(s), which are usually selected based solely on their physical properties. Kossack et al. [4] pointed out that solvent screening based on physical properties such as selectivity alone may result in unfavorable entrainer choice. It is necessary to take separation process properties such as energy consumption, number of stages, etc., into account during solvent selection. A short-cut evaluation model which can quickly evaluate both the solvents' physical/mixture and process properties is important in designing an optimal separation-based process. In other words, with the assistance of the short-cut evaluation model, different solvent-based separation alternatives can be quickly generated and evaluated.

This work focuses on developing a hybrid separation synthesis framework for the separation of nonideal mixtures. First, solvents/entrainers that can break the azeotropic barrier are designed by a computer-aided molecular design method or identified via literature search. In this step, the solvent/entrainer physical and mixture properties are applied as constraints. Then, the solvents' process properties, like energy consumption and the number of stages, are calculated using the developed short-cut evaluation model. In the short-cut evaluation model, for extractive distillation systems, the extractive column is divided into three sections (rectifying, extractive, stripping), and the Underwood and Fenske equations are applied in each section separately. The analysis of each section separately enables consideration of variations in the relative volatility across the whole column, thereby providing accurate estimates of the minimum reflux ratio for the column and the number of stages in each section. The minimum reflux ratio is used to estimate the minimum reboiler duty through enthalpy balance. The short-cut evaluation model employs a scoring approach, which considers physical properties, mixture properties, and process parameters to select the optimal solvent-based separation process. Finally, the top-ranked alternatives are verified using rigorous models.

The developed hybrid separation synthesis framework is applied to different case studies, such as dimethyl carbonate and isobutanol production. For all the examples, the framework was able to generate multiple process alternatives that are quickly evaluated to identify the optimal separation process.

References

[1] Shen, W., Dong, L., Wei, S. A., Li, J., Benyounes, H., You, X., & Gerbaud, V. (2015). Systematic

design of an extractive distillation for maximum-boiling azeotropes with heavy entrainers. AIChE Journal, 61(11), 3898-3910.

[2] Luyben, W. L. (2008). Control of the maximum-boiling acetone/chloroform azeotropic distillation system. Industrial & engineering chemistry research, 47(16), 6140-6149.

[3] You, X., Rodriguez-Donis, I., & Gerbaud, V. (2015). Improved design and efficiency of the extractive distillation process for acetone–methanol with water. Industrial & Engineering Chemistry Research, 54(1), 491-501.

[4] Kossack, S., Kraemer, K., Gani, R., & Marquardt, W. (2008). A systematic synthesis framework for extractive distillation processes. Chemical Engineering Research and Design, 86(7), 781-792.

628766 Discrete-Steepest Descent: A Solution Method for Process Synthesis Generalized Disjunctive Programs

Abstract:

The optimal process design and operation challenge chemical engineers, specifically in the Process Systems Engineering(PSE) community. Chemical processes need to satisfy operational, economic, and environmental goals, which can be integrated through optimal design, i.e., an area studied within Process Synthesis [1,2,3]. A promising method that tackles the design of processes with structural decisions is the superstructure approach, where different possible process flowsheets are combined into a single structure. Hence, the objective is to select which units and interconnections lead to the optimal process, considering both operating conditions and desired specifications [4]. This superstructure method leads to a mathematical optimization problem that can be tackled using tools borrowed from Mathematical Programming. Moreover, due to the usual presence of both nonlinear terms and discrete variables, these problems can be posed as Mixed-Integer Nonlinear Programs (MINLP). The solution to these optimization problems is challenging given their combinatorial and nonconvex nature. Therefore the Generalized Disjunctive Programming (GDP) framework has been proposed to resolve the challenges of MINLP solution strategies [5]. Accordingly, the main contribution of this work is to consolidate the Discrete-Steepest Descent Algorithm (D-SDA) as an efficient GDP solution method through its implementation and automatization in Pyomo. The D-SDA exploits a mathematical structure often present in GDP superstructures, i.e., the inherent positional order present in superstructure design problems.

A common objective in process design is to designate discrete locations to units or streams over a superstructure, e.g., unit operations can be included in series, such as trays in distillation columns or positions in reactor networks. When posing these problems as GDP, Boolean variables represent the discrete decisions relevant to a given superstructure choice together with all the constraints associated with it. These Boolean variables belong to an ordered set representing positions of possible locations within the superstructure. Furthermore, Boolean variables are usually subject to knapsack equality constraints, where only a fixed number of the Boolean variables within that ordered set need to be set to True, e.g., three trays need to contain catalyst in a reactive distillation column [6]. Knowledge of this mathematical structure is integrated into the proposed algorithm to efficiently explore the discrete solution space through a variable reformulation technique.

We propose the Discrete-Steepest Descent Algorithm (D-SDA) as a systematic algorithm tailor-designed to solve GDP superstructure problems with cardinality or knapsack equality constraints defined over ordered Boolean variables. The D-SDA relies on a reformulation of the Boolean variables in the GDP problem into integer choices named external variables[7]. The external variables provide a succinct representation of locations within the superstructure by exploiting its characteristic of sequentially ordered positions. This variable selection reduces the problem's dimensionality, as it integrates several Boolean variables into just one external variable and allows the method to efficiently explore the combinatorial space of the discrete variables, referred to as the neighborhood. The algorithm uses search criteria such as neighborhood, and line searches to perform a systematic scouting of the discrete feasible region. This exploration leads to an efficient choice of external variables resulting in Nonlinear Programming (NLP) subproblems that are solved sequentially until a local optimum is found. The D-SDA uses as termination criterion the integrally local optimality [8],allowing it to find local optima that are not necessarily considered by other MINLP and GDP solution algorithms. Motivated by this technique's early success for complex and highly nonconvex MINLPs [7,6,9], we present a formal description of the algorithm within the scope of GDP, together with an open-source implementation of the method.

The D-SDA is presented as a tailored solution method for GDP. The specialized solution methods for GDP are usually based on generalizing algorithms for MINLP, where the optimization problems are decomposed, so the discrete variables are fixed and allow to solve the problem only in terms of the continuous variables. Furthermore, we show that the D-SDA is related to other existing GDP solvers such as Logic-based Branch-and-Bound (LBB) and the Logic-based Outer-Approximation (LOA) [10]; that is, the D-SDA reduces to these algorithms in limiting cases. The D-SDA does not provide a dual bound for the optimal solution because it does not entail the optimal solution of a relaxation of the original problem. Thus, the D-SDA's optimality criterion does not rely on computing the optimality gap as LOA and LBB. Instead, building upon the developments of [8], the D-SDA's stopping criteria are modified to guarantee local integral optimal convergence.

We implemented the D-SDA method in an extensible Python implementation compatible with Pyomo.GDP [10] to address discrete optimization problems in Process Synthesis [11]. The performance of this algorithm is compared against MINLP and GDP solvers by addressing three process superstructure problems: A GDP reformulation of the volume minimization of a series of Continuously Stirred Tank Reactors (CSTR) considered by Liñán et al.[7], the optimal economic design of a tray-distillation column proposed by Ghouse et al.[12], and a GDP reformulation of the optimal design for chemical batch processing presented by Kocis and Grossmann[13]. For the CSTR superstructure, the proposed D-SDA framework resulted in a more efficient strategy to obtain the globally optimal solution in computational time than state-of-the-art MINLP solvers. Furthermore, even with the problem's GDP reformulation, the D-SDA performed faster than other tailored GDP solvers such as LOA and LBB. The D-SDA proved to be more efficient in the distillation column example than both MINLP and GDP solvers and improved the best-known GDP solution to this challenging optimization problem. For the batch processing problem, MINLP solvers performed slightly faster than the proposed D-SDA method; nevertheless, it converged to the optimum within a few seconds, a time within the same order of magnitude as state-of-the-art nonlinear discrete optimization solvers.

The examples mentioned herein, together with a general D-SDA implementation, were written in Python and made compatible with the open-source algebraic modeling language Pyomo.GDP [10], and can be found in an openly-available GitHub repository (https://github.com/bernalde/dsda-gdp).

References

[1] E. Pistikopoulos, A. Barbosa-Povoa, J. H. Lee, R. Misener, A. Mitsos, G. Reklaitis, V. Venkatasubramanian, F. You, and R. Gani, "Process systems engineering–the generation next?" Computers & Chemical Engineering, p.107252, 2021.

[2]Q. Chen and I. Grossmann, "Recent developments and challenges in optimization-based process synthesis," Annual review of chemical and biomolecular engineering, vol. 8, pp. 249–283, 2017.

[3]M. Rafiei and L. A. Ricardez-Sandoval, "New frontiers, challenges, and opportunities in integration of design and control for enterprise-wide sustainability," Computers & Chemical Engineering, vol. 132, p. 106610, Jan. 2020.

[4]H. Yeomans and I. E. Grossmann, "A systematic modeling framework of superstructure optimization in process synthesis," Computers & Chemical Engineering, vol. 23, no. 6, pp. 709–731, 1999.

[5]I. E. Grossmann and F. Trespalacios, "Systematic modeling of discrete-continuous optimization models through generalized disjunctive programming, "AIChE Journal, vol. 59, no. 9, pp. 3276–3295, 2013.

[6]D. A. Liñán, D. E. Bernal, L. A. Ricardez-Sandoval, and J. M. Gómez, "Optimal design of superstructures for placing units and streams with multiple and ordered available locations. Part II: Rigorous design of catalytic distillation columns, "Computers & Chemical Engineering, vol. 139, p. 106845, 2020.

[7]——, "Optimal design of superstructures for placing units and streams with multiple and ordered available locations. Part I: A new mathematical framework, "Computers & Chemical Engineering, vol. 137, p. 106794,2020.

[8] K. Murota, "Discrete convex analysis, "Mathematical Programming, vol. 83, no. 1, pp. 313–371, 1998.

[9]D. A. Liñán, D. E. Bernal, J. M. Gomez, and L. A. Ricardez-Sandoval, "Optimal synthesis and design of catalytic distillation columns: A rate-based modeling approach, "Chemical Engineering Science, vol. 231, p. 116294, 2021.

[10]Q. Chen, E. S. Johnson, D. E. Bernal, R. Valentin, S. Kale, J. Bates, J. D. Siirola, and I. E. Grossmann, "Pyomo.gdp: an ecosystem for logic based modeling and optimization development."

[11]L. Biegler, I. Grossmann, and A. Westerberg, Systematic Methods of Chemical Process Design, ser. Physical and Chemical Engineering Sciences. Prentice Hall PTR, 1997.

[12]J. H. Ghouse, Q. Chen, M. A. Zamarripa, A. Lee, A. P. Burgard, I. E. Grossmann, and D. C. Miller, "A comparative study between GDP and NLP formulations for conceptual design of distillation columns," in Computer Aided Chemical Engineering. Elsevier, 2018, vol. 44, pp. 865–870.

[13]G. R. Kocis and I. E. Grossmann, "Global optimization of nonconvex mixed-integer nonlinear programming(MINLP) problems in process synthesis, "Industrial & engineering chemistry research, vol. 27, no. 8, pp. 1407–1421,1988.

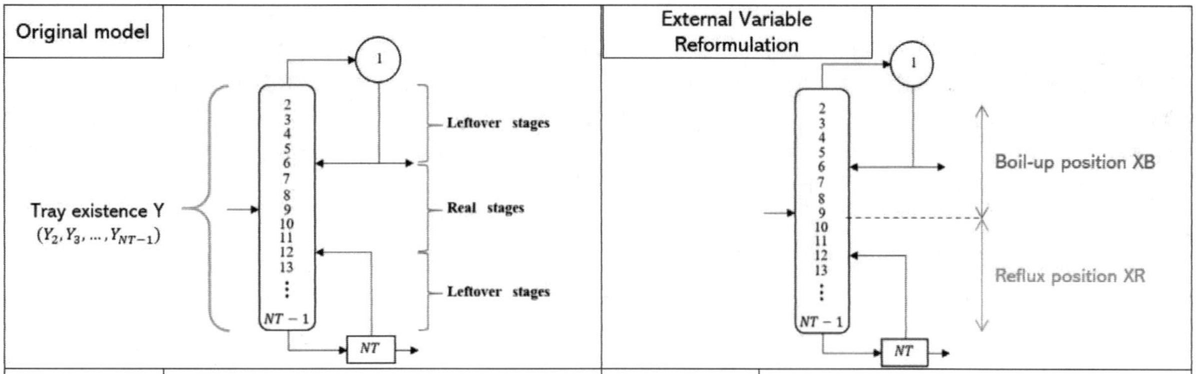

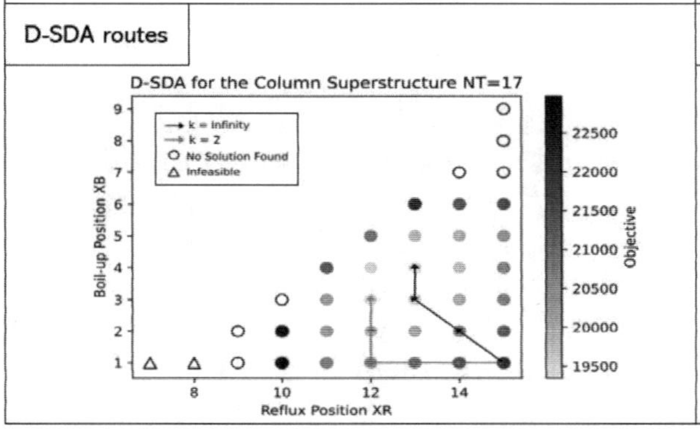

Results

Method comparison using Knitro		
Method	**Execution time [s]**	**Gap [%]**
D-SDA-Infinity	**8.58**	**0.00**
D-SDA-2	**6.34**	0.54
MINLP-BigM	65.66	0.54
MINLP-Hull	116.26	-
GDPopt-LOA	19.12	8.12
GDPopt-GLOA	161.64	**0.00**
GDPopt-LBB	900.0	15.12
Complete enumeration	67.38	**0.00**

621549 Optimization of Natural Gas Liquefaction Process Using Detailed Differential Algebraic Equation Based Multi-Stream Heat Exchanger Design

Abstract:

Natural Gas has become an increasing source of energy around the world. It is seen as a transition fuel from traditional energy sources like coal and oil to renewable energy sources like solar and wind. Worldwide trade of natural gas has also increased with many nations using it as the fuel in their power generating plants. Transporting natural gas overseas requires liquefaction under high pressure and cryogenic temperatures.

The liquefaction of natural gas is an energy intensive process where a mixture of hydrocarbon and nitrogen is used as the mixed refrigerant to cool the gas inside a multi-stream heat exchanger. Most optimization studies considering natural gas liquefaction process do not include the design effects of the multi-stream heat exchanger. This could lead to sub-optimal or infeasible solution to the real world flowsheet optimization problem. To the best of author's knowledge, only Tsay et al. (2017) implemented detailed design of multi-stream heat exchanger inside natural gas flowsheets.

In this study, we formulate a DAE based model for multi-stream heat exchanger design which incorporates phase change and heat exchanger between the natural gas and mixed refrigerant streams. The design equations of a spiral wound coil heat exchanger are used for the detailed design of the multi-stream heat exchanger. The phase change is modeled using complementarity constraint based flash equations and the heat transfer is modeled based on first principles heat partial differential equation.

We use the trust-region filter based method (by Eason and Biegler, 2018) to embed the detailed DAE model into the natural gas liquefaction process flowsheet. Reduced models are used to represent the DAE model inside the flowsheet models which are updated after each trust region iterations using the solution from the DAE design model. The filter is used to guarantee convergence to both feasible and optimal point of the overall design problem of natural gas liquefaction process with detailed heat exchanger design. We will present the DAE model formulation and the trust-region algorithm along with results from a real world natural gas liquefaction flowsheet optimization problem.

References:-

1. Eason, J.P. and Biegler, L.T. (2018), Advanced trust region optimization strategies for glass box/black box models. AIChE J, 64: 3934-3943
2. Tsay, C., Pattison, R.C. and Baldea, M. (2017), Equation-oriented simulation and optimization of process flowsheets incorporating detailed spiral-wound multistream heat exchanger models. AIChE J., 63: 3778-3789.

632331 Ensuring Process Safety of Heat Exchanger Networks Under Uncertainty

Abstract:

Heat exchanger network synthesis (HENS) offers a cost effective means of achieving a plant's energy demand [1,2]. Often overlooked, however, are the process safety metrics, including the potential for overpressure. One possible consequence is a tube rupture, an overpressure event whereby the low pressure shell side is pressurized by the high pressure tube side [3,4,5]. If the pressures are not considered, the chances of selecting unsafe exchanger configurations that are susceptible to tube rupture increase. To overcome this, prior work has incorporated a Safety Rating (SR) for individual heat exchangers [6,7]. This metric captures the severity of a tube rupture for a single exchanger and can serve as a benchmark when comparing the safety of different exchanger configurations. The specific challenges of safety-ensured HENS, however, include the complexity in incorporating complicated SR calculations, ensuring safety of connected exchangers in a network setting, and the difficulty in imposing the safety constraints under uncertainty. The deterministic SR is obtained via dynamic simulation that requires specifications and highly nonlinear thermodynamic property predictions for liquid-liquid, vapor-liquid, and flashing liquid-liquid systems. We tackle this challenge by developing a data-driven piecewise linear (PWL) underestimation of SR, which is then incorporated in the HENS formulation that retains the MINLP structure of the synthesis problem. Originally proposed by Rebennack and Kallrath [8], the PWL formulation ensures an efficient and tight but, at the same time, a conservative underestimation (vs. approximation) of SR. In the context of process safety, this is important because of the severity of erroneous approximation. The simplicity of PWL allows a minimum SR constraint to be applied to all exchangers to obtain a final HEN configuration that meets a plant's level of risk.

We further extend the HEN synthesis to account for the uncertainty of SR predictions. This uncertainty is assumed to be in the form of pressure variations across a HEN's streams. These uncertainties, which are common during normal operation, can lead to lower SR values. If left unaccounted for, the SR minimum threshold originally proposed may not perform as expected in a plant setting and instead give a false sense of security. As the SR is itself a function of pressure, we use a robust counterpart of the safety constraint to tackle the worst-case scenario and ensure that the designed HEN can tolerate a certain level of variance that may be present in a plant. One benefit of this approach is that plants could use historical operating data to assume a level of variance and subsequently determine the most cost effective HEN configuration. We will demonstrate the overall safety-ensured HEN synthesis framework using several case studies.

References:

[1] Yee, T. F., & Grossmann, I. E. (1990). Simultaneous optimization models for heat integration—II. Heat exchanger network synthesis. *Computers & Chemical Engineering*, *14*(10), 1165-1184.

[2] Ponce-Ortega, J. M., Jiménez-Gutiérrez, A., & Grossmann, I. E. (2008). Optimal synthesis of heat exchanger networks involving isothermal process streams. *Computers & Chemical Engineering*, *32*(8), 1918-1942.

[3] Grim, L. (2017). CSB Investigation: Williams Geismar Olefins Plant Reboiler Rupture and Fire. In *ASSE Professional Development Conference and Exposition*. American Society of Safety

Engineers.

[4] API Standard 521. (2014). Pressure-Relieving and Depressuring Systems.

[5] Hellemans, M. (2009). *The safety relief valve handbook: design and use of process safety valves to ASME and International codes and standards*. Elsevier.

[6] Harhara, A., & Hasan, M. F. (2020). Dynamic modeling of heat exchanger tube rupture. *BMC Chemical Engineering*, *2*(1), 1-20.

[7] Harhara, A., & Hasan, M. F. (2019). Incorporating Process Safety into Heat Exchanger Network Synthesis and Operation. In *Computer Aided Chemical* Engineering (Vol. 47, pp. 221-226). Elsevier.

[8] Rebennack, S., & Kallrath, J. (2015). Continuous piecewise linear delta-approximations for univariate functions: computing minimal breakpoint systems. *Journal of Optimization Theory and Applications, 167(2)*, 617-643.

630439 Use of Dimensionality Reduction and Transfer Learning in Deep Reinforcement Learning Controller for Hydraulic Fracturing

Abstract:

Reinforcement Learning (RL) is an area of machine learning where an agent learns optimal control actions by directly interacting with the environment in order to maximize the cumulative sum of rewards. However, the application of RL to process control has been recent and limited [1]. There are multiple approaches to solving the RL problem in the literature but recent successes in RL has been due to combining concepts from deep learning with RL algorithms. The idea of leveraging deep neural networks (DNNs) as function approximators in RL has delivered tremendous success in video games like Atari games [2] etc. On the contrary, in process control, a Deep RL (DRL) controller was recently proposed to control discrete-time nonlinear processes [3]. The DRL controller is a model-free off-policy actor-critic algorithm in which learning is based on temporal difference (TD) error [1] and Deterministic Policy Gradient (DPG) theorem [4]. The actor and the critic are represented using DNNs so that the policy and the value functions can be generalized to the continuous state and action spaces. Ideas like replay memory, target networks, and gradient clipping were used in the DRL controller to make learning appropriate for process control applications. The DRL controller was able to track setpoints for single-input-single-output (SISO), multi-input-multi-output (MIMO), and a nonlinear system with external disturbances [3].

Hydraulic fracturing is the extraction of oil and gas from rocks with low porosity and permeability. In order to extract oil and gas from such rocks, an artificial medium of proppant (sand) is created within the fractures using controlled explosions followed by the pumping of fracturing fluids at high pressures. The efficiency of the extraction process is dependent on the final fracture geometry and proppant concentration along the fracture. In order to achieve the desired objectives with respect to these process variables, model predictive controllers (MPC) have been designed recently [5]. But these controllers (a) require an accurate model which is difficult to obtain in the case of hydraulic fracturing given its complexity, (b) involve exorbitant computational costs, and (c) require regular re-tuning of controller parameters. Therefore, in this work, we propose to design a model-free DRL controller for hydraulic fracturing. Despite its success, the DRL controller has a few limitations which include the requirement of long training times, and careful initialization of hyperparameters for fast convergence [6]. In order to overcome these challenges, we propose to implement transfer learning wherein the controller learns a suboptimal policy offline using a data-based reduced-order model (ROM) before learning online from the actual process. In this work, we used a high-fidelity model of hydraulic fracturing process as a virtual numerical experiment of the actual process. Additionally, we used principal component analysis (PCA) to reduce the dimension of the RL state before using it for learning. Consequently, the actor and the critic were trained in the reduced-PCA space. With these proposed steps included in the DRL controller training, it shows convergence to an optimal policy with the objective of obtaining uniform proppant concentration along the fracture.

Literature cited:

[1] Sutton, R.S., Barto, A.G. Reinforcement Learning: An Introduction. *Cambridge: MIT Press*, 1998.

[2] Mnih, V., Kavukcuoglu, K., Silver, D., Graves, A., Antonoglou, I., Wierstra, D., Riedmiller, M. Playing Atari with Deep Reinforcement Learning. *arXiv Preprint, arXiv:13125602*, 2013.

[3] Spielberg, S., Tulsyan, A., Lawrence, N.P., Loewen, P.D., Bhushan Gopaluni, R. Toward self-driving processes: A deep reinforcement learning approach to control. *AIChE J*, **65**(10), 2019.

[4] Silver, D., Lever, G., Heess, N., Degris, T., Wierstra, D., Riedmiller, M. Deterministic policy gradient algorithms. *In ICML*, 2014.

[5] Siddhamshetty, P., Yang, S., Kwon, J. S.-I. Modeling of hydraulic fracturing and designing of online pumping schedules to achieve uniform proppant concentration in conventional oil reservoirs. *Comput. Chem. Eng.*, **114**:306 – 317, 2018.

[6] Shin, J., Badgwell T.A., Liu, KH., Lee, J.H. Reinforcement Learning – Overview of recent progress and implications for process control. *Computer Aided Chemical Engineering*, **44**:71-85, 2018.

632115 Harnessing Cognitive AI Technology in Refining to Enhance Operator Decisions

Abstract:

This talk will explore a case study on how one of the largest refiners in the US is leading the way to implement Cognitive AI technologies that digitize refinery campaign processes from planning to operator action. We'll explain how this technology is helping operators to choose a judgment about what action to take at any given time, leading to better coordination and higher refinery margin through intermediate upgrades, product giveaway, and unit optimization.

Digital technologies and AI are becoming crucial for refiners to improve margins and reduce costs. There is a desire to move quickly so that you can be the company that achieves higher profitability before it becomes the standard. Yet we are observing inefficiency and resistance to digital adoption. Resistance and inefficiency persist at the business level from uncertainly in how to define the most impactful use case, and at a personnel level from a lack of trust in novel digital technologies. In this talk, we'll discuss how operations teams can begin building trust in AI technologies and share a roadmap of how a refinery launched a major initiative to optimize refinery processes and improve profitability by digitizing the refinery operating plan and leveraging best practices and knowledge to enhance operator decision making abilities.

Commercial optimization of a refinery is a unique exercise in many respects, providing both challenge and opportunity. These include crude slates that change according to market demand, variability in feedstock quality, and the continually changing operating capability of equipment from degradation and many other factors. What makes one refinery perform better than its competitors is often its ability to execute a set of objectives consistently while navigating through this uncertainty. According to Accenture, 61% of refiners believe process optimization technologies are a primary benefit driver to production planning and execution. Refinery campaigns are fundamental to improving margins but using a traditional problem-solving workflow between engineers and operators can be ineffective resulting in a very transactional execution process, relying heavily on individual interpretation of the guidance. The burden is on the operator to fully understand context, leading to inconsistencies over time and between different teams.

To close the talk, we'll highlight primary benefits from Cognitive AI technologies that are being used in refining today, including providing context to operators, quantifying decision confidence, and allowing for accelerated improvement across campaign plans, ultimately closing the profitability gap.

3 main takeaways:

1. How refiners can build trust in novel digital technologies including artificial intelligence and define the more impactful use cases for their refinery.
2. How one of the largest refineries in the US implemented Cognitive AI technologies to optimize refinery processes and improve profitability by digitizing refinery campaign processes from planning to action in operations.
3. New insights into how Cognitive AI technologies are helping operators to make a judgment about what action to take at any given time, leading to better coordination and helping refineries to better capture the full refinery margin.

629822 Regression Model for Tool Wear Monitoring in Precision Machining

Abstract:

In metal cutting operations, tool wear must be monitored either by operators or by a tool wear monitoring system to avoid possible damage to products and machine tools. In precision machining specifically, the naturally occurring and progressing flank wear leads to unwanted vibration, which deteriorates the surface finish and results in dimensional inaccuracy of the machining part. In the worst case, the cutting tool can break while it is engaged with the workpiece. With the ever-increasing demand for cost savings and improvements in quality, tool condition monitoring is becoming increasingly important in modern manufacturing [1].

Modern tool wear monitoring systems leverage data-driven models applying machine learning (ML) to predict tool wear and detect machine abnormalities [2]. Many studies have involved ML algorithms that use techniques such as K-nearest neighbors [3], support vector machine [4], and convolutional neural networks [5]. Although these methods have shown good performance, the fundamental methodology and complete toolchain that is suitable to predict tool wear still need to be investigated. Moreover, regression models that predict tool wear as a continuous process state/variable are seldom. The often-applied classification aims to predict discrete class labels, but tool wear is a continuously progressing system disturbance. Therefore, it is important to explore regression models to predict tool wear of continuous variables. The objective of this work is to propose a regression-based framework for predicting tool wear in CNC systems. We use support vector regression (SVR) as the ML algorithm, to provide insights into and quantify tool wear prediction in the machining process.

As shown in Fig.1, in the proposed framework, we collect, tabulate, and label audio and vibration data of the CNC process under varying tool wear and explore the performance of the proposed framework using regression. First, audio and vibration signals are converted to frequency domain through Fast Fourier Transform (FFT) to reduce the complexity of the audio and vibration time waveforms collected. Then, Principal Components Analysis (PCA) is used to eliminate features of small variance, increase the information content, reduce overfitting and computational effort. The selected features of audio and vibration signals are modeled and analyzed with an SVR model. The framework of Fig.1 was applied to a Mazak Variaxis 630-5X II T Mill/Turn Machine 40 hp 5-axis CNC machine. The experimental facility was instrumented to acquire process data for machine power, spindle power, spindle vibration, and to record audio/video and is MTConnect capable. Moreover, the machine was controlled to perform milling at various spindle speeds, widths of cut, depths of cut, and feed rates. Application of the framework of Fig. 1 for tool wear monitoring will be presented. In terms of the accuracy in tool wear prediction, it will be shown that the audio and vibration signals are sufficiently rich in information about tool wear so that a very accurate model (0.013 and 0.007 in RMSE, respectively) could be developed with those signals. Computation of the relative error between real tool wear measurement and predicted tool wear showed 1.54% error in audio signal-based regression and 3.1% error in vibration signal-based regression. Fusion of these signals further improved the accuracy of the regression model. In conclusion, the acoustic signals that often lead engineers and operators identify issues and anomalies in the operation of precision machining equipment were used successfully to learn ML models to predict tool wear in a very accurate manner. Extrapolation of these models in time to be useful in machine prognostics is currently underway and will be discussed in this presentation.

Acknowledgment: This material is based upon work supported by the U.S. Department of Energy's Office of Energy Efficiency and Renewable Energy (EERE) under the Advanced Manufacturing Office Award Number DE-EE0007613.

makes any warranty, express or implied, or assumes any legal liability or responsibility for the accuracy, completeness, or usefulness of any information, apparatus, product, or process disclosed, or represents that its use would not infringe privately owned rights. Reference herein to any specific commercial product, process, or service by trade name, trademark, manufacturer, or otherwise does not necessarily constitute or imply its endorsement, recommendation, or favoring by the United States Government or any agency thereof. The views and opinions of authors expressed herein do not necessarily state or reflect those of the United States Government or any agency thereof."

References

[1] H. Shao, H. L. Wang, and X. M. Zhao, "A cutting power model for tool wear monitoring in milling," *Int. J. Mach. Tools Manuf.*, vol. 44, no. 14, 2004, doi: 10.1016/j.ijmachtools.2004.05.003.

[2] Z. Li, R. Liu, and D. Wu, "Data-driven smart manufacturing: Tool wear monitoring with audio signals and machine learning," *J. Manuf. Process.*, vol. 48, 2019, doi: 10.1016/j.jmapro.2019.10.020.

[3] A. Jegorowa, J. Górski, J. Kurek, and M. Kruk, "Use of nearest neighbors (K-NN) algorithm in tool condition identification in the case of drilling in melamine faced particleboard," *Maderas Cienc. y Tecnol.*, vol. 22, no. 2, 2020, doi: 10.4067/S0718-221X2020005000205.

[4] D. Shi and N. N. Gindy, "Tool wear predictive model based on least squares support vector machines," *Mech. Syst. Signal Process.*, vol. 21, no. 4, 2007, doi: 10.1016/j.ymssp.2006.07.016.

[5] F. Jia, Y. Lei, J. Lin, X. Zhou, and N. Lu, "Deep neural networks: A promising tool for fault characteristic mining and intelligent diagnosis of rotating machinery with massive data," *Mech. Syst. Signal Process.*, vol. 72–73, 2016, doi: 10.1016/j.ymssp.2015.10.025.

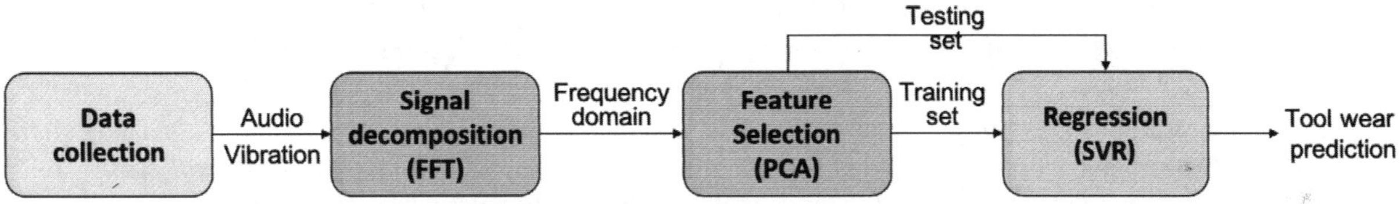

Figure 1. Flow diagram of a regression-based framework for tool wear prediction.

626215 Selection of Combined Index Weights to Optimize Anomaly Detection in Big Area Additive Manufacturing

Abstract:

As additive manufacturing (AM) becomes more prominent in our society, there are many questions about AM's repeatability and reliability. Today, AM is a process that involves constructing an object layer-by-layer; although this method allows the realization of complex computer-aided design (CAD) files into physical objects, a layer-by-layer process can also be more prone to error. Therefore, each layer has to be qualified so that the whole part can have a functional, rather than a purely aesthetic, purpose. Although AM has been used for several decades, most of the final part quality can be attributed to extensive operator knowledge and experience, rather than to the use of data-informed approaches. To assist operators, particularly those who may not be as knowledgeable about the process, and to monitor part quality, it is important to adapt data-driven approaches to warn operators of impending anomalies in the process. To achieve this, it is important to develop robust algorithms to interpret the vast quantities of collected data, especially when a priori knowledge of the data structure is nonexistent.

One traditional approach to monitor a manufacturing process is statistical process control (SPC). Before SPC is applied, a model that can adequately describe source of variation must be identified. Principal component analysis (PCA) [1] is one technique available to this end. The main idea is to describe normal variations by means of a number of principal components (PCs). This number (K) is smaller than the number of original variables (J). If there is no a priori knowledge about the data structure, it can be difficult to automatically calibrate a PCA model by selecting the correct number of PCs to retain. To overcome this issue, an existing proposed technique that involves the calculation of an ignorance score [2] to calibrate a probabilistic PCA model (PPCA) [3] is used in this study. An optimal model is selected by choosing the number of PCs for which the ignorance score achieves the minimal value.

After a model is selected using the ignorance score, SPC monitoring statistics such as Hotelling's T^2 statistic (H) and the squared prediction error (Q) can be used to detect anomalous data during process operation [4]. Anomalous data is identified when the H statistic and/or the Q statistic exceed their respective theoretical upper control limits (UCLs). While most applications make use of two charts, one for Q and one for H, a combined index () has been proposed more recently [5]. The combined index is a weighted sum of H and Q as seen in Eq. (1). The two weights, w_1 and w_2, are typically thought of as separate parameters [5]. So far, the weights of the combined index are calculated from UCLs for H and Q [5,6]. For example, the UCL of the H statistic relies on the assumption on how the H statistic is distributed (the chi-squared distribution was assumed in the original work [5]) and on selected confidence limits. Using UCLs as weights in the combined index is not motivated theoretically, only practically. Consequently, these UCL-based weights may render results obtained with the combined index difficult to translate from one application to the next. Today, there is no procedure available to set the weights (w_1 and w_2) so that optimal results can be assured, either theoretically or empirically. In this work, we present the first results of a systematic study where the chosen weights are varied deliberately in order to explore their effects on the combined index as a classifier during process monitoring.

Importantly, the combined index and its use for process monitoring is invariant to scale. Multiplying w_1 and w_2 by the same factor results in proportional change of the combined index and the upper control limit obtained for this combined index. To standardize our approach, we always chose the weights such that the sum of the eigenvalues of the empirical covariance matrix equals the following weighted sum of the same eigenvalues as seen in Eq. (2). The range of w_2 is chosen so that w_1 remains positive. Then, w_1 is calculated based on the above constraint.

Since the weights affect the estimated combined index, there likely exists an optimal set of relative weights that can optimize the performance of the combined index across numerous data types. The hypothesis is that the relative PPCA weights w_1=1 and w_2= (where) give the optimal anomaly detection performance. When these weights are applied, the combined index becomes equal to the squared Mahalanobis distance using the covariance matrix corresponding to the PPCA model [2,3] as seen in Eqs. (3-5). In essence, we are suggesting that the new optimal weights found in this study should be used consistently across literature to provide the best anomaly detection with the combined index.

The effect of the relative weights on the combined index is explored with simulated data sets [2]. Data types B1, B2, and C with noise to signal ratios up to 55% are used to generate simulated data sets with 1024 samples. The calculated combined indices are averaged over the number of samples. As seen in Figs. 1-2, the value of the combined index for a particular data set is a function of the selected weights. One should note that the vertical lines visually represent the locations of select weight ratios, such as the PPCA weights and UCL weights, where the chi-squared UCL has a specified confidence limit (i.e. 90%, 95%, and 99%). The simulated data sets allow the exploration of the effect of the noise to signal ratio and a nonoptimal number of retained PCs on the identification of the best performing relative weights. In order to compare the performance of each set of relative weights, the area under the curve (AUC) metric is used [7]. The AUC of the receiver operator characteristic curve (ROC) essentially describes the effectiveness of the combined index, given a set of relative weights, as a classifier to detect the true and false anomalies. The set of relative weights that lead to the maximum AUC (above 0.5 which is a random classifier) is the set of relative weights that produce the best classifier performance.

[1] Jolliffe, I.T. and Cadima, J., "Principal Component Analysis: A Review and Recent Developments." *Philosophical Transactions of the Royal Society A: Mathematical, Physical and Engineering Sciences*, 374(2065), 2016, pp. 20150202.

[2] Russo, S., Li, G., and Villez, K., "Automated Model Selection in Principal Component Analysis: A New Approach Based on the Cross-validated Ignorance Score," *Industrial & Engineering Chemistry Research,* 58(3), 2019, pp. 13448-13468.

[3] Tipping, M.E. and Bishop, C.M., "Probabilistic Principal Component Analysis," *Journal of the Royal Statistical Society: Series B (Statistical Methodology)*, 61(3), 1999, pp. 611-622.

[4] Montgomery, D., "Introduction to Statistical Quality Control", *Wiley*, 8th ed., 2019.

[5] Yue, H.H. and Qin, S.J., "Reconstruction-based Fault Identification Using a Combined Index," *Industrial & Engineering Chemistry Research*, 40(20), 2001, pp. 4403-4414.

[6] Jackson, J.E. and Mudholkar, G.S., "Control Procedures for Residuals Associated with Principal Component Analysis," *Technometrics*, 21(3), 1979, pp. 341-349.

[7] Fawcett, T., "An Introduction to ROC Analysis," *Pattern Recognition Letters,* 27(8), 2006, pp. 861-874.

$$\gamma = \frac{Q}{w_1} + \frac{H}{w_2} \tag{1}$$

where γ is the combined index, Q is the squared prediction error, H is Hotelling's T^2 statistic, and w_1 and w_2 are weights.

$$\sum_{j=1}^{J} \lambda_j = w_1 \sum_{j=1}^{K} \lambda_j + w_2 \sum_{j=K+1}^{J} \lambda_j = w_1 \sum_{j=1}^{K} \lambda_j + w_2 \,(J-K) \tag{2}$$

where J represents the dimension of the data, K represents the number of retained PCs, $\sum_{j=1}^{K} \lambda_j$ is the sum of the retained eigenvalues, $\sum_{j=K+1}^{J} \lambda_j$ is the sum of the residual eigenvalues, and $\sum_{j=1}^{J} \lambda_j$ is the sum of all eigenvalues from the empirical covariance matrix.

$$\gamma_i = {y_i}^T (\Sigma)^{-1} y_i \tag{3}$$

$$\Sigma = w_1\, \overline{P\,\Lambda P}^T + w_2\, \underline{P P^T} = P\Lambda^* P^T \quad (4)$$

$$\Lambda^* = \begin{bmatrix} w_1\overline{\Lambda} & 0 \\ 0 & w_2 I \end{bmatrix} \quad (5)$$

where γ_i represents the combined index of the ith row, y_i represents the ith row in the mean centered data matrix (n samples x J variables), Σ is the covariance matrix corresponding to the PPCA model, $\overline{P}$ are the retained PCs (J x K), $\overline{\Lambda}$ is a diagonal matrix of retained eigenvalues (K x K), $\underline{P}$ are the residual PCs ((J-K) x K), P are all of the PCs (J x J), and Λ^* is a diagonal matrix of all of the weighted eigenvalues (J x J).

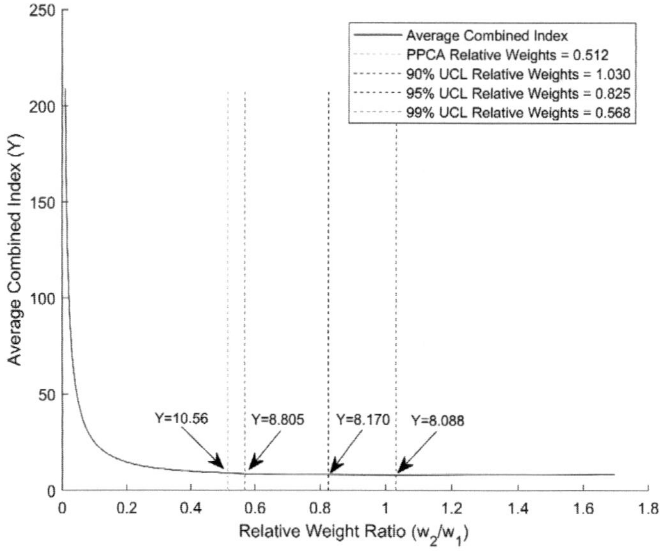

Figure 1. The combined index of data type B1.04 with a noise to signal ratio of 25% (calibrated with 3 PCs for a 9 dimensional data set). The vertical lines represent the locations of select relative weights. The theoretical UCL confidence level corresponds to the chi-squared control limit for the H statistic. The specified values on the plot represent the value of the combined index at the intersection of the selected relative weights indicated by the vertical lines.

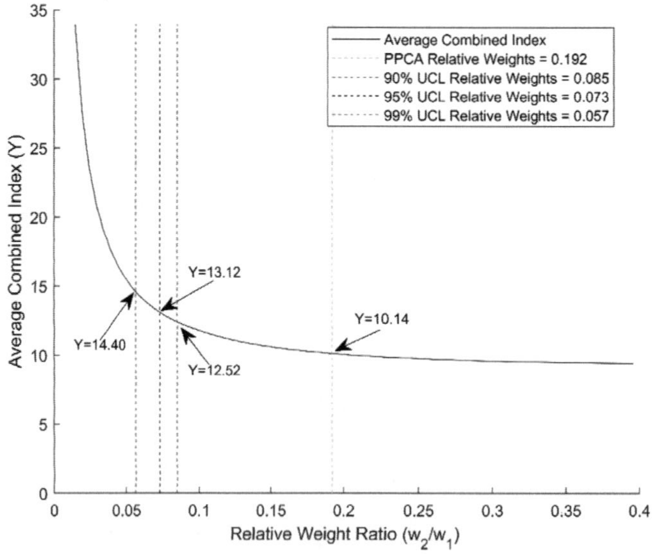

Figure 2. The combined index of data type C.02 with a noise to signal ratio of 25% (calibrated with 8 PCs for a 10 dimensional data set). The vertical lines represent the locations of select relative weights. The theoretical UCL confidence level corresponds to the chi-squared control limit for the H statistic. The specified values on the plot represent the value of the combined index at the intersection of the selected relative weights indicated by the vertical lines.

634105 Keynote Talk - Topological Data Analysis: Concepts, Computation, and Applications in Manufacturing

Abstract:

A key hypothesis that drives engineering practice is that data has structure. The dominant paradigms for describing such structure are statistics (e.g., moments, correlation functions) and signal processing (e.g., convolutional neural nets, Fourier series). Topological Data Analysis (TDA) is a field of mathematics that analyzes data from a fundamentally different perspective. TDA represents datasets as geometric objects and provides dimensionality reduction techniques that project such objects onto low-dimensional descriptors. The key properties of these descriptors (also known as topological features) are that they provide multiscale information and that they are stable under perturbations (e.g., noise, translation, and rotation). In this talk, we review key mathematical concepts and methods of TDA and present different applications in chemical engineering. Specifically, we discuss how to use TDA techniques to conduct real-time process monitoring, to design chemical sensors, and to quickly screen solvents for biomass processing.

634106 Keynote Talk - Platforms and Algorithms for Digitally-Enabled Next-Gen Manufacturing

Abstract:

Current initiatives calling for a significant increase in the degree of computing and communications networking technologies used to automate and optimize manufacturing can be traced to the 2011 – 2015 time period. The phrase “Industrie 4.0” was coined Wolfgang Wahlster and first used at the 2011 Hannover Fair in Germany (Kagermann, et al. (2011)). The phrase “Fourth Industrial Revolution” was defined in 2015 by Klaus Schwab (Schwab (2015)) and was used as the theme for the 2016 World Economic Forum Annual Meeting at Davos.

During the same time period, operating companies, notably ExxonMobil, issued a call for fundamental changes to the state-of-the-art of industrial control systems to address “pain points” including (1) a lack of interoperability between systems from different suppliers, (2) unacceptably high
costs for upgrades or replacements compared to other computing systems, and (3) restrictions on business value generation caused by barriers to insertion of third-party technology innovations (Bartusiak, et al. (2015)). This call to action resulted in the establishment of the Open Process Automation Forum of The Open Group in 2016 that is working to establish the industry standards for an open, secure, interoperable process automation architecture (https://www.opengroup.org/forum/open-process-automation-forum). The intent of the Open Process Automation Standard is to solve the identified industrial control system pain points and to enable realization of the vision of Industrie 4.0 and the Fourth Industrial Revolution for manufacturing.

Central to the Open Process Automation and Industrie 4.0 visions is a step-change increase in the pace of innovation with technologies including applied artificial intelligence, big data, and data analytics applied to manufacturing compared to the *status quo* with current industrial control system products that are of a closed, proprietary nature. Explicitly, one of the intentions of Open Process Automation is to increase opportunities for academic researchers and other technology providers to get their research transferred to industrial practice by reducing commercial barriers and the time it takes to deploy new technology.

In this talk, we will provide updates on the status of the Open Process Automation Standard (O-PAS), industrial prototypes of O-PAS based systems (Bartusiak, et al. (2021)), O-PAS conformance certification, and the state of Open Process Automation business ecosystem building. This will include guidance to academic researchers on how they can leverage the O-PAS Standard to facilitate the adoption of their research. We will comment on the relationships among Open Process Automation, Industrie 4.0, and similar initiatives -- such as the Clean Energy Smart Manufacturing Innovation Institute (CESMII) (Edgar and Pristikopoulis EN. (2018)), Manufacturing x Digital (MxD) (https://www.mxdusa.org), NAMUR Open Architecture (Klettner, et al. (2017)), and NAMUR Module Type Package (Bernshausen, et al. (2016)). Finally, we will comment on our outlook for new process systems engineering technologies -- including applied artificial intelligence, machine learning, big data, and data analytics -- to enable next-gen manufacturing efficiency.

References

Bartusiak RD, Bitar S, DeBari DL, Patin DL, Balsubramanian M, Foltz D, Caro RL, Brandl DL. (2015), "Functional Requirements for an Open Architecture Industrial Control System: Part 2 – What We Want". *ExxonMobil Research and Engineering Technical Report.*

Bartusiak RD, Bitar S, DeBari DL, Houk BG, Stevens D, Fitzpatrick B, Sloan P. (2021). "Open Process Automation: A standards-based, open, secure, interoperable process control architecture". *Proc. 2021 IFAC World Congress.*

Bernshausen J, Haller A, Holm, T, Hoernicke M, Obst, M, Ladiges J. (2016). "Namur Modul Type Package -- Definition". *atp-magazin* 58(1-2):72-81.

Edgar TF, Pistikopolous EN. (2018). "Clean Energy Smart Manufacturing Innovation Institute", Comp. Chem. Eng. 114:130-144.

Kagermann H, Lukas W-D, Wahlster W. (2011). "Industrie 4.0". VDI nachrichten. https://web.archive.org/web/20130304101009/http://www.vdi-nachrichten.com/artikel/Industrie-4-0-Mit-dem-Internet-der-Dinge-auf-dem-Weg-zur-4-industriellen-Revolution/52570/1

Klettner C, Tauchnitz T, Epple U, Northdurft L, Diedrich C, Schroeder T, Grossmann D, Banerjee S, Krauss M, Iatrou C, Urbas L. (2017). "Namur Open Architecture". *atp-magazin* 59(1-2):20-37.

Schwab K. (2015). "The Fourth Industrial Revolution". *Foreign Affairs.* https://www.weforum.org/agenda/2016/01/the-fourthindustrial-revolution-what-it-means-and-how-to-respond/

623538 High-Throughput Measurement of a Machine Learning Model for Polyester Biodegradation

Abstract:

The U.S. generates more than 35 thousand tons of plastic waste every year and less than 20% is recycled, leading to an excessive amount of plastic accumulating in landfills. Many used plastics never make it to controlled end-of-life locations and are left in the environment to degrade, which can take more than a century. There is a dire need for degradable alternatives that can compete with current everlasting plastics. Currently, the study of compostable and biodegradable candidates is a low-throughput process, in part due to the present methods and standards to study biodegradation which require testing times of multiple months and capital-intense measurement equipment. These methods substantially reduce the size of data sets on polymer biodegradation, limiting our ability to develop group contribution theories and data-driven models to predict biodegradation performance.

Here, we combine high throughput synthesis, high throughput biodegradation assays, and machine learning to produce a predictive model for polyester biodegradability. This study utilized the clear zone test to obtain biodegradation rates of hundreds of polyesters using a sample of different bacterial strains. For the clear zone test, polymer is homogeneously dispersed in media, and its degradation during bacterial growth is monitored optically. This biodegradation test was applied to a library of hundreds of polyesters representing diverse synthetic routes and monomer functionalities, including biobased and synthetic systems to generate a large data set for machine learning. Different representations for polymer chemical structure and machine learning algorithms were explored in order to identify key chemical features that promote degradation and to identify models with the greatest predictive capacity.

632907 Buffering Effects on the Solution Behavior and Hydrolytic Degradation of Poly(β-amino ester)s

Abstract:

Poly(b-amino ester)s (PBAEs), cationic polymers originally developed as gene therapy vectors, have found their way into increasingly diverse applications, owing to their simple and efficient synthesis, facile incorporation of a variety of functional groups, and highly tunable hydrolytic degradation. The protonation of tertiary amines on each repeat unit along the backbone yields pH-responsive solution properties and degradation profiles, as well as a high buffering capacity. However, the opposing effects of pH on PBAE solubility and hydrolysis have led to conflicting reports of acidic conditions accelerating degradation in some cases and slowing degradation in others. Therefore, to unify understanding of PBAE degradation and solution properties, we studied PBAE hydrolysis as a function of pH and buffer concentration. We observed that, at high buffer concentrations, well above the molar concentration of PBAE amines, the solution pH remains constant throughout the degradation process. In these high buffer concentration conditions, increasing pH accelerated hydrolytic degradation. Yet, it is common to employ PBAE-based materials in 1X phosphate-buffered saline (PBS) to mimic physiological solutions, and 1X PBS contains just 10 mM phosphate buffer. When the concentration of PBAE amines exceeds that of buffer, the polymer and its hydrolytic degradation essentially control solution pH. Regardless of the original pH of a 10 mM buffer, dissolution of 20 mg/mL PBAE (corresponding to 60 mM amines) afforded a solution pH of 8, which decreased as PBAE hydrolysis evolved acid. These studies showcase the role of buffering on the pH-dependent degradation and solution properties of PBAEs, providing guidance for programming degradation of these polymers for applications ranging from drug delivery to thermosets.

623563 Reducing Moisture Sensitivity of Protein-Based Thermosets through Protein Charge Modification and Melt Polymerization with Hydrophobic Monomers

Abstract:

Proteins are promising renewable feedstock for materials manufacture due to their large volumes produced in agricultural waste streams. The extensive intermolecular interactions and highly hydrogen bonded nature of proteins can be leveraged to provide mechanical reinforcement in engineering plastics. Proteins have a wide range of reactive groups amenable to chemical modification, enabling strong and tough protein copolymers to be prepared via various graft polymerization strategies. However, as proteins have high softening temperatures and limited solubility in organic solvents, many strategies were limited to aqueous-based chemistries and water soluble copolymers. Previously, this challenge was addressed by using a surfactant as both compatibilizer and plasticizer, allowing protein macromonomers to be blended with hydrophobic acrylate comonomers and copolymerized solvent-free using thermoplastic processing techniques. However, proteins are hydrophilic and are easily plasticized by water, leading to materials with low extensibility in dry conditions and low mechanical strength at high humidity levels. This critically limits the application of protein-based materials to a limited humidity range where mechanical properties are optimal.

Here, we investigate the influence of charged protein functional groups on humidity sensitivity, and demonstrate the reduction of water uptake through the copolymerization of modified whey proteins with the hydrophobic n-butyl acrylate. A combination of esterification and acetylation reactions superneutralize the protein by installing uncharged ethyl esters and amide groups onto carboxylic acids and amines respectively. Superneutralization reduced water solubility of the protein, and resulted in a 46% reduction in water absorption at 90% relative humidity, compared to unmodified protein. Incorporation of the charge modified proteins into a copolymer is enabled by the flexible nature of the employed surfactant compatibilization strategy, where a range of ionic surfactants were shown to be effective at compatibilizing the otherwise immiscible protein-monomer mixtures. Subsequent methacrylation allows the protein to be incorporated into protein-polyacrylate network, where the hydrophobic polyacrylate further reduced moisture absorption. The lowered water uptake of superneutralized proteins is reflected in the copolymer mechanical properties, which remain stable across a wider humidity range compared to non-charge modified and supercharged protein copolymers. Structure-property relationships were further studied using Fourier-transform infrared spectroscopy and atomic force microscopy. Overall, this work provides a strategy to address the performance variability of protein-based materials due to changes in humidity.

628603 Growing Silk Fibroin in Advanced Materials for Food Security

Abstract:

In this presentation we provide an overview of the micro-/nanofabrication methods developed for structural biopolymers, highlighting recent advances in the rapid and ease construction of complex and multifunctional silk fibroin-based devices by integrating top-down manufacturing with bottom-up molecular self-assembly. Of particular interest is the development of a new nanofabrication strategy that employs templated crystallization to direct silk fibroin folding and assembly from a suspension of disordered, random coil molecules to ordered, hierarchical mesostructured materials. Such advancements in structural biopolymers fabrication provide the basis for engineering a new generation of technical materials that can be interfaced with food and plants. In particular, we will present newly developed techniques to direct the assembly of structural proteins into nanostructured materials that can serve as: edible coatings to prolong the shelf-life of perishable food, microenvironments to boost seed germination in saline soil and injectors to deliver payloads in plant vasculature. These examples will provide an opportunity to discuss how the establishment of a successful interface between biomaterials and plants tissues requires the development of a basic scientific knowledge on: mechanics of disorder to order transitions in proteinaceous materials during condensation phenomena, fluid mechanics and transport phenomena in plants vasculature, and swelling of porous materials exposed to plant fluids.

631671 Engineering Charged Silk Conjugates for Ionically Crosslinked Hydrogels

Abstract:

Silk fibroin extracted from the fibers of the *Bombyx mori* silkworm's cocoon is a high molecular weight protein whose versatility has encouraged its application to address a wide variety of biomedical challenges. *B. mori* fibroin consists of hydrophilic regions (blocks) that can be modified using a variety of aqueous and organic phase chemistries, as well as repeating hydrophobic regions (blocks) that associate by hydrogen bonding. This hydrogen bonding serves to physically cross-link the polymer chains, results in the formation of beta sheet secondary structures, and provides unique opportunity to tune the mechanical and degradation properties of the protein. In addition to its tunable properties, silk fibroin is biocompatible in a variety of biomedical applications in formats that include films, scaffolds, and hydrogels. The existing silk hydrogels are often formed by covalent cross-linking through dityrosine bond formation or by induction of the beta sheets, neither of which are reversible under cell compatible conditions. The goal of this work is to synthesize electrostatically crosslinked hydrogels using charged silk conjugates and to determine the extent to which the cross-linking process is reversible. This work utilizes an efficient synthetic route to carboxylate silk fibroin, a reaction that generates negatively charged silk conjugate materials. The reaction employed succinic anhydride in the presence of urea to modify the hydroxyl groups of the serine and threonine residues and amine groups of the lysine, arginine, and histidine residues, resulting in the enrichment of carboxylic acid functional groups. The addition of urea during the reaction led to a two-fold increase in the degree of protein carboxylation compared to existing methods, which is attributed to the disruption of hydrogen bonding in the hydrophobic sections of the protein. The increase in charged groups enabled the preparation of hydrogels via electrostatic interactions with oppositely charged molecules. Amine containing molecules (Jeffamine D and ED series) with varying molecular weights were mixed with carboxylated silk in a 1:1 molar charge ratio, which resulted in the formation of hydrogels via electrostatic crosslinking. The gelation kinetics were very slow for 1 w/v% carboxylated silk solution but were found to increase with increasing silk concentration up to 8 w/v%. Because the electrostatic interactions are non-covalent, this mode of crosslinking also offers the opportunity to prepare silk-based hydrogels that are reversible. The reversibility of gels was investigated by adding excess amine solution to hydrogels. The shear thinning behavior of the electrostatically crosslinked hydrogels was quantified by using dynamic shear rheology. The viability of human mesenchymal stem cells (hMSCs) encapsulated within the hydrogel was >90% as determined by Nucblue live reagent staining under fluorescence microscopy. The high viability indicated that the electrostatic network formed between carboxylated silk and Jeffamine ED showed no toxicity toward cells. Our ongoing work focuses on investigating reversibility through rheology and measuring degradation behavior of the silk hydrogels. Tuning the electrostatic charges on silk fibroin thus provides a viable route to synthesize electrostatically cross-linked hydrogels suitable for encapsulating hMSCs for soft tissue engineering applications.

627175 Renewable Polymers Via Direct Functionalization of Lignocellulosic Sugars

Abstract:

The need for sustainable polymers has driven the development of an array of chemical and biological pathways to potential monomers derived from renewable sources.[1-5] Despite decades of innovation, low conversion efficiencies from raw lignocellulosic feedstocks persist.[6] Here we present a new class of renewable polymer precursors that can be produced from the hemicellulose fraction of lignocellulosic biomass in high yields using a previously described aldehyde-facilitated biomass pretreatment.[7–9] By leveraging the use of aldehydes with different functionalities, we can install targeted functional handles onto both sides of the hemicellulose-derived xylose—creating a tricyclic, difunctional structure that presents as a potential renewable monomer for high-glass transition bioplastic synthesis. Here, we demonstrate the use of an easily bio-sourced carboxylic acid-functionalized aldehyde (glyoxylic acid) to produce a novel, tricyclic, diacid precursor from xylan in unmodified biomass in 83% yield and directly from commercial xylose in up to 95% yield. Polycondensation of the corresponding diester with a range of diols yielded a family of amorphous polyesters with tough mechanical properties, high glass transition temperatures and good gas barrier properties. Selective chemical recycling of the polyesters was possible at high yields even from a mixed plastic waste stream via alcoholysis and hydrolytic stability studies reveal that the polyesters will break down in neutral water at room temperature after an initial lag period of 77 days, during which time the polyesters remain mechanically useful. Due to the transformation's simplicity, techno-economic analysis shows that even production from commercial xylose could be cost competitive with fossil alternatives. Finally, we explore polymer chemistries beyond polyesters using alternative functionalities to carboxylic acids to expand material production opportunities from lignocellulosic biomass.

References:

1. Zhu, Y., Romain, C. & Williams, C. K. Sustainable polymers from renewable resources. *Nature* **540**, 354–362 (2016).

2. Zhang, X., Fevre, M., Jones, G. O. & Waymouth, R. M. Catalysis as an Enabling Science for Sustainable Polymers. *Chem. Rev.* **118**, 839–885 (2018).

3. Vilela, C. *et al.* The quest for sustainable polyesters – insights into the future. *Polym. Chem.* **5**, 3119–3141 (2014).

4. Isikgor, F. H. & Becer, C. R. Lignocellulosic biomass: a sustainable platform for the production of bio-based chemicals and polymers. *Polym. Chem.* **6**, 4497–4559 (2015).

5. Grignard, B., Gennen, S., Jérôme, C., Kleij, A. W. & Detrembleur, C. Advances in the use of CO_2 as a renewable feedstock for the synthesis of polymers. *Chem. Soc. Rev.* **48**, 4466–4514 (2019).

6. Hillmyer, M. A. The promise of plastics from plants. *Science* **358**, 868–870 (2017).

7. Shuai, L. *et al.* Formaldehyde stabilization facilitates lignin monomer production during biomass depolymerization. *Science* **354**, 329–333 (2016).

8. Questell-Santiago, Y. M., Zambrano-Varela, R., Talebi Amiri, M. & Luterbacher, J. S. Carbohydrate stabilization extends the kinetic limits of chemical polysaccharide depolymerization. *Nature Chemistry* **10**, 1222–1228 (2018).

9. Lan, W., Amiri, M. T., Hunston, C. M. & Luterbacher, J. S. Protection Group Effects During α,γ-Diol Lignin Stabilization Promote High-Selectivity Monomer Production. *Angewandte Chemie International Edition* **57**, 1356–1360 (2018).

630752 Grazing-Incidence Diffraction Reveals Cellulose Lattice Contraction with Dehydration of Plant Primary Cell Wall

Abstract:

Cellulose, a major structural component of plant cell wall, is an essential feedstock biopolymer in industries such as paper, textiles, and renewable energy. It natively exists in the form of semi-crystalline cellulose microfibrils, which are important in the mechanics of primary cell wall expansion. Characterization of the crystalline properties of cellulose is crucial in understanding plant growth, as well as the degradation of cellulosic biomass for industrial applications. The plant primary cell wall is highly hydrated in its native state, yet most X-ray scattering studies of this system have been conducted in the dried state. To study primary cell walls under hydrated conditions, we employ grazing-incidence wide-angle X-ray scattering (GIWAXS) with a humidity chamber. GIWAXS near the critical angle for total external reflection provides enhanced scattering and signal-to-noise, while a humidity chamber keeps the sample hydrated with minimal interference from water. Onion outer epidermal peels were used as model system for primary cell walls. After sample dehydration, we observed cellulose crystal lattice contraction and changes in the relative intensities of diffraction peaks. We attribute these changes to the removal of water that, while in the hydrated state, interacts with the hydrogen bonding network within the cellulose crystal. Density functional theory (DFT) models of hydrated and dry cellulose microfibrils qualitatively support our hypothesis. These results elucidate the influence of water on native cellulose crystal structure and by extension, the material properties of primary cell walls.

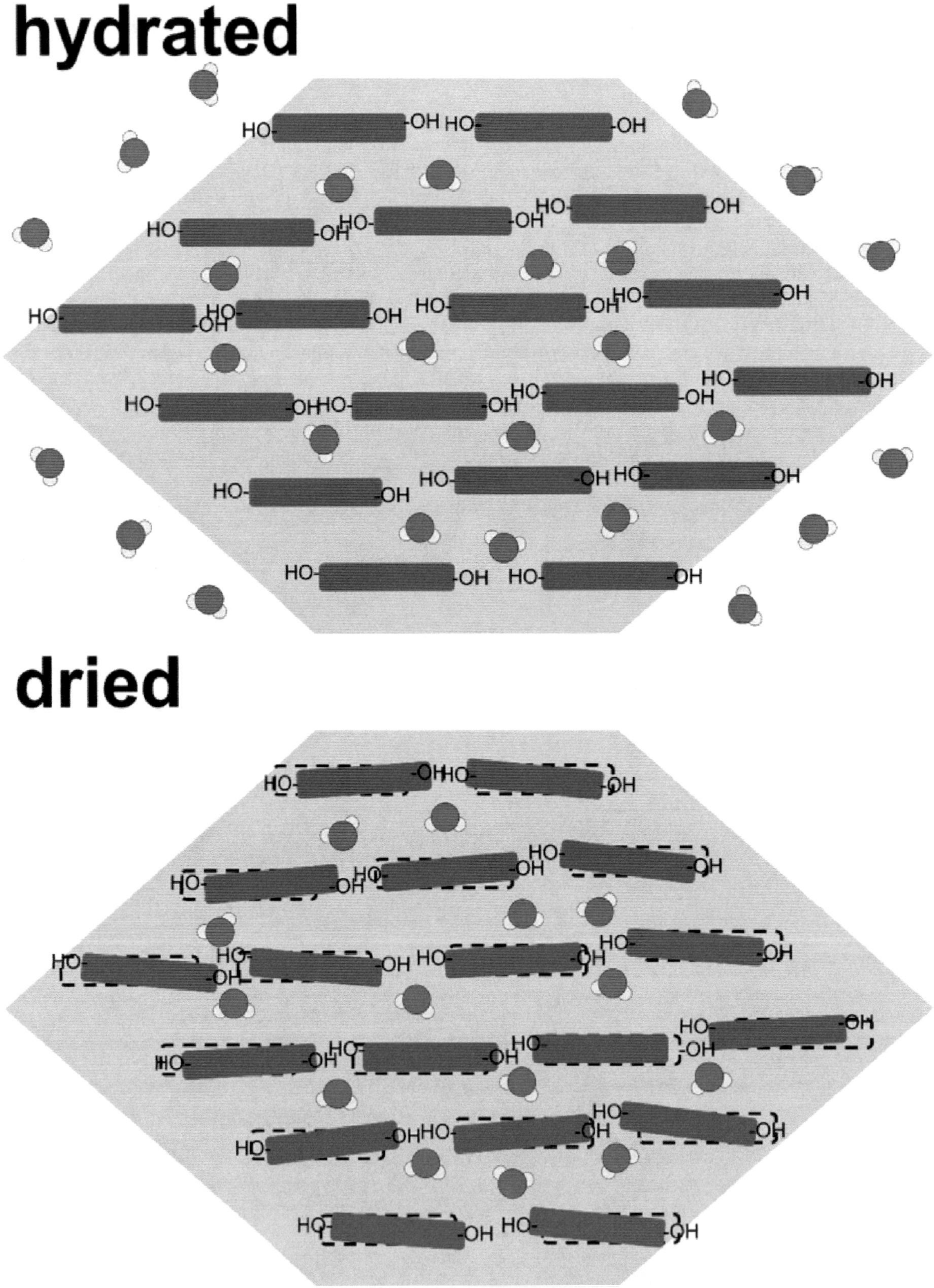
hydrated
HO-
-OH
dried

626851 Golem: A Probabilistic Approach to Robust Experiment and Process Optimization Based on Regression Trees

Abstract:

Numerous challenges in science and engineering can be framed as optimization tasks, including the maximization of reaction yields, the optimization of molecular and materials properties, and the fine-tuning of automated hardware protocols. Data-driven optimization strategies are often adopted to solve these tasks efficiently. Increasingly, these strategies are coupled with automated hardware to enable autonomous experimental platforms. The majority of the sequential model-based strategies used, however, do not consider robustness against the variability of experiment and process conditions. In fact, it is generally assumed that these parameters are exact and reproducible. Yet some experiments may have considerable noise associated with some of their conditions, and process parameters optimized under precise control may be applied in the future under variable operating conditions. In either scenario, the optimal solutions found might not be robust against input variability, affecting the reproducibility of results and returning suboptimal performance in practice. In this talk, I will introduce *Golem*, an algorithm that is agnostic to the choice of experiment planning strategy and that enables robust experiment and process optimization. *Golem* identifies optimal solutions that are robust to input uncertainty, thus ensuring the reproducible performance of optimized experimental protocols and processes. It can be used to analyze the robustness of past experiments, or to guide experiment planning algorithms toward robust solutions on the fly. We assess the performance and domain of applicability of *Golem* through extensive benchmark studies and demonstrate its practical relevance by optimizing an analytical chemistry protocol under the presence of significant noise in its experimental conditions.

Reference: Aldeghi, M., Häse, F., Hickman, R. J., Tamblyn, I. & Aspuru-Guzik, A. Golem: An algorithm for robust experiment and process optimization. *arXiv:2103.03716* (2021).

628299 Learning Based Scheduling of Industrial Hybrid Renewable Energy Systems

Abstract:

In order to reduce the global CO_2 emissions for low-carbon development while improving the efficiencies of operation, energy systems have started undergoing major transitions to facilitate the broad integration of distributed renewable energy technologies. The inclusion of renewable energy technologies to a typical fossil-fuel based energy source have gradually transformed the concept of centralized generation into distributed energy systems. The choice of optimal power contribution among the available hybrid sources often depends on several factors such as the time of use (TOU) price of the grid electricity, renewable energy generation forecast, electricity demand estimation etc.

Due to the high intermittent nature of renewable energy sources, the system complexity increases, thereby making the optimal decision making more challenging using the existing physics-based model techniques. The inevitable uncertainties related to energy sector such as climate changes, energy market fluctuations, and energy technology developments often demand complex physics-based model and excessive computational time especially for control applications. The physics-based energy scheduling approach generally solves an optimization problem in background to decide the optimal power mix among the available power sources. However, model dependencies, parameter uncertainties and multi-dimensional impact of the embedded uncertainties impose the decision making to be more problematic, especially when such an approach is extended to large-scale and complex energy systems.

The utilization of machine learning techniques in the energy sector has gradually attracted researchers' attention for the effective management of the aforementioned complexities due to their purely data-driven nature. Reinforcement learning (RL), which is a branch of machine learning, can adopt model-free approaches for scheduling under uncertainties by employing agents to participate in the process of interest. The RL agents make control decisions and maximize its total reward by interacting with the simulation environment and evaluating the series of actions taken.

In this study, a data-driven approach that adopts RL techniques to determine the schedule of power mix in a hybrid renewable energy system (HRES) is presented, with a case study on the highly energy intensive chlor-alkali process. This process involves electrolysis of sodium chloride solution to produce chlorine as the primary product together with hydrogen and sodium hydroxide (caustic soda) solution as the secondary products. Almost 50-70% of the overall cost of operating a chlor-alkali process comes from electricity usage, thereby making the process highly sensitive to electricity price trends. To combat the increased grid dependency of such a highly energy intensive process and to achieve profitability, a grid connected hybrid energy system comprising of solar PV panels, wind turbines and fuel cells are considered that supplies power to the chlor-alkali process.

We implement Proximal Policy Optimization (PPO), an Advantage Actor-Critic RL method that enable samples to be trained for multiple epochs of mini-batch updates, to effectively distribute energy among the various available sources without having access to the dynamics of the system. In our exclusive experiments, the RL based scheduling provides efficient operational strategies with saving in cost and show nearly on-par performance with the optimal solution. For stochastic tasks under uncertainties, our RL scheduling approach demonstrates strong superiority over the conventional optimization. This framework can generally be used in any data-driven scheduling problems taking the advantages of its model-free property, and can be generalized to multiple task scenarios via transfer learning.

634107 Keynote Talk: Modelling and Monitoring with Dynamic Auto-Regressive Latent Variable Methods

Abstract:

With the advent of Industry 4.0, industrial data can be collected at very high sampling rates, leading to more complicated relation between process and quality variables[1, 2]. To extract valuable information from these variables and avoid the curse of dimensionality, multivariate latent variable methods (LVM) are widely used, such as partial least squares (PLS)[3], canonical correlation analysis (CCA)[4] and latent variable regression (LVR)[5]. PLS and CCA are employed to construct the relations between process variables **X** and quality variables **Y** by maximizing their covariance and correlation respectively, while LVR aims to maximize the projection of **Y** in the latent space. Several variants of LVM are proposed to achieve better modeling and monitoring performance, including concurrent ones[6-8] and nonlinear versions[9, 10].

The aforementioned models obtain satisfactory modeling and monitoring performance; however, only static variations are considered in these models, and this static assumption constrains their applicability in actual dynamic systems. Several dynamic variants were proposed to extract temporal information from the data. For instance, augmented matrices were formed to include lagged process and quality variables in the work of Ku et al.[11] and Qin and McAvoy[12]. However, it is hard to interpret their latent structures because of the augmentation matrices. To address this issue, dynamic inner LVM was proposed to construct dynamic outer model by maximizing the relations between current samples and a linear combination of past samples, and a dynamic inner model that is consistent with the dynamic outer model[13-16]. Dynamic weighted variants were also proposed in [1, 17] to reduce the redundant information between adjacent historical samples by introducing basis functions. For aforementioned dynamic variants of LVM, only dynamic cross-correlations between the current quality variables and past process and quality variables are considered, leaving large portion of valuable quality information unexploited.

In this work, a series of dynamic auto regressive LVM (DALVM) algorithms are proposed to fully exploit the valuable information of both process and quality variables. DALVM aims to maximize the dynamic relations between current quality samples and past process and quality samples, and to achieve consistency, an auto-regressive exogenous (ARX) model is designed in the inner model. In DALVM, in addition to the auto-regressive relations of quality variables, the cross-correlations between **X** and **Y** are also constructed. Further, aiming for comprehensive modeling and monitoring, subsequent decompositions are proposed to decompose the process and quality variables concurrently into seven dynamic and static subspaces, and the corresponding monitoring scheme is designed for all subspaces. Two industrial dataset, Tennessee Eastman process and the three-phase flow facility, are used to illustrate the advantages of the proposed algorithms.

References

[1] Zhu, Q., 2020. Auto-regressive modeling with dynamic weighted canonical correlation analysis. Journal of Process Control, 95, pp.32-44.

[2] Xu, B. and Zhu, Q., 2020. Online quality-relevant monitoring with dynamic weighted partial least squares. Industrial & Engineering Chemistry Research, 59(48), pp.21124-21132.

[3] Geladi, P. and Kowalski, B.R., 1986. Partial least-squares regression: a tutorial. Analytica chimica acta, 185, pp.1-17.

[4] Hardoon, D.R., Szedmak, S. and Shawe-Taylor, J., 2004. Canonical correlation analysis: An overview with application to learning methods. Neural computation, 16(12), pp.2639-2664.

[5] Zhu, Q., 2020. Latent variable regression for supervised modeling and monitoring. IEEE/CAA Journal of Automatica Sinica, 7(3), pp.800-811.

[6] Qin, S.J. and Zheng, Y., 2013. Quality-relevant and process-relevant fault monitoring with concurrent projection to latent structures. AIChE Journal, 59(2), pp.496-504.

[7] Zhu, Q., Liu, Q. and Qin, S.J., 2017. Concurrent quality and process monitoring with canonical correlation analysis. Journal of Process Control, 60, pp.95-103.

[8] Zhang, H. and Zhu, Q., 2021. Multi-layer fault monitoring with concurrent kernel latent variable regression. Submitted to Industrial & Engineering Chemistry Research.

[9] Melzer, T., Reiter, M. and Bischof, H., 2003. Appearance models based on kernel canonical correlation analysis. Pattern recognition, 36(9), pp.1961-1971.

[10] Zhang, Y., Zhou, H., Qin, S.J. and Chai, T., 2009. Decentralized fault diagnosis of large-scale processes using multiblock kernel partial least squares. IEEE Transactions on Industrial Informatics, 6(1), pp.3-10.

[11] Ku, W., Storer, R.H. and Georgakis, C., 1995. Disturbance detection and isolation by dynamic principal component analysis. Chemometrics and intelligent laboratory systems, 30(1), pp.179-196.

[12] Qin, S.J. and McAvoy, T.J., 1996. Nonlinear FIR modeling via a neural net PLS approach. Computers & chemical engineering, 20(2), pp.147-159.

[13] Dong, Y. and Qin, S.J., 2018. Dynamic latent variable analytics for process operations and control. Computers & Chemical Engineering, 114, pp.69-80.

[14] Dong, Y., Liu, Y. and Qin, S.J., 2019. Efficient dynamic latent variable analysis for high-dimensional time series data. IEEE Transactions on Industrial Informatics, 16(6), pp.4068-4076.

[15] Zhu, Q., Qin, S.J. and Dong, Y., 2020. Dynamic latent variable regression for inferential sensor modeling and monitoring. Computers & Chemical Engineering, 137, p.106809.

[16] Qin, S.J., Dong, Y., Zhu, Q., Wang, J. and Liu, Q., 2020. Bridging systems theory and data science: A unifying review of dynamic latent variable analytics and process monitoring. Annual Reviews in Control.

[17] Xu, B. and Zhu, Q., 2020. Online Quality-Relevant Monitoring with Dynamic Weighted Partial Least Squares. Industrial & Engineering Chemistry Research, 59(48), pp.21124-21132.

622388 Phase Separation Kinetics in Liquid-Liquid Extraction and Settler Design

Abstract:

The kinetics of droplet swarm sedimentation and coalescence in liquid/liquid phase separation systems was investigated. Target of this study was to develop an appropriate rate model for settler design. In a first series of batch phase separation experiments the system MIBK/water was investigated. The system properties density, viscosity and surface tension were adjusted by adding PEG 4000 (5 wt% to 40 wt%) and sodium chloride (0.05 mol L^{-1} to 1 mol L^{-1}) to the aqueous carrier phase. Additionally, the system cyclohexanone/water was investigated. Latter system differs from the first series as far as interfacial tension and viscosity increase with increasing amount of PEG 4000, while in the first series the interfacial tension decreases with increasing amount of PEG 4000. Phase separation was monitored by an optical system and by ultrasonic scanning. Since any of the phase separation processes is nonlinear, a modified 1st order rate model, shown in equation 1, was applied to fit the experimental data.

From single droplet-interface coalescence experiments Cockbaine et al [1] already concluded that the coalescence rate follows 1st order decay. They assumed that coalescence delay is determined by the rate of film drainage. They proposed separation mechanisms but did not discuss settler design. Detailed investigation of droplet-droplet coalescence was performed by Kamp [2] and Villwok [3]. These studies mainly focused on droplet-droplet coalescence mechanisms, but also did not discuss settler design.

Based on the settler balance shown in equation 2, and the modified 1st order rate model for nonlinear droplet swarm sedimentation and droplet coalescence, the design of continuous settlers, actually the settler area, can be performed.

The modified 1st order rate model can address nonlinear droplet swarm sedimentation and droplet-bulk coalescence with two kinetic parameters, the preexponential (b) and the rate constant (k). Basically these parameters are just fit parameters, deduced from batch phase separation experiments. To find a link to the basics of phase separation in liquid-liquid extraction the parameters were compared with the physical properties of the investigated systems. The model parameters compare well with the density, viscosity and interfacial tension. Figure 1 representatively shows the phase separation results for the separation of the system cyclohexanone/water with 15 wt% PEG 4000. The figure shows the experimental data and the modeled separation curves. Figure 2 shows the correlation of the half-life time for coalescence of all experiments with the physical properties.

The results of this study, covering a total number of 21 experiments with a density span of the aqueous phase from 1013 to 1076 kg/m^3, a viscosity span from 3 mPas to 30 mPas and a span of the interfacial tension from 3 to 10 mN/m, show a simple correlation of the model parameters and the physical properties. The results also show a correlation of the rate of sedimentation and the rate of coalescence, confirming, that the modified 1st order rate model offers a simple and easy to apply tool in settler (area) design in liquid-liquid extraction. The results of droplet swarm sedimentation modeling have already been published [4].

Literature:

[1] Cockbain, E.G.; McRoberts, T.S. (1953), The stability of elementary emulsion drops and emulsions, J. Colloid Sci. 8, 440–451

[2] Kamp, J. (2017), Systematic coalescence investigations in liquid/liquid systems – From single drops to technical applications, PhD Thesis, Technische Universität Berlin

[3] Jörn Villwock (2019), Systematische Analyse des Koaleszenzverhaltens von zweiphasigen Flüssigsystemen bei Ionenzugabe, PhD Thesis, Technische Universität Berlin

[4] Pornprapa Bol, Georg Rudelstorfer, Annika Grafschafter, and Matthaeus Siebenhofer, (2021), The Kinetics of Droplet Sedimentation in Liquid-Liquid Extraction, Chem. Ing. Tech. 2021, 93, No. 1–2, 260-272

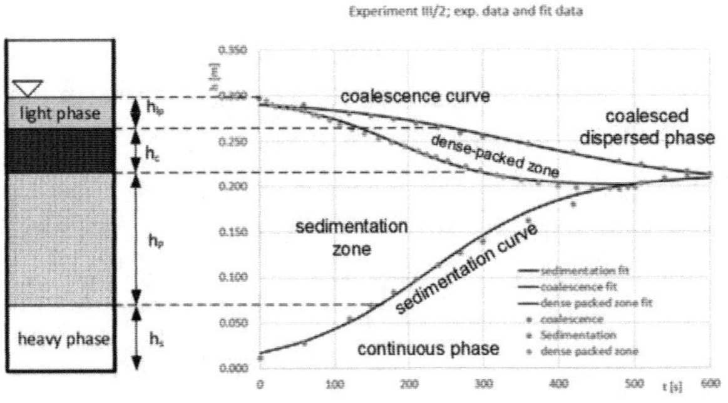

Figure 1: Height of the sedimentation zone, the dense packed zone and the coalescence zone for experiment III/2 (cyclohexanone/water plus 15 wt% PEG 4000; T = 293 K). The experimental data and the modeled data are compared. Abbreviations: h_S: settling height (clear continuous phase), h_P: dispersion height; h_C coalescence height and h_{lP}: height of the coalesced (clear) light phase.

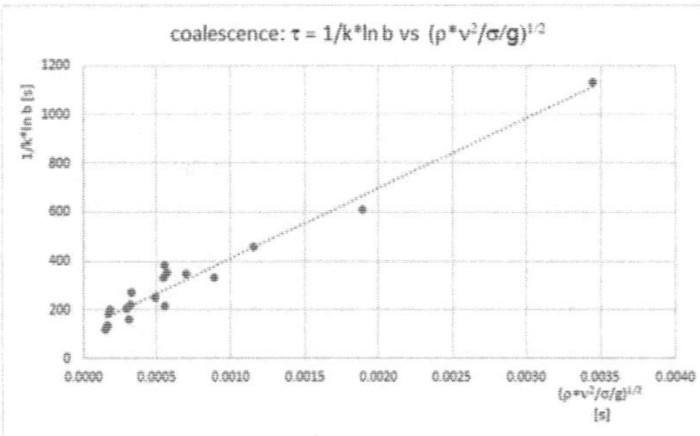

Figure 2: Correlation of the half-life time of coalescence of all phase separation experiments at T = 293 K. Abbreviations: ρ: density of the aqueous phase [kg/m³]; ν: kinematic viscosity of the aqueous phase [m²/s]; σ: interfacial tension [N/m] and g: acceleration due to gravity [m/s²].

$$w(h_S) = \frac{dh_{S(t)}}{dt} = \frac{h_{max}*b*k*\exp(-k*t)}{(1+b*\exp(-k*t))^2}$$

Equation 1 - ($w(h_S)$: settling rate [m s^{-1}]; h_{max}: height of the specific phase; b: fit parameter (with the half-life time τ = 1/k*ln b); k: rate constant [1 s^{-1}]; t: time [s];)

$$F_v * c_{dp} = (F_v + dF_v) * c_{dp} - w(h) * c_{dp} * dA$$

Equation 2 - F_V: volumetric flow rate [m³s^{-1}]; c_{dp}: dispersed phase fraction; A: settling area [m²]

626457 Three Phase Flow and Four Phase Flow in the Taylor Couette Disc Contactor for Combination of Liquid-Liquid Extraction and Chemical Reaction

Abstract:

Production processes of chemicals and drugs typically involve several synthesis steps with phase contact patterns of solids (S) gases (G) and liquids (S). The different phases can act as carrier, reactant, catalyst or solvent. Typically, continuously stirred tank reactors or batch reactors are used for multiphase processes. They are well investigated and adaptable to specific tasks, but also suffer from limited performance. For economic reasons the process design of continuous multiphase processes requires new flexible reactor concepts. The Taylor-Couette Disc Contactor (TCDC) [1] perfectly meets this requirements of intensive multiphase contact up to four phases (GLLS). The design of the TCDC allows continuous mixing and separation of four phases in one apparatus. Three phase operation was successfully validated with heterogeneously catalyzed esterification of acetic acid in aqueous solution with simultaneous extraction of the reaction product into a solvent phase. Simultaneous reaction and liquid-liquid extraction increases the total conversion of acetic acid from 15% (without extraction) to over 30%. The SLL concept may also be applied in hydrometallurgy for leaching combined with solvent extraction, or biotechnological applications. Chemical conversion plus liquid-liquid extraction in the TCDC is best modelled with a tank reactor (CSTR) cascade model. Meanwhile, four phase operation was successfully performed. The hydrodynamics, including dispersed gas phase holdup, solvent phase holdup, the flow regimes as well as axial backmixing and $k_L a$-determination, were investigated and compared with three phase contact hydrodynamics (LLS). The four phase process concept has not yet been confirmed with an appropriate experiment, but it may be useful in heterogeneously catalyzed gas-liquid reactions with phase transfer.

[1] A. Grafschafter, (2019), Simple and Effective: The Taylor-Couette Disc Contactor for liquid-liquid extraction (Design and Performance), Dissertation,

628004 New Developments in Membrane Solvent Extraction

Abstract:

Solvent extraction usually carried out by dispersing one phase as drops in the other phase needs phase coalescence after extraction; the devices used are mixer-settlers or tall columns in industrial operations. This dispersive operation requires a phase density difference and encounters problems of flooding, loading and low values of allowable phase flow rate ratios. Further, phase coalescence can be problematic especially if emulsion formation takes place. To bypass such problems, nondispersive solvent extraction via a porous hydrophobic membrane was developed sometime back. It is also called membrane solvent extraction (MSX). Here the organic phase flowing on one side of a porous hydrophobic membrane wetted the membrane pores whereas the aqueous phase flowing on the other side and not wetting the pores was maintained at the same or a higher pressure. There is a maximum allowable pressure difference after which aqueous phase breaks through and is dispersed in the organic phase; nondispersive operation is no longer possible. There are commercial units operating based on this concept. Countercurrent flow is used for high solute recovery in membrane solvent extraction as well. Inevitably there is significant pressure drop in the liquid flowing on each side of the membrane which may lead to or exceed maximum allowable phase pressure difference at both ends of the narrow flow channels encountered with hollow fiber membranes. To avoid such problems a new technique has been developed. We will describe this technique and illustrate membrane solvent extraction for different systems. A model developed will also be described.

629988 Oil Recovery from Dilute Oil-Water Mixtures Via Hydrophobic Hollow Fiber Membranes

Abstract:

Robust, high-efficiency oil-water separations are crucial to achieving the circular water economy. Previous investigation of oil permeation through hydrophobic hollow fiber membranes has demonstrated their ability to efficiently recover oil from oil-water solutions for extended durations. Furthermore, researchers have identified mechanistic competition between permeation and coalescence which results in transport phenomena that deviate from traditional pore flow models. These studies have generally focused on insoluble organic concentrations in excess of 1%. However, for many industrial wastewaters, oil concentrations may be substantially lower. This work investigates the oil flux and recovery behavior for dilute oil-water emulsions by examining the effects of both operating parameters (e.g., transmembrane pressure, influent flow rate) and solution properties (e.g., influent oil concentration, interfacial tension, viscosity, ionic strength). Of particular significance, we identify the presence of optimal and suboptimal operating ranges which ultimately highlights the competition between mass transfer resistances within the system.

630299 Cellulose Nanocrystals- and Lignin Magnetic Nanocomposites Enhance the Ethanol Extraction from Aqueous Solution Using Castor Oil As the Extractant

Abstract:

Ethanol use as a gasoline additive keeps rising in the USA, with approximately 15,000 million gallons consumed in 2019. Multiple separation processes are needed to achieve low-water content ethanol, and commonly distillation is performed, which requires multiple stages and high energy consumption. This work focuses on using liquid-liquid extraction at room temperature to separate ethanol from its aqueous solution using castor oil as the extractant phase. We chose castor oil due to its high ethanol solubility, immiscibility with water, low toxicity, low cost, high boiling point (~ 313 °C), and low heat capacity. Also, we investigated superparamagnetic iron oxide, Fe_3O_4, nanoparticles coated with kraft lignin or Fe_3O_4-coated cellulose nanocrystals, and used these as additives to investigate the potential enhancement of the ethanol extraction process by imposing external magnetic fields. Kraft lignin, a hydrophobic material with low molecular weight, high stability, and high strength material, was chosen to protect the Fe_3O_4 nanoparticles from oxidation and formed multicore magnetic structures. Similarly, cellulose nanocrystals (CNCs) provided a hydrophilic substrate with a unique aspect ratio that served as a platform capable of attaching multiple magnetic nanoparticles onto its surface. The Fe_3O_4 nanoparticles and the biobased nanocomposites were synthesized and coated using the co-precipitation method and characterized using Fourier Transform infrared spectroscopy, thermogravimetric analysis, X-ray diffraction, transmission electron microscopy, dynamic light scattering and zeta potential measurements, and vibrating sample magnetometry. For the liquid-liquid extraction experiments, different concentrations of ethanol aqueous solution (2%, 5%, 10%, 20%, and 30% wt.%) extracted by different ratios of castor oil (50%, 70% of total volume) were studied. Different amounts of nanoparticles (0, 0.15, 0.25, and 0.5 w/v%) added into the extraction system were studied. Ethanol aqueous solution and castor oil mixtures with nanoparticles were emulsified at 13,000 rpm for 5 minutes and kept at room temperature for 24 hours to reach equilibrium. A magnetic field (300 mT) was applied using a Neodymium rare-earth magnet to break-up the emulsions faster and separate the aqueous and the organic phase. After achieving phase separation, headspace gas chromatography (HS-GC-FID) was used to determine each phase's ethanol concentration. The liquid-liquid extraction experiments results showed that water/castor oil/ethanol emulsions were formed by adding modified-magnetic nanoparticles (lignin or CNCs). Due to the magnetic nanoparticles' presence, the emulsions were easier to separate into two phases under an external magnetic field. Our current studies revealed that the nanoparticles enhanced the extraction process by increasing the amount of ethanol transferred from the aqueous phase to the oil phase (50:50 v/v) by 33% compared to the liquid-liquid extraction, without the nanoparticles. Once ethanol transfers to the oil phase, and due to the boiling point difference between the castor oil and the ethanol, simple evaporation could lead to high ethanol purity levels.

624247 Replacement of Fluorinated Aqueous Fire-Fighting Foams (AFFF)

Abstract:

Fluorinated aqueous fire-fighting foams were developed to fight flammable liquid (Class B) fires. These foams contain per- and poly-fluoro alkyl substances such as perfluorooctanesulfonic acid (PFOS). The United States military was directed to replace these foams with non-fluorinated firefighting agents by October 1, 2023 as part of the 2020 National Defense Appropriations Act (NDAA). This presentation will provide an overview of the characteristics and performance of foams used to fight Class B fires under different conditions such as pooled liquids or tank fires, along with obstacles to the development of replacement foams matching the performance of fluorinated AFFF. Lastly, issues related to the environmental performance of the replacement foams will be discussed.

623036 Green and Sustainable Processes for Critical Metal Recovery Using Oxalate Chemistry

Abstract:

Green chemistry is the design of energy-efficient chemical processes and products with environmentally benign input materials while preventing waste generation. This presentation will describe sustainable metal extraction processes from three different types of mixed metal oxide materials: spent lithium-ion battery cathodes, bauxite, and ilmenite. Recycling of spent lithium-ion battery cathodes have the potential to meet the growing demand of two critical metals: lithium (Li) and cobalt (Co). The demand for both of these metals are exponentially increasing with the rapid growth in electric vehicles. A green, closed-loop, and sustainable process for recovery of both of these metals from lithium cobalt oxide ($LiCoO_2$) has been developed utilizing oxalate chemistry. In this process, the recovery and recycle of the aqueous acid through a novel ion-exchange process minimizes waste generation. The Li is separated in the form of lithium carbonate (Li_2CO_3) and the Co is separated either as cobalt oxide (Co_3O_4) or cobalt hydroxide ($Co(OH)_2$). The oxalate-based process has also been applied for refining of ores such as bauxite and ilmenite and recovery of commercially valuable metals such as aluminum (Al), iron (Fe) and titanium (Ti). These metals can be efficiently separated and recovered in the form of their metal hydroxides. This presentation will cover the experimental and mechanistic details along with a techno-economic analysis of our closed-loop processes with existing technologies.

626392 Evaluating Green Solvent Nanoparticle Production Via Supercritical Fluid Synthesis

Abstract:

Nanoparticles, materials with at least one dimension less than 200nm, have been the subject of increasing research interest due to unique nanoscale properties for a variety of applications. In addition to high surface-area-to-volume ratio, nanoparticle shape and nature of exposed surfaces impacts functionality with much research dedicated to controlling these nanoparticle features. Supercritical fluid (SCF) synthesis can utilize “green” (e.g., non-toxic, abundant, and renewably sourced) solvents, such as water or ethanol, by applying the unique properties of a solvent as it transitions from subcritical to supercritical state to specifically manipulate nanoparticle formation, with potential to control for size and shape. It is a relatively simple process with few steps and has been successfully scaled with supercritical water, enabling manufacture of 1000 tons of nanoparticles per year. While this water synthesis has shown to be successful, scaled synthesis with supercritical ethanol is nascent and may have potential benefits over water synthesis. For example, ethanol has a lower critical point than water so potentially lower energy requirements for similar SCF synthesis benefits. In addition, ethanol can act as a reducing agent at SCF conditions, enabling synthesis of pure metal nanoparticles, while water is oxidizing and mainly creates metal oxide nanoparticles unless substantial amounts of reducing agents are added. In this study, these potential benefits are evaluated with a life cycle perspective to weigh considerations with solvents, synthesis energy consumption, and nanoparticle quality. Here, knowledge of the full-scale supercritical water plant along with Aspen process models are applied to evaluate the life cycle assessment impacts of supercritical ethanol synthesis versus supercritical water synthesis at the lab scale and full production scale. This study informs solvent selection for supercritical fluid nanoparticle synthesis to make the best decision regarding environmental impacts and quality of nanoparticle product.

624346 Green Solvent Selection Using End-of-Life Metrics

Abstract:

Solvents are an often overlooked yet indispensable aspect of chemical reactions and processing. Improving solvent consumption and reuse/recycle has the opportunity to greatly decrease overall process waste production. Current green solvent selection guides either completely ignore solvent process impacts or lack a comprehensive metric for determining the role of solvent end-of-life options in their solvent ranking. End-of-life options include solvent recycle using separations to purify and reuse the solvent or disposal/incineration. In previous works the determination of solvent end-of-life impacts are either overly simplified or require rigorous computational analysis and/or significant process and thermophysical property data. The goal of this work is to provide a comprehensive yet simple way to compare solvent end-of-life options solely relying on basic chemical properties.

The end-of-life options that are compared in this work include recycling via distillation and membrane separations as well as incineration. For each option, the minimum energy of separation is determined and compared to the energy recovered via incineration as well as energy demand for solvent manufacture. This approach is used across a large range of solvents and solvent mixtures in order to help determine the lowest impact options for each solvent and help to evaluate green solvents for various process considerations.

624493 Emergy and SPI Assessment of Solvent Recovery Pathways

Abstract:

Introduction

Conventional waste solvent handling techniques such as incineration, onsite, and offsite disposal increase the process's ecological burden and carbon footprint (Cseri et al., 2018; Das et al., 2012; Oppelt, 1990; Riber, 2007). The solvent waste generation will continue to increase as global solvent consumption expands (Chea et al., 2020). Sustainability assessment has become fundamentally important to decision-makers due to the increased environmental concerns associated with waste solvent disposal. Therefore, the recovery and reuse of solvents from waste streams is paramount to improving the overall sustainability of industrial processes (Cavanagh et al., 2014; Slater et al., 2010, 2007).

Emergy and Sustainable Process Index (SPI) are sustainability metrics that have can be used to evaluate the energy efficiency and greenness of industrial processes (Brown and Ulgiati, 2002; Krotscheck and Narodoslawsky, 1996; Narodoslawsky, 2015; Ulgiati et al., 1994). Emergy, defined as the amount of available energy of one kind that is directly and indirectly used up in transformations to make a product, can help identify the major energy-based hotspots within a process. Emergy helps to account for all the work done by nature and human-economy using solar-emjoules (sej) as the basis for calculation. However, SPI is an ecological footprint that evaluates the environmental burden of processes. It estimates the total area needed by a process to produce a unit product or service. In sustainability assessment, it is important to have multiple indices, as a single metric is insufficient in capturing and addressing the various dimensions encompassing a process. Therefore, to offer a holistic sustainability analysis for solvent recovery processes, both SPI and Emergy, need to be integrated in the evaluations.

Methodology

In this work, the Emergy and SPI of solvent recovery framework has been evaluated using a superstructure-based optimization approach that captures all the complexities of the problem. Individual technologies were categorized into stages. The Emergy models include contents of the waste solvent, material of construction of technologies, labor, purchase cost of technologies, utilities, maintenance and overhead costs. The SPI models consist of areas needed for raw material consumption, energy consumption, infrastructure, staff accommodation, embedded air, water, and soil emissions into the ecosphere. The optimization was formulated as a mixed-integer non-linear programming (MINLP) problem. The objective was to simultaneously minimize the SPI, Emergy and Economics of the recovery pathway while meeting the exit stream's purity requirements.

Results & Discussion

The recovery framework was applied to an illustrative case study. In this case study the goal is to recover dimethoxyethane (DME) at a purity of 95% from a process waste stream consisting of 21.3% DME, 35.3% water, 41.3% toluene, and 1.3% 1-ethoxy-1-methoxy ethane (EME). The combined capital and operating cost of the recovery process is \$330,000 per annum (Chea et al., 2020). Figure 1 shows the results for the SPI and Emergy analysis for this representative case study. It can be observed that the major contribution to Emergy is the maintenance and overhead cost while the area needed to accommodate water emissions is the dominant footprint for the SPI analysis. The results support the need to have multiple metrics to help identify hotspots and implement corrective measures to reduce the impacts of the major contributing footprints. This was observed based on the representative case study

References

Brown, M.T., Ulgiati, S., 2002. Emergy evaluations and environmental loading of electricity production systems. J. Clean. Prod. 10, 321–334. https://doi.org/10.1016/S0959-6526(01)00043-9

Cavanagh, E.J., Savelski, M.J., Slater, C.S., 2014. Optimization of environmental impact reduction and economic feasibility of solvent waste recovery using a new software tool. Chem. Eng. Res. Des. 92, 1942–1954. https://doi.org/10.1016/j.cherd.2014.02.022

Chea, J.D., Lehr, A.L., Stengel, J.P., Savelski, M.J., Slater, C.S., Yenkie, K.M., 2020. Evaluation of Solvent Recovery Options for Economic Feasibility through a Superstructure-Based Optimization Framework. Ind. Eng. Chem. Res. 59, 5931–5944. https://doi.org/10.1021/acs.iecr.9b06725

Cseri, L., Razali, M., Pogany, P., Szekely, G., 2018. Organic Solvents in Sustainable Synthesis and Engineering, in: Green Chemistry. Elsevier, pp. 513–553. https://doi.org/10.1016/B978-0-12-809270-5.00020-0

Das, A., Gupta, A.K., Mazumder, T.N., 2012. A comprehensive risk assessment framework for offsite transportation of inflammable hazardous waste. J. Hazard. Mater. 227–228, 88–96. https://doi.org/10.1016/j.jhazmat.2012.05.014

Krotscheck, C., Narodoslawsky, M., 1996. The Sustainable Process Index a new dimension in ecological evaluation. Ecol. Eng. 6, 241–258. https://doi.org/10.1016/0925-8574(95)00060-7

Narodoslawsky, M., 2015. Chapter 3 - Sustainable process index, in: Klemeš, J.J. (Ed.), Assessing and Measuring Environmental Impact and Sustainability. Butterworth-Heinemann, Oxford, pp. 73–86. https://doi.org/10.1016/B978-0-12-799968-5.00003-8

Oppelt, E.T., 1990. Air emissions from the incineration of hazardous waste. Toxicol. Ind. Health 6, 23–51.

Riber, C., 2007. Evaluation of waste specific environmental impacts from incineration. Kgs. Lyngby.

Slater, C.S., Savelski, M.J., Hounsell, G., Urbanski, F., Geiger, J., Knoechel, D., 2010. Green Engineering Analysis of a Multi-use Solvent Recovery System for Small Volume Waste Steams in the Pharmaceutical Industry. Presented at the 2010 Meeting of the American Institute of Chemical Engineers, Salt Lake City, UT.

Slater, C.S., Savelski, M.J., Taylor, S., Kiang, S., LaPorte, T., Spangler, L., 2007. Pervaporation as a green drying process for solvent recovery in pharmaceutical production, in: Amer. Institute Chem. Engng. Annual Meeting.

Ulgiati, S., Odum, H.T., Bastianoni, S., 1994. Emergy use, environmental loading and sustainability an emergy analysis of Italy. Ecol. Model. 73, 215–268. https://doi.org/10.1016/0304-3800(94)90064-7

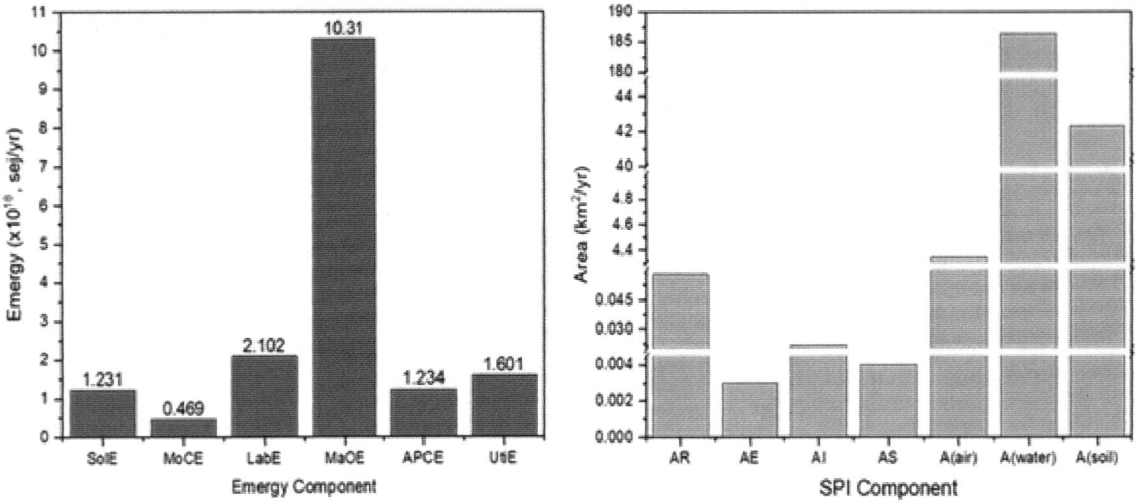

Figure 1. SPI and Emergy analysis for optimal pathway of sedimentation-pervaporation-ultrafiltration. [AR - Area for raw material consumption; AE - Area for energy consumption; AI - Area for installation; AS - Area for staff usage; A(air) - Area needed to embed air emissions; A(water) - Area needed to embed water emissions; A(soil) - Area needed to embed soil emissions; SolE – emergy content of waste solvent; MoCE – emergy associated with material of construction of selected technologies; LabE – emergy associated with labor; MaOE – emergy due to maintenance and overhead cost; APCE – emergy due to annualized purchase cost of technologies; UtiE – emergy associated with utilities]

629170 Modeling and Comprehensive Analysis for an Industrial Complex Coupling Allam Power Cycle, Air Separation Unit, and Ammonia Manufacturing Process

Abstract:

The sustainable development of global economy requests the effective carbon capture, utilization, and storage (CCUS) for all business sectors. The most cost-effective way for current chemical/petrochemical industries with fossil fuels as main energy/material sources to perform CCUS should focus on cleaner productions instead of end-of-pipe processing. One of such promising technologies is to employ the Allam cycle for power production, which bears a high production efficiency without caron emissions but the commercial-grade CO2 generation instead [1]. In this paper, a new and highly material-and-energy integrated industrial complex, i.e., A3 complex, has been developed, modeled, and comprehensively analyzed. It is a natural gas empowered complex, which includes an Allam power cycle power to generate electricity and pipeline-grade CO2 for commercial utilization; an air separation unit (ASU) powered by the Allam cycle while simultaneously supply the pure oxygen to the Allam cycle as feedstock; as well as an ammonia plant to utilize the nitrogen from ASU to produce ammonia. Compared with standalone ASU, Allam cycle, and ammonia plant, the integrated A3 complex has improved their strengths and mitigated their weaknesses simultaneously. The efficacy of the developed A3 complex has been quantitively demonstrated by rigorous large-scale modeling and simulations with Aspen Plus. Meanwhile, the economic and environmental analyses including production, energy consumptions, and emission comparisons between the developed A3 complex and ordinary standalone processes have also been conducted, which systematically exemplify the excellence of the developed A3 complex.

References

Allam, R.; Martin, S.; Forrest, B.; Fetvedt, J.; Lu, X.; Freed, D.; Brown, G.W., Jr.; Sasaki, T.; Itoh, M.; Manning, J. Demonstration of the Allam Cycle: An update on the development status of a high efficiency supercritical carbon dioxide power process employing full carbon capture. In Proceedings of the 13th International Conference on Greenhouse Gas Control Technologies, GHGT-13, Lausanne, Switzerland, 14–18 November 2016. Energy Procedia 2017, 114, 5948–5966.

632437 Application of Mathematical Algorithms for Real Time Optimization of Continuous Flow Extraction Processes

Abstract:

Contemporary chemical engineering approaches to the separation and extraction of material from one solvent into another are dominated by the use of batch techniques. These batch methods, though ubiquitous in both research and industrial applications, have labor and safety issues that can be difficult to address, and often prove to be challenging to optimize. Continuous flow extraction provides a platform to develop and design extraction processes in safe handling conditions, such as low control volume microfluidic extractions, while also allowing for rapid optimization and subsequent scale up. When used in tandem with inline analytics tools and techniques, paired with centralized controls and processing software, continuous flow extractions provide a rapidly optimizable alternative with broad applicability in both research and industrial contexts. This presentation will discuss the development and demonstration of such processes, where inline tools combine with mathematical algorithms and remote process controls to design a real time optimized and chemistry agnostic method for continuous extraction in both single and multistage setups.

624382 Advanced Operating Strategies to Enhance the Performance of Chemical Looping Natural Gas Reforming Processes

Abstract:

Natural gas is a crucial feedstock for chemical production in modern industry, and the reforming of natural gas into syngas, a mixture of H_2 and CO, is often an essential first step to convert natural gas into chemicals. However, conventional natural gas reforming processes, mainly including steam methane reforming (SMR) and autothermal reforming (ATR), suffer greatly from efficiency losses induced by the endothermic reactions and the need for air separation, respectively. Moreover, the efficiency of these processes become even lower when they are integrated with CO_2 capture. Chemical looping reforming (CLR) is a novel process that can address these challenges faced with conventional natural gas reforming. CLR performs the reforming reaction in two separate reactors, namely a reducer (fuel reactor) and a combustor (air reactor). In the reducer, natural gas is partially oxidized by metal oxides into syngas. In the combustor, the reduced metal oxides are regenerated using air. Unlike SMR, the CLR system can operate under autothermal conditions without external fuel combustion. Unlike ATR, the CLR system does not require air separation. These advantages allow the CLR system to achieve a much higher energy efficiency and much lower capital and operating costs.

By producing the syngas and regenerating the oxygen carriers in separate reactors, the CLR process offers significant flexibilities in process design. Many advanced process design strategies can be adopted in CLR processes to further improve the product yield, the energy efficiency, and the economic benefits. This work focuses on analyzing these different strategies and quantify their benefits. The following strategies are studied in detail: (1) separating the reducer into multiple modules operated at different conditions; (2) operating the reducer and combustor at different pressures by using solid pressure changing device; and (3) recycling downstream flue gas and purge gas back to the reducer. The CLR system as well as its integration with downstream chemical production are simulated in ASPEN Plus. Material, heat, and power balances are established in the process models to achieve autothermal operation in the overall process systems, and their performance is compared with SMR and ATR. The results indicate that the CLR processes can improve the product yield by 5-15% over conventional technologies. Moreover, using the advanced operating strategies can further improve the energy efficiency of those CLR processes by ~5%, which is an additional benefit granted by the flexibility of the CLR processes.

624269 Data-Driven Predictive Model and Optimization Based on Machine Learning on Steam Reforming Process

Abstract:

Hydrogen has recently been in the spotlight as its potential as a future eco-friendly mobility energy source. Although the on-site steam methane reforming (SMR) process is widely used for hydrogen refueling stations, the difference between the thermal efficiencies of commercial products and the theoretical maximum is very large. To maximize thermal efficiency, numerous studies regarding mathematical modeling have been conducted. However, mathematical modeling has a disadvantage in that it is difficult to build high-fidelity models of complex equipment and processes. Besides, the on-site SMR process has complicated reactors and processes. Therefore, in this study, a predictive model based on machine learning is developed to overcome the difficulty. Firstly, a data pre-processing step is performed to improve the data quality, removing outliers and noise of 485,710 actual operation datasets. Next, the optimal number of hidden layers, number of hidden nodes, optimizer, and activation function are determined to develop a high accuracy deep neural network model (R^2 value of 0.9987). The predictive model estimates six variables including steam temperature, reformate gas flow rate, and its CO, CO_2, CH_4, and H_2 compositions. To determine the optimal operating conditions, a total of 387,420,489 cases are calculated, considering both nine operating variables and five constraints. In conclusion, the thermal efficiency is maximized to 85.6%, higher than those of commercial products.

625783 Use of Bayesian Modeling for Failure Risk Analysis and Control Strategy Design

Abstract:

Process characterization and definition of appropriate control strategies is critical to ensure quality in the manufacture of active pharmaceutical ingredients. According to the principles of quality by design (QbD), process development should include identification of critical quality attributes (CQAs), characterization to understand the relationship between process parameters and these CQAs, and an evaluation of risk for the process. This presentation provides a case study wherein Bayesian probabilistic modeling was used to evaluate process risk for a specific critical impurity. Formation and purging of this impurity was studied via DoE and mechanistic studies spanning the proven acceptable ranges (PAR) for reaction and crystallization parameters. Probabilistic models were developed to predict process performance and assess and visualize predicted failure rates across the multivariate parameter space. This analysis informed assignment of critical process parameters and appropriate multivariate PARs to ensure robust processing. The probabilistic models also enabled estimation of the impact of common cause plant variability on the anticipated failure rate against specification levels of the critical impurity. Overall, these various failure rate analyses, enabled by the Bayesian models, helped to ensure implementation of an effective control strategy and will aid in future process transfers and process monitoring upon commercialization.

626120 Accelerated Process Design and Optimization for a Small Molecule COVID-19 Therapeutic

Abstract:

Molnupiravir is an investigational oral antiviral treatment for SARS-COV-2 that works as a ribonucleoside analog to inhibit replication of RNA viruses. The proposed commercial manufacturing route is a five step synthetic route from the regulatory starting material. This presentation will discuss the optimization of the final transformation on a highly accelerated timeline to ensure metric-ton supply of high purity molnupiravir during the coronavirus pandemic. This step begins with an acid-catalyzed deprotection reaction to form molnupiravir from the penultimate intermediate which also forms impurities due to acid-sensitive hydrolysis. Small-scale high throughput experimentation was used to build kinetic models which were used to optimize impurity generation and yield productivity. The workup of the reaction also introduces the risk of producing lightly purple colored active pharmaceutical ingredient (API) through trace metal coordination, which was mitigated through the addition of a chelating agent. The chelated metal and other aqueous-soluble impurities are rejected through extractions. Remaining impurities are then rejected into the mother liquor during the azeotropic distillative crystallization. Because of the potential variability of the distillation across several equipment trains and scales, the crystallization was designed for better control and less sensitivity to these variations. The resulting process performance has been monitored at four clinical and commercial manufacturing sites. In summary, this talk will discuss the development of control strategies across the last step of the molnupiravir synthesis to ensure the critical quality attribute of purity is met.

628959 A Digital Twin of Flexible Modular Continuous API Manufacturing Process

Abstract:

Advanced continuous manufacturing (CM) is emerging as a preferred platform for the production of active pharmaceutical ingredients (APIs). The future continuous pharmaceutical manufacturing is desired to be based on modular plug and play concept. In this approach, mobile modules consisting of a group of unit operations have been developed that can be easily reconfigured to assemble a scalable pharmaceutical manufacturing process for different products. It is also suitable for on-demand manufacturing such as in the case of COVID-19. However, because of different levels of complexity, the development and adaptation of flexible and modular CM plant is still an open area of research. Therefore, a systematic framework of methods and tools are needed to develop, adapt and evaluate the flexible modular CM plant for API manufacturing with reduced time and resources. In particular, a digital twin is critical for quick design, adaptation, optimization, scale up, and control of continuous APIs manufacturing process, allowing for virtual modeling of operations without extensive process development.

In this work, a digital twin for continuous APIs manufacturing process using modular components has been demonstrated. The process flowsheet model is the heart of 'digital twin' and consists of mathematical representation of three modules with the options of adding more as needed. The first module is for feeding and dispensing. It consists of a refill unit, feed tank, pump, mixing, and preheater. The second module is for performing the chemical reactions required to produce the target APIs. This module consists of tubular reactors placed inside a heating and cooling jacket. The third module is focused on separation of API from impurities. A continuous chromatographic model is currently used for separation purposes that can be easily replaced with any other type of separation techniques such as continuous crystallization. The applications of the digital twin have been demonstrated for design and development of suitable control architecture for a continuous APIs manufacturing process. The developed control architecture is also modular in nature and can be easily adapted for different manufacturing processes of APIs that may have different material and information flows, dead time variations, and tuning parameters. The proposed digital twin can save the time and resources needed for continuous APIs manufacturing and improve the product quality significantly.

The objective of this presentation is two-fold; first to highlight the developed digital twin and then demonstrate its application for design of process and control architecture of continuous API manufacturing processes.

624677 Quality-By-Control of a Novel Unit for Continuous Integrated Filtration-Drying of Drug Substances

Abstract:

Quality-by-Control (QbC) is a recent evolution [1] of the Quality-by-Design [2] initiative, in which active process control represents the core feature of the control strategy. With an active process control system in place, the critical process parameters of the process are automatically manipulated in response to disturbances, so that the target critical quality attributes of the product are ensured. The QbC framework has acquired even more importance within the recent interest in continuous integrated pharmaceutical manufacturing [3]. Continuous manufacturing (CM) guarantees greater product consistency compared to batch processing, as continuous processes present inherently greater process controllability. QbC has emerged as a very useful tool for the transition to CM, which requires a systematic framework for process control.

In this study, we develop a QbC framework for an intensified unit for continuous integrated filtration-drying of crystallization slurries. Continuous and integrated solutions for filtration and drying in upstream manufacturing are not much studied in the literature, although these processing steps represent a main obstacle for the implementation of end-to-end CM. The unit studied in this work is a novel carousel technology [4]. The carousel presents multiple processing stations located within a main cylindrical body. In the first station, the crystallization slurry is loaded. In the subsequent stations, slurry filtration, cake washing, cake deliquoring and thermal drying are carried out. The dry cake is discharged from the last station through the action of a pneumatic piston. All stations operate simultaneously, processing different batches of material: every fixed cycle duration the carousel rotates, enabling pseudo-continuous operation. Filtration, washing and drying operations present high coupling *per se*, and are even more interacting within the rotation mechanism of the carousel, which fixes the residence time of each processing step.

To aid the development of the control strategy for the unit, we develop a comprehensive digital twin of the carousel. The digital twin is based on detailed first-principles models of the processing steps [5], combined together for simulating real time operation of the carousel. The developed framework represents a unique and effective environment for testing and comparing different control strategies, since: *i)* time-consuming experiments with different control strategies can be quickly simulated on the digital twin, and *ii)* the critical quality attributes of the product, such as the solvents and impurities content, are always accessible in the *in-silico* simulations, while they cannot be measured in real time on the physical unit.

After having developed the digital twin, we validate it with filtration and drying experimental data collected on a physical carousel. Then, we conceive and implement on the digital twin a three-level control system for the carousel, based on a recently proposed approach [1] that follows the ISA-95 Enterprise-Control System Integration Standard. The Level 0 of the control system is made up of the built-in controllers of the carousel and of the accessory equipment, such as the pumping system. End-point control is resorted to at Level 1, for automatically setting the duration of each cycle to ensure that the discharged cake meets the target residual solvents and impurities content. Within the end-point control loop, the available measurements from the process are used to infer in real time the time instant at which the cake can be discharged because the desired quality has been met, and the carousel rotation routine is consequently automatically run. We also implement a model-based Level 2 control system, which includes a state estimator for continuous monitoring of the critical quality attributes and real time optimization. We successfully test the proposed control system in normal operating conditions and in presence of filter mesh fouling, which is known to be the main disturbance occurring during carousel operation.

References

[1] Su, Q., M. Moreno, C. Laird, Z. Nagy and G. Reklaitis (2019). A perspective on Quality-by-Control (QbC) in pharmaceutical continuous manufacturing, *Comput. Chem. Eng.*, **125**, 216–231.

[2] Food & Drug Administration, 2004. Guidance for industry, PAT-A Framework for Innovative Pharmaceutical Development, Manufacturing and Quality Assurance. http://www. fda. gov/cder/guidance/published. html. Accessed on 03/29/2021.

[3] Lee, S. L., T. F. O'Connor, X. Yang, C. N. Cruz, S. Chatterjee, R. D. Madurawe, C. M. V. Moore, L. X. Yu and J. Woodcock (2015). Modernizing Pharmaceutical Manufacturing: from Batch to Continuous Production. *J. Pharm. Innov.,* **10,** 191–199.

[4] Liu, Y.C., A. Domokos, S. Coleman, P. Firth, Z.K. Nagy (2019). Development of continuous filtration in a novel continuous filtration carousel integrated with continuous crystallization, *Org. Proc. Res. & Dev.*, **23** (12), 2655-2665.

[5] Destro, F., V. Wang, M. Abdi, X. Feng, E. Wood, E., S. Coleman, P. Firth, A. Barton, M. Barolo and Z. K. Nagy, (2021). Continuous integrated filtration, washing and drying of aspirin: digital design of a novel intensified unit. Proc. of the 11th IFAC Symposium on Advanced Control of Chemical Processes, Venice (Italy), June 13-16 2021. In press.

630559 Data Pre-Treatment Analysis of Residence Time Distribution (RTD) Profiles for Pharmaceutical Manufacturing Applications

Abstract:

Given the numerous advantages of continuous manufacturing (CM) [1] in the pharmaceutical industry, there has been an immense research thrust aimed towards understanding process dynamics [2] for accurate process prediction and efficient implementation of CM. Amongst the various techniques [2, 3] implemented for this end, the characterization of residence time distribution (RTD) has been performed for various unit operations as well as entire manufacturing lines, to capture the overall system dynamics followed by a change in the processing conditions [4]. RTD profile of any unit operation in the pharmaceutical manufacturing lines, can be obtained by adding a pulse or step change of the selected tracer at the inlet of the unit operation and measuring the tracer concentration at the outlet as a function of time using Process Analytical Technology (PAT).

However, the tracer concentration profile can include significant intrinsic and extrinsic noises from various sources. The different sources of "extrinsic noise" include environment and process conditions. The sources of "intrinsic noise", on the other hand, refer to the internal fluctuations of the system, which can be classified into the sensor, equipment, and material variability related to the inherent variability of the sensor during RTD measurement, equipment fluctuations during operation, and differences of tracer and bulk material properties, respectively [5, 6]. It is important to measure, quantify and remove the external noise as well as measurement variability from the RTD profiles obtained experimentally, as they can lead to inaccuracies in developing predictive models for evaluating system dynamics. However, the available literature focusing on the application of RTD for pharmaceutical manufacturing vaguely mentions enough details on this aspect of data pre-treatment of RTD profiles.

In this work, we aim to present and discuss various data pre-treatment and noise handling strategies aimed towards identifying effective strategies for denoising RTD profiles. The first aspect of the proposed work focuses on the quantification of the noise associated with the experimental RTD measurements. Following this, the second aspect focuses on handling the measurement variability and external noise for reliable RTD measurements. To this end, strategies similar to those available in image processing [7] and signal processing [8] are implemented. Some examples of noise filtering include linear and non-linear filters like adaptive filter, Kalman filter, recursive least-square, least mean square filter, and median filter. To reduce the variability, signal transforms, such as Fourier transform and Wavelet transform [9], are used for noise reduction, followed by low-pass or high-pass filtering. In the proposed work, the various data pre-treatment and noise handling strategies including time and frequency averaging, and exponential smoothing techniques, are evaluated and compared, to obtain a good signal-to-noise ratio (SNR) of the experimentally obtained RTD profile. Lastly, the effectiveness of different averaging and denoising techniques is analyzed to ensure that the important information pertaining to the RTD profile is retained. Thus, the proposed work aims towards developing guidelines for data pre-treatment strategies of RTD, which would be helpful towards effective utilization of RTD studies for predictive modeling of manufacturing lines.

References:

1. Schaber, S.D.e.a., *Economic Analysis of Integrated Continuous and Batch Pharmaceutical*

Manufacturing: A Case Study. . Ind. Eng. Chem. Res. , 2011. **50**: p. 10083–10092.
2. M. Ierapetritou, F.J.M., G. Reklaitis, *Perspectives on the continuous manufacturing of powder-based pharmaceutical processes.* AIChE Journal, 2016. **62**: p. 1846-1862.
3. V. Vanhoorne, C.V., *Recent progress in continuous manufacturing of oral solid dosage forms.* International Journal of Pharmaceutics, 2020. **579**: p. 119-194.
4. Danckwerts, P.V., *Continuous flow systems: Distribution of residence times.* Chemical Engineering Science, 1953. **2**: p. 1-13.
5. Kirsch, J.D. and J.K. Drennen, *Near-infrared spectroscopy: applications in the analysis of tablets and solid pharmaceutical dosage forms.* Applied Spectroscopy Reviews, 1995. **30**(3): p. 139-174.
6. Blanco, M. and M. Alcalá, *Content uniformity and tablet hardness testing of intact pharmaceutical tablets by near infrared spectroscopy: a contribution to process analytical technologies.* Analytica chimica acta, 2006. **557**(1-2): p. 353-359.
7. Brailean, J.C., et al., *Noise reduction filters for dynamic image sequences: A review.* Proceedings of the IEEE, 1995. **83**(9): p. 1272-1292.
8. Vaseghi, S.V., *Advanced digital signal processing and noise reduction*. 2008: John Wiley & Sons.
9. Xu, Y., et al., *Wavelet transform domain filters: a spatially selective noise filtration technique.* IEEE transactions on image processing, 1994. **3**(6): p. 747-758.

629294 Methanol Synthesis Processes

Abstract:

Hydrogenation of carbon dioxide may contribute to the reduction of industrial carbon dioxide emissions. The bulk chemical methanol is a suitable hydrogenation product. It is used for paints, plastics, and adhesives, as feedstock for the production of chemicals such as formaldehyde, ethylene, propylene, methyl tertiary-butyl ether, and acetic acid, and as fuel additive or alternative fuel. It provides promising properties for energy storage, as it can raise the energy density of hydrogen-based energy carriers by one order of magnitude. Low temperature level in reforming also suggests methanol for fuel cell applications. State-of-the-art methanol production is based on syngas, a mixture of carbon monoxide, carbon dioxide, and hydrogen, and preferably $Cu/ZnO/Al_2O_3$ catalysts. State of the art reactor- and process design suggest looping systems with sophisticated internal heat energy management. Huge efforts have meanwhile been made to directly process carbon dioxide instead of syngas. To our knowledge neither of the carbon dioxide based processes has established on the market, mainly because of catalyst life time limitations. To sum up the present status, the establishment of direct methanol synthesis from carbon dioxide suffers from limitations in feed gas preparation and/or appropriate catalysts. Appropriate high carbon dioxide concentration sources are available in huge quantities in the iron and steel industry or cement industry, but appropriate solutions are still missing.

When processing carbonate based iron ores (siderite) in hydrogen atmosphere our group registered that within a temperature window of 600 to 800 °C and ambient pressure it is possible to alter the carbon monoxide/carbon dioxide ratio of the exiting gas. The final solid product from ore reduction still maintains this "catalytic" property, suggesting it for syngas preparation from carbon dioxide with a cheap and robust catalyst which can easily be processed after deactivation. The process was tested and validated in a lab size plug flow reactor setup.

For experimental simulation/validation of the methanol synthesis in lab scale a CSTR loop reactor system with external cooling loop for condensing the gaseous products and pressure drop controlled gas feed was constructed. The operation window of the setup is 500 °C at maximum and 10 MPa. At present the reactor setup is equipped with two gas supply lines equipped with mass flow controllers, limiting operation flexibility to fixed stoichiometry of the carbon dioxide/carbon monoxide source. The methanol synthesis process was validated with a Cu/ZnO/MgO catalyst prepared by wet incipient impregnation and stoichiometric CO_2/H_2 feed at 250 °C and 5 MPa. Whether this catalyst can withstand long term operation with CO_2/H_2 feed with acceptable loss of activity, or whether an upgrade of the feed gas by pretreatment in the syngas reactor is needed, cannot be concluded from present experience. However, either route, dual step synthesis with feed gas pretreatment or direct synthesis shows promising features, since catalyst life span is not a crucial issue.

630139 Systematic Approaches for Discovering Innovations to Enable a Sustainable Circular Economy

Abstract:

The recent push towards Sustainable Circular Economy (SCE) needs to be substantiated with modelling and systems design of value-chains. In our previous work, we have built a general framework to assess and design life-cycle value chains within a superstructure network for establishing SCE [1]. This SCE framework was applied to a carrier bags case study with various plastic and paper alternatives and multiple End-of-Life (EoL) options in the superstructure [2]. Due to the possibility of multiple stakeholders, multi-objective optimization was performed within the framework for objectives from the three domains: Environmental, Economic and Circularity. The trade-offs between these three objective types were quantified using pareto surfaces, and compromise solutions were proposed from the pareto optimal set. This study led to insights about currently available alternatives of the value-chain, e.g., quantifying the paper-plastic dilemma, probing effect of re-use, selecting best EoL strategy etc.

However, this approach is limited by the expanse of the constructed superstructure network. The proposed designs can only be derived from the set of currently available technologies and product systems in the life cycle of products. Evidently, win-win solutions are likely to be obtained by substituting conventional technologies with innovations and emerging technologies. This work is aimed at (i) guiding innovation in value chains and (ii) discovering novel products and pathways. For the first goal, we use hotspot and sensitivity analysis, while the second goal relies on superstructure-free design methods.

First, we identify activities in the carrier bags value-chain which act as hotspots of emissions, cost and 'loss of circularity' using life-cycle allocation and displacement methods. For instance, the highest amount of circularity loss in the optimal carrier bags value-chain occurs at the segregation process. These hotspots are obtained for various scenarios of technological advancements and recycling consciousness in society. Through these hotspots, we gain knowledge about where the highest scope of improvement in all the three objective domains lie. This is followed by sensitivity analysis using Monte-Carlo simulations to find effect of perturbations in the parameter space on the pareto surface. We also explore prospects of performing more comprehensive yet tractable sensitivity analysis using (i) analytical approaches: contingent derivatives of perturbation maps [3]; and (ii) stochastic approaches: multi-objective evolutionary algorithms (MOEA)[4]. Sensitivity analysis allows us to rank value-chain activities based on pareto dominance that can be achieved by enhancing their respective efficiencies through innovation . In turn, this helps in guiding innovations towards most promising sectors for SCE. For example, such an approach can point toward the need for innovation in chemical recycling technologies of plastic carrier bags.

For discovering innovations, we explore the use of superstructure-free methods. As stated earlier, the SCE framework only evaluates technologies and connections specified within the superstructure. We expect that novel products and pathways can be found if a more flexible design space is used. In order to do so, we evaluate various PSE methods such as State Task Network (STN), General Modular Framework (GMF) and Unit, Port, Conditioning Stream (UPCS) representation [5] to best describe the nodes and connections within the cradle-to-cradle value chain. We scan the method pool for prevalent analogies between units and life-cycle activities, possibility of directed multigraph required for circular flows and flexibility in designing node connections. However , we do not define or generate the superstructures beforehand. Instead, we let the connections and amounts transferred be decision variables of the design problem. UPCS defines the 'minimal' and 'feasible' rule s for nodes which ensure that no redundant flows enter a unit, and the necessary flows for transformations do enter, respectively. In addition to these, we add the following rules: (i) each product flow is associated with both input and output ports of UPCS, and (ii) there must exist a unit between every output and input port. The streams linking the ports are captured through an adjacency matrix with binary decision variables in each cell. Each cell of this matrix corresponds to a transformation and refers to a sub-matrix of tuples containing (i) scaling factor of transformation (decision variable), (ii) enthalpy and exergy changes (parameters), (iii) yield per unit input (parameter). Within the manufacturing sector, these transformations can be modelled as reaction-separation networks (e.g., from Reaxys), whereas value-chain activities can be modelled using exergy changes. Physico-chemical constraints such as flow conservation are added as a function of adjacency matrix and scaling factors. We demonstrate this method using an illustrative example and find optimal nodes and connections for a small set of products and elementary flows (to and from nature). We also probe the scalability of this method for larger number of products, and to this effect, consider using techniques like Bender's decomposition and evolutionary algorithms (like NSGA-II) for superstructure-free design.

Through this presentation we aim to convey further developments in application of our SCE framework [1,2] for guiding innovation and emerging technologies towards activities with most scope and highest impact, using hotspot and sensitivity analyses, respectively. Additionally, we create a highly flexible design space for value chains using superstructure and superstructure-free methods. With a large enough sub-matrix for underlying transformations, we expect that this new methodology can yield optimal solutions which haven't yet been considered in academia or industry. We also probe the scalability of these methods for large and practically relevant problems using advanced mathematical modeling techniques.

References

1. Thakker, Vyom, and Bhavik R. Bakshi. "Toward sustainable circular economies: A computational framework for assessment and design." *Journal of Cleaner Production*295 (2021): 126353.
2. Thakker, Vyom, and Bhavik R. Bakshi. "Towards Sustainable Circular Economy: Design Framework and Application to Grocery Sacks". *2020 AIChE Annual Meeting*. Video available at: https://youtu.be/dWj6R0UXMjM
3. Tanino, Tetsuzo. "Sensitivity analysis in multiobjective optimization." *Journal of Optimization Theory and Applications*3 (1988): 479-499.
4. Deb, K., Pratap, A., Agarwal, S., and Meyarivan, T. A. M. T. "A fast and elitist multiobjective genetic algorithm: NSGA-II". *IEEE transactions on evolutionary computation*, *6*(2) (2002): 182-197.
5. Wu, WenZhao, Carlos A. Henao, and Christos T. Maravelias. "A superstructure representation, generation, and modeling framework for chemical process synthesis." *AIChE Journal*9 (2016): 3199-3214.

SHAPING THE FUTURE WITH A HYDROGEN VALUE CHAIN SIMULATION PLATFORM

Ian Willetts, AVEVA Software LLC, Carlsbad, CA
Cal Depew, AVEVA Software LLC, Lake Forest, CA

Abstract

Today's top international oil companies (IOCs) are responding to the climate crisis and setting goals to be net zero carbon emission companies by 2050 or sooner. They are on a journey to transition from being international oil companies to becoming international energy companies (IECs). These companies will implement near term, medium term, and longer-term initiatives for decarbonization as they progress along their energy transformation journey. These steps will include:

- Reducing methane emissions
- Reducing carbon dioxide emissions
- Carbon Capture, Utilization and Storage (CCUS)
- Accelerating decarbonization

A key accelerator all companies are embracing as part of their strategy to drive to their net-zero goals is digital transformation which involves identifying and implementing methods to bring digital tools and work processes closer together to drive more informed business decisions. Many industry stakeholders see the Digital Twin as the most important building block for the digital transformation of the process industries and digital transformation of standard work processes is required to accelerate today's energy transition.

While legacy simulators are well-suited to accurately simulate conventional processes, their decades-old architectures mean they are not ideally suited to serve the entire plant lifecycle nor to support the digital transformation journey for the nascent hydrogen or low-carbon economy.

Digital transformation is undoubtedly a critical lever that both EPCs and Owner Operators can use to deliver both short term sustainability gains and to elevate the quality of capital project execution while minimizing the environmental impact of new capital investments for traditional oil and gas processes, CCUS projects or low or zero carbon energy projects.

Process simulation is accepted as a required tool for the process design of traditional oil and gas and chemical plant facilities, from conceptual design, through engineering and on to operations. However, with the energy transition underway, the landscape of the process industries will change dramatically over the next three decades. As such, a process simulator must be able to simulate both existing and new energy processes as well as support new and improved process engineering practices. The paper discusses the challenges of the current simulation tools from a process modeling perspective when it comes to the need for:

- Improved process equipment models suitable, for example, for modeling gray and blue hydrogen generation including steam methane reforming, absorption processes to capture carbon dioxide and adsorption processes to create high purity hydrogen

- Environmental and economic KPI calculations that allow process engineers to design for production, and quality goals while balancing economic and environmental objectives.
- New equipment models to support the modeling and optimization of the full hydrogen value chain including green hydrogen generation including, for example, renewable power sources, electrical grid models and electrolyzers
- Accurate thermodynamics for key sustainable processes including, for example, carbon capture and hydrogen liquefaction
- Capabilities to simulate liquid organic hydrogen carrier (LOHC) processes that increase the energy density of hydrogen to facilitate easier transportation

The paper also discusses the challenges of today's tools from a platform perspective when it comes to updating the tools for the hydrogen value chain:

- Legacy process simulators are overloaded with tightly woven niche features and functions only used by specialists. The complexity of the code makes it difficult for software vendors to adapt them fast enough for rapidly changing industry trends.
- Legacy process simulators are typically poorly integrated into engineering workflows beyond the process world, and if so, only with a single directional information flow. This leads to poor FEED quality and subsequent capital project cost and schedule overruns.

To address these challenges, the next generation process simulator needs the following, and other functionality, to be true enablers of digital transformation for the new lower carbon economy:

- Emission calculations linked with capital and operating cost models to support multi-constraint optimization of new process designs
- An open modeling framework to be able to quickly add new process models as they are rapidly developed to address the evolving hydrogen economy
- The ability to take Hydrogen value chain models used for process design and reuse them for dynamic simulation, control strategy design and for operator training simulators thereby shortening project schedules and lowering the total cost of simulation
- The ability to put those same models on-line for real time performance monitoring, process optimization and predictive analytics for incremental profitability, all in a cost-effective manner

The authors will show that the above identified obstacles can be readily overcome today with a next generation process simulator - AVEVA Process Simulation – that has been prepared with all the necessary capabilities and architecture needed to be able to support EPCs and Owner Operators as they go through their energy transition journey.

Beginning an Energy Transition Journey

Today's top international oil companies (IOCs) are responding to the climate crisis by setting goals to be net zero Carbon companies by 2050 or sooner. They are inevitably on a journey to transition from being international oil companies to being international energy companies (IECs).

This journey will include deploying near term sustainability solutions for current assets as well as investing in lower carbon energy projects in the future. TotalEnergies, for example, plans to ramp its capital investment in clean energy projects from 3% of total capital investment today to 20% over the next 20 years.

A Deloitte Insights article titled "The future of work in oil and gas" [1] published in Oct 2020 outlined one potential energy transition pathway that oil and gas companies will likely undertake to achieve their net-zero goals. This is shown in Table 1.

Table 1. Energy Transition Pathways for Oil and Gas Companies

Energy Transition Pathway	Theme
First Steps Foundational sustainability measures and HSE requirements	Water, safety, health, land etc.
Building Blocks Advanced emissions, electrification, and energy efficient measures	Asset electrification
	Energy efficiency and management
	Fugitive emissions
New Blueprints Portfolio of lower-carbon energy sources, greener fuel mix, and carbon capture solutions	Energy from low-carbon sources
	Greener end-product mix (natural gas)
	New revenue models (renewables, biofuels, EV stations etc.)
	Carbon capture solutions and services
Winning the Future License to lead in a clean-energy economy	Net zero emissions from purchases and operated assets
	End-of-life treatment of sold products to create a circular economy
	Shape and influence paths to net-zero of consumers and societies

While Deloitte is looking at the world from the Owner Operator perspective, the building blocks and new blueprints above will rely on EPCs to align with Owner Operators to deliver on those initiatives. As such this transition will equally impact both EPCs and OO's and both will have to rethink the way they do business today.

This view by Deloitte is well aligned with the viewpoint of the Oil and Gas Climate Initiative (OGCI) who published a progress report "Delivering on a Low Carbon Future"[2] in December 2020. In the report the OGCI argue that it is a commercial imperative for oil and gas companies to implement a broad and ambitious portfolio of near, mid, and long-term low carbon solutions to meet their net-Carbon zero goals. The focal areas they outline include:

- Reducing methane emissions
- Reducing carbon dioxide emissions
- Carbon Capture, Utilization and Storage (CCUS)
- Accelerating decarbonization

Table 2 captures some of the recommended levers that oil and gas companies should deploy in these four areas:

Table 2. Levers that Oil and Gas Companies Should Deploy

Focus	Primary Levers	Additional Comment
Reduce methane emissions	Detect, measure, and mitigate fugitive emissions	Must be cost effective and scalable to be deliver substantive emissions reductions
	Replace or upgrade high-emitting equipment	
Reduce carbon dioxide emissions	Improve energy efficiency	Focus on reduction in LNG and Refineries. Consider CO2 emissions from production facilities as well as from emissions associated with the required imports of electricity and steam
	Reduce and ultimately eliminate flaring	
	Electrify operations with renewables where possible	
	Cogenerate electricity and useful heat	
Carbon Capture, Utilization and Storage (CCUS)	Captured CO2 can be stored underground with or without enhanced oil recovery. Captured CO2 can also be used for producing chemicals and concrete	Scaleup is the biggest challenge. Per the IEA CCUS in Clean Energy Transitions report (Sept 2020) a 38-fold increase in CCS is needed to meet the 2020 1.5 Gt goals.
Accelerate decarbonization	Requires investment in low and zero carbon energy sources including green hydrogen	11% of GHG emissions come from energy production including oil and gas. The percentage is much higher when considering Scope 1,2 and 3 emissions

Confluence of Energy Transition and Digital Transformation

A key accelerator all companies are embracing as part of their strategy to drive to a net-zero goal is digital transformation. Digital transformation involves bringing together today's digital tools and work processes to drive more informed business decisions.

Figure 1. Digital Transformation Merges Digital Tools and Work Processes

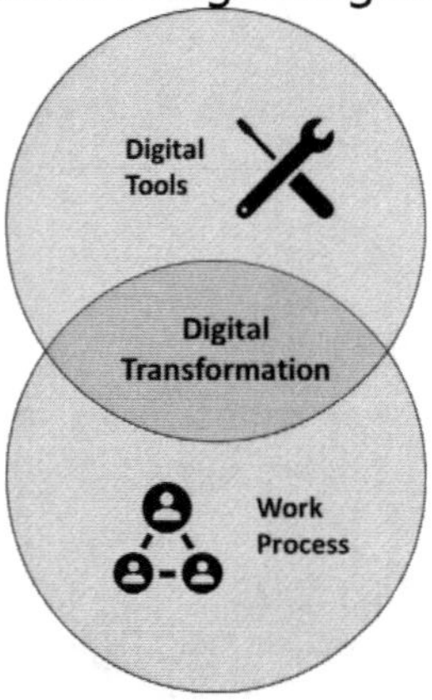

Per Covestro[3], new process technology must outperform existing technology to be acceptive in a competitive environment. But the existing technology and processes have been optimized for decades. We now have only a short time for the introduction of new sustainable processes. To achieve the same level of maturity, we need innovation for both the processes and the design workflows to accelerate the new process technology. This need for innovation applies to both chemical companies addressing the circular economy and oil and companies in their energy transition to become energy companies.

As oil and gas companies begin to execute their energy transition strategies, they must still deliver profitability in what remains a highly competitive industry. More informed business decisions can drive many traditional key performance indices (KPIs) of both EPCs and Owner Operators, across greenfield and brownfield facilities including:

- Higher quality FEED design leading to lower capital cost and shorter schedules
- Engineering efficiency improvements across disparate multi-disciplinary teams
- Improved safety
- Less unplanned downtime
- Higher margins and profitability

With the drive for a sustainable future, additional metrics come into consideration, including but not limited to:

- Project emission factors
- Overall CO2 intensity of products produced
- Resource-energy efficiency
- Specific Energy Intensity
- Carbon efficiency

One key accelerator of digital transformation is the Digital Twin, which is a digital representation of a real-world entity, a model that mirrors an actual asset to be built or an existing brownfield asset. In the Process Industries there are many different views of what represents a Digital Twin. One common viewpoint mirrored by many of the largest oil and gas companies is shown in the following image from AVEVA Group Plc, a leading provider of digital transformation solutions.

Figure 2. The Digital Twin

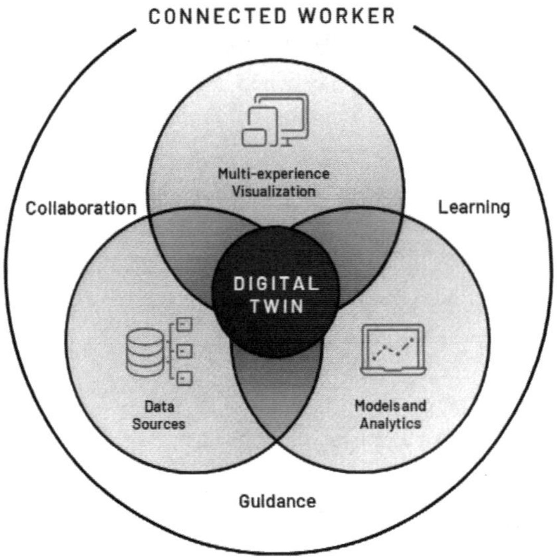

The Digital Twin includes:

- Data sources: the engineering design data and real time plant data
- Simulation models: first principles, physics-based models, data analytic models and hybrid models
- Visualization layer: the tools necessary for the people to be able to access and visualize data so that they can make appropriate and informed decisions to drives KPIs

The simulation model is often referred to as the behavioral digital twin as it can provide knowledge and insight into how the process will behave. Knowledge and insight provided by such a model cannot be gained in any other way other than by operating the plant. However, in the nascent hydrogen economy there are many projects planned but almost all are in the early conceptual or the early FEED project phase where there is no opportunity to gather plant operating data.

Simulation Throughout the Capital Project Lifecycle

Incorporating a process simulation model as part of the digital twin at every phase of a CAPEX project reduces risk and ensures projects are much more likely to be delivered on schedule and within budget. Successfully developing and leveraging the Process Digital Twin throughout a project lifecycle requires both the EPC and the Owner Operator to think differently and change the traditional business models. The following highlight the value a digitally transformed workflow can provide a Capital project from initial concept through design, commissioning, and startup and into on-going operations:

From concept to FEED

The Construction Industry Institute (CII) report number RR331-11[4] that a project with a high FEED maturity and accuracy outperformed projects with a low FEED maturity and accuracy by 24 percent in terms of cost growth, and 12 percent in terms of change order performance.

A process simulation model is a key enabler to achieve a highly mature FEED by using the insight and knowledge provided by the simulation model to eliminate errors and mistakes in the design before capital is committed and when the cost to correct an error is the lowest.

Detailed design

As a project moves toward detailed design some of the equipment in the FEED phase will change. It is important to keep the simulation in lockstep with design changes to ensure that as the physical asset takes shape the digital twin remains an exact mirror image. The simulation can be used continually to ensure that the latest detailed design changes do not negatively impact the controllability and operability of the asset3.

Commissioning and start-up

Armed with a highly responsive digital twin, the EPC and the Owner-Operator can turn their attention to preparing for start-up and ensuring their teams have the right skills to safely start up and operate the plant. Training operations staff using the behavioral digital twin allows operations staff to practice startup repeatedly and even fail repeatedly in a completely safe environment. Practice makes perfect and the operations staff, over time, develop the required skills and competency to ensure a smooth and error free startup.

Ongoing operations

The value of a simulation does not end with startup. It can be used to support on-going operations and drive efficiencies. This can be in the form of connecting to plant data to provide inferential measurements that can support decision making for operational excellence. Models can also be used for troubleshooting to investigate the cause of abnormal operations. Staff turnover is also inevitable, and the Process Digital twin can be a backbone of a plant's workforce competency development program.

Simulation Challenges for Today's Industry

Legacy Code

Perhaps the most known simulation platforms in the Process Engineering domain are PRO/II, Aspen Plus, HYSYS and UniSim.

- PRO/II is the flagship steady state simulation package of AVEVA, formerly Simulation Sciences Inc who were incorporated in 1967. Today the product is known as AVEVA PRO/II Simulation.
- The Aspen Project started in 1976 and this project in turn led to the creation of Aspen Technology in 1981. Today Aspen Plus is one of the two the flagship steady state simulation software packages of Aspen Technology.
- HYSYS was first conceived and created by the Canadian company Hyprotech. The HYSYS Version 1.1 Reference Volume was published in 1996. In 2002 Aspen Technology acquired Hyprotech and today this product is known as Aspen HYSYS.
- Following a 2004 ruling by the United States FTC, Aspen Technology was forced to divest its Hyprotech assets, including HYSYS source code, selling these to Honeywell who

ultimately produced the UniSim product. UniSim is built on the same foundation as Aspen HYSYS.

These software programs have evolved and morphed over the years as they have been used by customers who require new features and updates as requirements change. Inevitably, such legacy software becomes more challenging to adapt as industry needs change. Today we are in an unprecedented industrial transformation and in need of a new and modern simulation platform built from the ground up, to serve the needs of the next generation, digitally native, process engineers.

Modeling Requirements

Whilst the existing legacy software platforms are very capable and certainly road tested, the energy transformation will drive a new set of modeling requirements to be fit for purpose including but not limited to:

- Improved process equipment models for gray and blue hydrogen generation such as reactor models for steam methane reforming, absorption processes to capture carbon dioxide and adsorption processes to create high purity hydrogen.
- Environmental calculations to allow process engineers to design processes that meet production and quality goals with minimum environmental impact.
- Equipment models for green hydrogen generation including renewable power sources, electrical grid models and electrolyzers to support the modeling and optimization of the full hydrogen value chain.
- Accurate thermodynamics for modeling hydrogen liquefaction and other sustainable processes.
- Capabilities to simulate liquid organic hydrogen carrier (LOHC) processes that increase the energy density of hydrogen and facilitate easier transportation.

Platform Requirements

Today's workers are digital natives and have very different expectations when it comes to technology. Unfortunately, legacy process simulators are overloaded with niche features and functions that have been added over the years and are often only usable by experts. This in turn makes it difficult for users to adapt them fast enough for today's rapidly changing industry trends.

A standalone simulation package is not an enabler of digital transformation. Legacy process simulators are typically poorly integrated into the engineering workflows beyond the process world. Even where some integration exists today, information exchange is only a single directional information flow. This leads to poor FEED quality and subsequent capital project cost and schedule overruns.

Hydrogen Value Chain Simulation Platform

As of August 2020, a Wood MacKenzie research report[5] highlights that the green hydrogen project pipeline has swelled from 3.5 GW to more than 15 GW just over the previous 10 months. The success of these and other future multibillion dollar capital projects will be measured against two key metrics, cost, and schedule. The traditional capital project execution methodologies of today result in projects being delivered, on average, 20 months late and 80% over budget.

It is paramount that this does not happen with these new renewable energy capital projects and a Digital Twin strategy involving simulation can materially impact these looming investments, provided there is a suitable Hydrogen value chain simulation platform available.

AVEVA Process Simulation

Unlike the legacy simulation platforms discussed earlier, AVEVA Process Simulation, conceived less than ten years ago, was developed from the ground up to deliver the process digital twin to the next generation of process engineers. AVEVA Process Simulation is differentiated from legacy platforms in three ways:

1. One process model can be used throughout the project lifecycle fully supporting a company's digital transformation goals.
2. It is built with next generation capabilities reflecting its conception and development into today's computing environment.
3. It is applicable to the full hydrogen value chain including gray, blue and green hydrogen processes.

One Model for the Process Lifecycle

Unlike any other simulator in the market today, AVEVA Process Simulation supports multiple simulation modes without the need to reconfigure the simulation and allows an engineer to simply toggle from one mode to another depending upon the task at hand. The following table highlights different ways that AVEVA Process Simulation may be used throughout the process lifecycle and the associated benefits for each use.

Table 3. Uses of Simulation throughout the Process Lifecycle

Project Phase	Requirements / Goal	Simulation Mode	Use Case	Benefit
Conceptual Design	Efficient and accurate optioneering	Steady State & Rating	Heat and Material Balance	Rapid optioneering on conceptual design supporting economic assessment for FID
			De-pressuring	Correct material selection for future integrity
			Safe Design	Ensure License to Operate will be granted and maintained
			Reactive Hazards Assessment	Design the process to mitigate the risk of reactive hazards to As Low As Reasonably Practicable (ALARP)
			Minimum Environmental Footprint	Includes GHG calculations to quantify environmental impact and allow most sustainable design to be built
FEED	High quality & maturity	Steady State and Dynamic Simulation	Most accurate design	Use the simulation to ensure a high quality or mature FEED which on average leads to 12% fewer change orders and 30% schedule reduction
			Equipment Sizing	Up to 3% saving in Total Installed Cost by eliminating overdesign or incorrect sizing leading to startup and commissioning delays
			Process Safeguarding	Implement a design that reduces the likelihood of loss of containment and that minimizes consequences should one occur
Detailed Engineering	Efficient and accurate design	Rating and Dynamics	Engineering Efficiency	Improved workflows from the digital transformation of the engineering processes can lead up to 30% fewer man-hours of engineering & 50% reduction in simulation effort
			Control and Operability Analysis	Ensure final design is easy to build and operate eliminating need for overly complex control strategies impacting safety and operating costs of the asset
			HAZOP Analysis	Support the HAZOP team to identify risk events or potential caused by deviations from design or operating intentions
			Flaring	Design so that flaring is not required during startup and CO2 emissions are minimized, e.g., include flare gas recovery in the initial design and not as a retrofit

Project Phase	Requirements / Goal	Simulation Mode	Use Case	Benefit
Detailed Engineering	Efficient and accurate design	Rating and Dynamics	Minimum Environmental Footprint	Includes GHG calculations to quantify environmental impact of standard operating procedures and abnormal operating conditions
			Energy Efficiency	Process intensification for lowest operating costs and lowest environmental impact
Commissioning and Startup	Controls Checkout and Operator Training	Rating and Dynamics	Controls Checkout	Verify and Validate controls database on the simulation model to eliminate issues and shorten commissioning time
			Development of Standard Operating Procedures	Establish SOPs on the simulator prior to live commissioning
			Start up and Shut Down	Practice start-up and shut-down procedures repeatedly in a safe environment to minimize commissioning and start up times
			Operator Training	Ensure the operations staff are ready for safe operations on Day One and ensure on-going
Operations	Process Improvement	Data Reconciliation, Steady State and Dynamics	Identify Integrity Operating Window Limits	Identify and eliminate conditions where equipment degradation is likely to occur ensuring minimum downtime
			Environmental Compliance	Maintain License to Operate
			Flaring	Reduced or zero flaring and lower environmental impact
			Process Upsets	Prepare operators to be able to manage abnormal operating conditions safely
			Troubleshoot Operations	Diagnose and correct operational issues using insight provided by the model
			Energy Efficiency	Identify and implement ways to minimize energy consumption
			Prescriptive Maintenance	intelligently schedule and plan asset maintenance with insight provided by the model
			Equipment and Asset Performance Monitoring	Monitor the technical and economic performance of the asset to identify opportunities for improvement

Next Generation Capabilities

The following table highlight some additional key features that differentiates AVEVA Process Simulation from other legacy systems.

Table 4. Key Features of AVEVA Process Simulation

Key Features	Advantages / Benefits	
Modern Graphical User Interface	Contemporary user interface in line with the expectations of today's younger process engineers	
Flexible Open Modeling	Easy to Customize and build libraries for new industries Supports traditional process modeling as well as the modeling needs of the lower carbon future process industries	Process Library Electrical Library Utilities Library Flare Library Renewables Library Economics Library Controls Library Pressure Swing Adsorption (PSA) Library Rate Based Library Membrane Library Environmental KPI calculations Others
Equation Oriented	Fast to solve and handles complex recycles very efficiently	
Cloud Enabled	Supports collaboration across multiple disciplines	
Optimization	Incorporates the industry leading AVEVA Process Optimizer (ROMeo) optimizer which provides data reconciliation and multi-constraint optimization capabilities	
Multi-Core/Threaded	Speed and performance factored into the original design	
Integrated with OSI PI	Supports real time performance monitoring and is integrated with the number one historian in the process industries today	
Integrated with AVEVA Engineering	Improves engineering efficiency by keeping process and data digital twin in lock step through the process lifecycle	
Integrated with AVEVA PRISM and Microsoft Bonai AI Platforms	Full support for hybrid modeling combining first principles and data-based analytical models	
Python Scripting Interface	Digitally native engineers can enhance and integrate third party applications with contemporary programming languages	

Hydrogen Value Chain Models

The following flowsheets demonstrate the usage of AVEVA Process Simulation for modeling gray, blue and green hydrogen production.

Figure 3. Steam Methane Reforming (Gray Hydrogen)

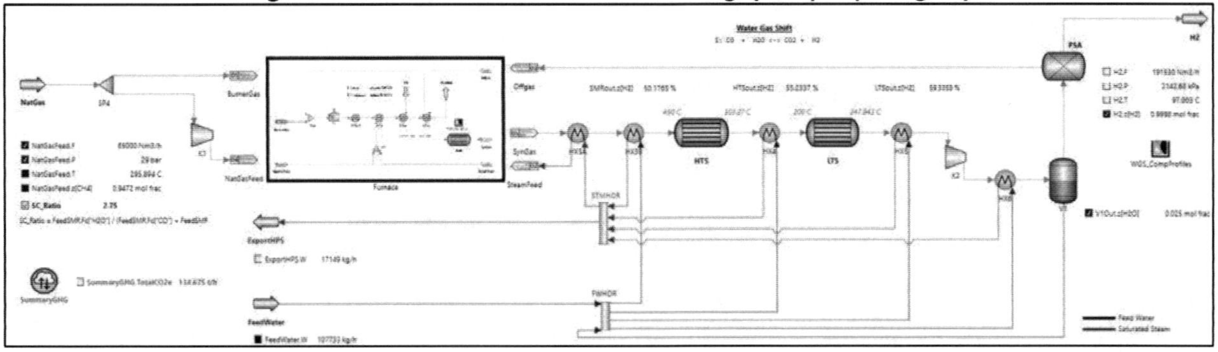

This example demonstrates a typical process to produce Hydrogen via Steam Methane Reforming (SMR). Natural Gas and high-pressure steam are reacted over a Ni / MgAl2O4 catalyst to produce H2, CO, and CO2. The CO and H2O in the outlet stream are then further refined in high-temperature and low-temperature water gas shift reactors. These reactors convert much of the remining CO and H2O into H2 and CO2. The final step of the process (not modeled here) separates H2 from the remaining components using Pressure Swing Adsorption (PSA).

Figure 4. Carbon Capture, Utilization and Storage (Blue Hydrogen)

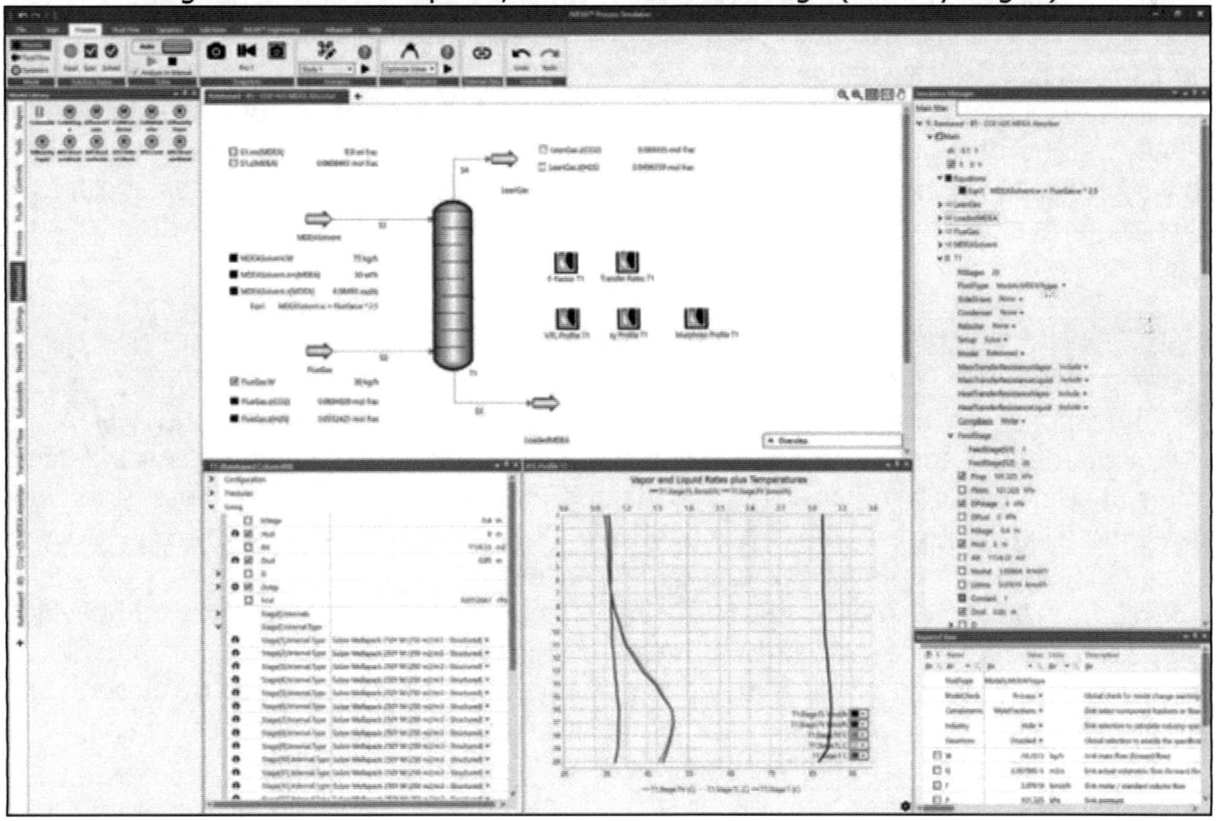

CO2 capture in absorbers is a not an equilibrium process but is mass transfer limited process. AVEVA Process Simulation includes a rate-based distillation column accurately modeling diffusion fluxes, diffusivity in the vapor and mass transfer calculations through the film between the vapor and liquid phases.

Figure 5. Sustainable Hydrogen Production (Green Hydrogen)

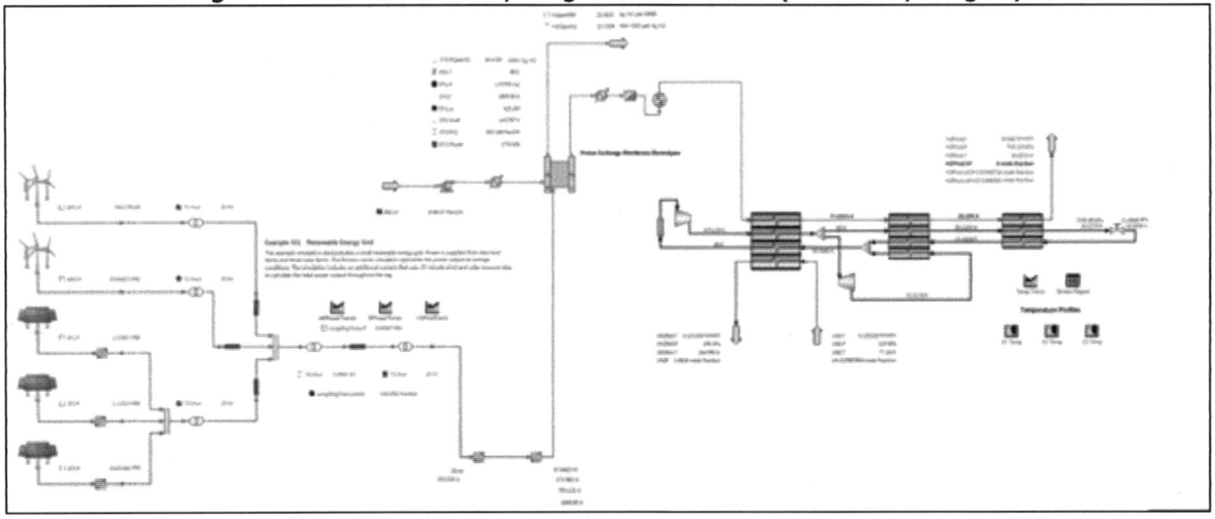

This simulation demonstrates a small renewable energy grid including renewable energy supplied from two wind farms and three solar farms. The model also includes both the process elements (pumps, heat exchangers, electrolyzers etc.), the electrical elements (transformers, power lines etc.) and a hydrogen liquefaction facility. The simulation runs at steady state to calculate average power output. The simulation also runs dynamically to predict the variability of power produced during the day as wind conditions change and the sun rises and falls. The model also looks at electricity supply and demand considering importing or exporting to the grid, spot prices and hydrogen requirements.

Conclusions

In response to climate change the process industries are today and will continue tomorrow to adapt to ensure a sustainable future. At the same time, oil and gas companies are catching up with the other industries and digitally transforming their businesses to operate smarter, more efficiently and more competitively. This digital transformation will be one key enabling building block for a successful energy transition.

AVEVA is strongly committed to a new sustainable future and has invested heavily over the last several years to develop AVEVA Process Simulation, a next generation simulation tool. AVEVA Process Simulation is ready to be the foundation of the digital twin, throughout the project lifecycle, for the new lower carbon projects that will deliver a more sustainable future.

Today's digitally native process engineers will be the front line when it comes to driving this change and the authors of this paper strongly encourage these change agents not to look at the simulation tools of the past but to look towards the future with AVEVA Process Simulation.

References

1. Dickson, D, Tilghman, N, Bonny T, "The Future of Work in Oil, Gas, and Chemicals", Deloitte (https://www2.deloitte.com/us/en/insights/industry/oil-and-gas/future-of-work-oil-and-gas-chemicals.html) 20 October 2020.
2. Oil and Gas Climate Initiative (OGCI) Report "Delivering on a Low Carbon Future" https://www.ogci.com/wp-content/uploads/2020/12/OGCI-Progress-Report-2020.pdf Dec 2020
3. Redepenning, C, Hanak, D. Broeker, S, Depew, C, "Building a Sustainable Chemical Industry Through Process Simulation" Chemweek Webinar, 11 May 2021
4. Construction Industry Institute Report No. RR331-11 - The Maturity and Accuracy of Front End Engineering Design (FEED) and Its Impact on Project Performance https://www.construction-institute.org/resources/knowledgebase/knowledge-areas/general-cii-information/topics/rt-331/pubs/rr331-11 2018
5. Hydrogen production costs to 2040: Is a tipping point on the horizon?", Wood MacKensie (https://www.woodmac.com/our-expertise/focus/transition/hydrogen-production-costs-to-2040-is-a-tipping-point-on-the-horizon/) 2020

622681 Nutrient Circularity to Abate Nitrogen Pollution: Techno-Economic Assessment of Nitrogen Recovery Systems for Livestock Facilities

Abstract:

Since the XIX century, the agricultural sector has experienced an accelerated industrialization progression pursuing its intensification and satisfy the demand of food production and agricultural products, i.e., increasing the agricultural production per unit of input resources (land, labor, time, fertilizer, seeds, and investment) (FAO, 2004). During the last decades, the agricultural intensification is driven by a sustained population increment, a worldwide average income growth, and the trade liberalization and logistics advancements leading the transnational trade of products (Baker and Da Silva, 2014). However, multiple environmental challenges must be faced as a consequence of the industrialization of the agricultural and farming activities: depletion of soil nutrients and organic matter due to mono-cropping practices, excessive and inefficient use of synthetic fertilizers to maintain high cropping yields, spatial concentration and inappropriate management of livestock manure, biodiversity loss, ecosystems degradation, etc. Focusing in the context of current nutrient management, the unbalanced use of synthetic fertilizers and the excessive organic waste generation from livestock facilities have decoupled a previous balanced scenario where the organic waste from livestock activities was used as a nutrient and organic matter supply for growing crops. This has created a dependency on mineral (phosphorus) and synthetic (nitrogen) fertilizers, and the efficient management of animal manure has become a critical need for livestock operations. Therefore, there is a disconnection between locations with excessive nutrient content from organic waste (livestock facilities), and croplands demanding nutrients to keep high yield rates. Also, due to the sparse location between facilities with in-excess nutrient content organic waste (including manure and digestate) and croplands with nutrient needs, the transportation of these wastes for nutrient redistribution is challenging and expensive. Therefore, the implementation of technologies and processes for the recovery of nutrients in a cost-effective form suitable for easy transportation and use in croplands is of utmost importance. Also, this would aid to restore the circularity of agricultural nutrients usage, disrupted by industrial practices (van der Wiel et al., 2020).

In this work, a systematic review and techno-economic analysis of the state-of-the-art technologies for nitrogen recovery from livestock waste is carried out through the development of modules for the calculation of the energy and mass balances, nutrient recovery yield, and techno-economic evaluation. The aim is to identify the most adequate technology to close the nutrient loop from livestock to crops, and the optimal design and operating conditions for its implementation. The processing and management of livestock waste involve all stages from waste collection to the production of the nutrient-rich product: manure preconditioning, optional anaerobic digestion of manure and biogas valorization, solid-liquid separation of manure or digestate, and nutrients recovery. The assessment of technologies is carried out with special emphasis on the resiliency of the technology selected for each scenario. This is because of the existence of multiple variable factors affecting the performance of nutrient recovery facilities, e.g., the variability of waste composition, the prices of the products obtained, and the cost of the utilities required by each unit. Therefore, a flexible framework capable of assessing changes in input parameters is developed, allowing the study of the impact of the fluctuation of these variables on the results, the techno-economic performance of the different nutrient recovery technologies, and the selection of the optimal configuration. In addition, the application of potential government incentives is evaluated, for designing and determining the most cost-effective incentive policies to stimulate the implementation of nitrogen recovery systems at livestock facilities.

The obtained information from the systematic design and evaluation of nitrogen recovery systems is essential for performing a multicriteria decision-making approach to deploy the optimal systems for nitrogen recovery considering the livestock facility size, type of livestock, its economic performance, and the availability of effective incentive policies to encourage the deployment of these facilities. Additionally, the obtained results are very valuable for the development of supply-chain studies to determine the optimal spatial distribution of nutrient recovery facilities, and the potential exchanges of waste and products across a multi-product coordinated network (Sampat et al., 2017).

References

Baker, D., and Da Silva, C., Trends in agri-food systems: Drivers, changes, impacts and overall assessment, FAO Policy Learning Programme (2014).

FAO, The ethics of sustainable agricultural intensification. FAO, (2004)

Sampat, A.M., Martin, E., Martin, M., and Zavala, V.M., Optimization Formulations for Multi-Product Supply-Chain Networks, Computers & Chemical Engineering, 104C, 296-310 (2017)

van der Wiel, B. Z., Weijma, J., van Middelaar, C. E., Kleinke, M., Buisman, C. J. N., and Wichern, F., Restoring nutrient circularity: A review of nutrient stock and flow analyses of local agro-food-waste systems. Resources, Conservation and Recycling, 160, 104901, (2020)

622575 Optimal Design of an Open-Cycle Ocean Thermal Energy Conversion System for Energy and Water Supply Considering Multiple-Objectives

Abstract:

Ocean thermal energy conversion systems (OTEC) represent an attractive economic alternative in communities where a high percentage of electricity production is based on oil, where the water is scarce and, where the communities are close to the ocean. This power generation technology uses the temperature difference between the deep cold water and warm surface water of the ocean to produce electricity and desalinated water using the principles of the Rankine cycle. The objective of this work is to present a rigorous model for the optimal design of the open cycle ocean thermal energy conversion system. This work addresses a multi-objective strategy for defining the relationships between design and utility policy prices in water-energy off-grid systems. The formulation allows obtaining the design and operation of the system. In addition, levels of subsidies and impacts social, energy, and economic policies are considered for defining the utility price for the end-user. The solution to the problem consists of minimizing the total annual cost while social welfare is maximized.

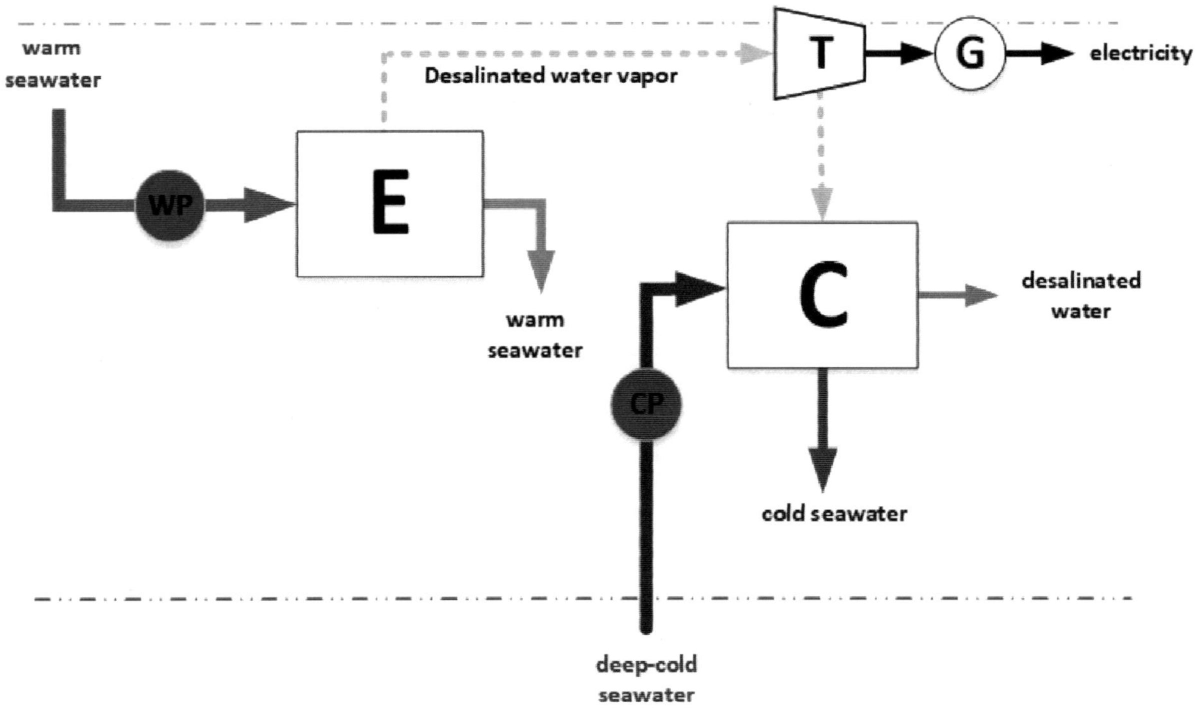

625525 High Yield and Economical Extraction of Rare Earth and Critical Elements from Coal Ash

Abstract:

Abstract:

Rare earth elements (REEs) are strategically important minerals with very limited domestic production. Coal and coal byproducts are a potential source of rare earth elements (REEs) with significant heavy rare earth element (HREE) content. Technologies enabling the environmentally benign recovery of REE, yttrium and scandium (REYSc) from coal combustion byproducts (coal fly ash) would provide the United States with a domestic supply of these strategically important materials while generating jobs and supporting economic growth. Under sponsorship from the Department of Energy, the team of Physical Sciences Inc., University of Kentucky/Center for Applied Energy Research, and Winner Water Services has developed and demonstrated a pilot scale plant to economically produce salable REYSc-rich concentrates and commercially viable co-products from coal ash feedstock using environmentally safe and high-yield physical and chemical enrichment/recovery processes. The pilot plant operates at the scale of approximately 0.4-1 tons per day (tpd) ash throughput for physical processing and about 0.5 tpd for chemical processing, producing at least 50 - 500 g of dry REYSc nitrates concentrate containing at least 10-20 wt.% REYSc (elemental basis). Various ash feedstocks have been qualified, and all contain between 400-500 ppm of REYSc. This presentation will describe our approach and key results for current work at the pilot scale in process optimization and efforts to increase the purity of the extracted rare earth elements to above >85 wt.% REYSc. The talk will also cover characterization of additional processes for the recovery and purification of critical materials as co-products from waste process streams. Target critical materials include aluminum (>50 wt.% purity) and scandium (>85 wt.% purity). The data obtained from the pilot plant optimization and the addition of aluminum and scandium co-products will be used to enhance and validate a commercial scale techno-economic analysis to enable the design of a commercial scale (hundreds of tpd) REYSc plant.

Acknowledgement: This material is based upon work supported by the U.S. Department of Energy under Award DE-FE0027167.

632229 *Gambling on Innovation*

Abstract:

The purpose of this talk is to provide a method for choosing an innovation portfolio, which is based on a simple and fundamental result from information theory. Given a set of innovation projects, estimated payoffs and probabilities of success for each, and an established capital fund, the question is "How much should one invest in each project, in order to maximize growth and minimize the risk of going bust?" That is, how should innovation leaders gamble on their innovation projects? The question relates to a type of process design for innovation processes. The concept in this article is especially important for early-stage innovation projects, where the probability of success is often <50%. As one heuristic, for $p = 50\%$, one needs a payoff $b = 4$ to achieve a 25% median compound annual growth rate. This article provides an algorithm to guide investment decisions in innovation, plus some heuristics that can be used for guidance. It also emphasizes the critical importance of estimating probabilities of success for innovation processes.

PROCESS DEVELOPMENT OF ORGANOSILICON COMPOUNDS

MIKE DEPIERRO AND KALYN FROESCHLE

AIChE Annual Meeting
Boston, Massachusetts

November 9, 2021

SILICON ENABLES AN EXTENSIVE FAMILY OF MATERIALS

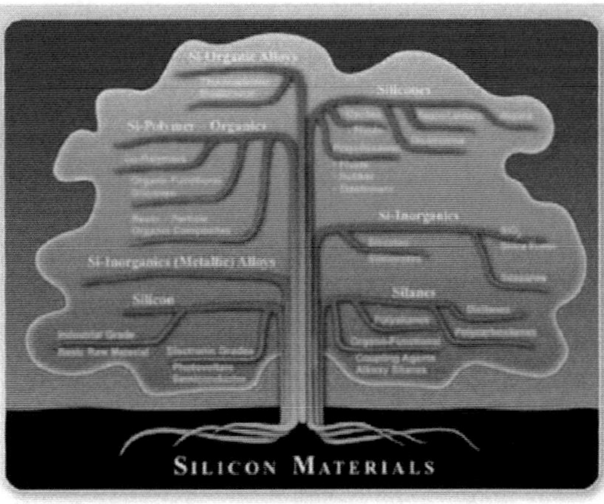

Si Based Chemistry Building Blocks

Silicon	Raw Metal: analogous to coal in the carbon framework.	
↓ Silane	Silanes and silanols are the monomers or building blocks used to make longer chain polymers.	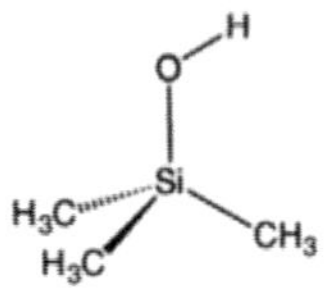
↓ Silanol		
↓ Siloxane (Silicone)	End product is a long-chain polymer with a silicon-oxygen backbone, usually with other molecules bonded on for a desired purpose.	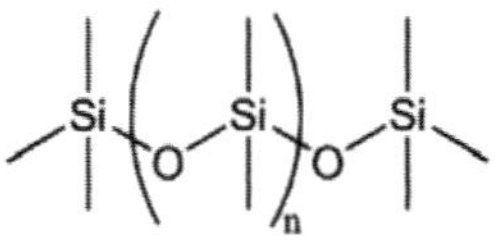

Silicone Applications

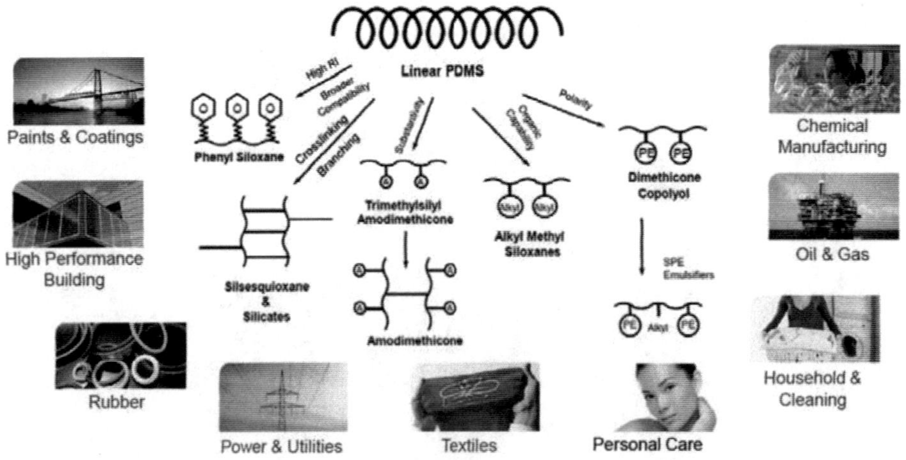

Functionalized Silicones

Siloxanes Functionalized with Pendant Groups

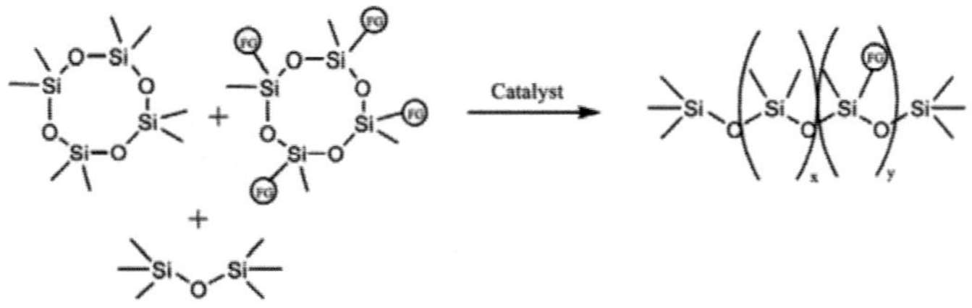

Siloxanes Functionalized with Terminal Groups

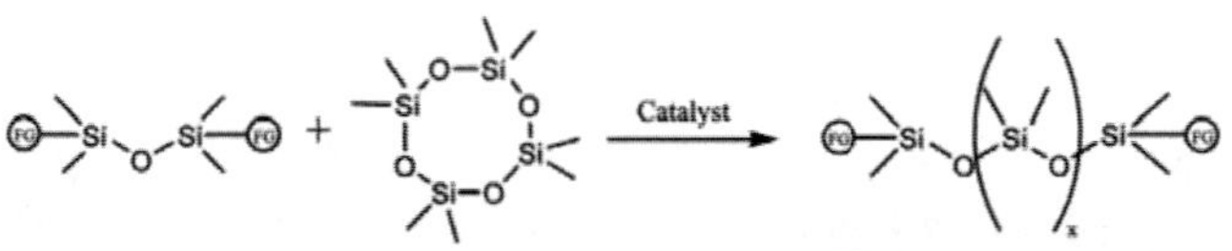

FG = Si-H, Si-Vinyl, Si-Phenyl, etc.

Crosslinking Via Hydrosilylation of Functionalized Silicones

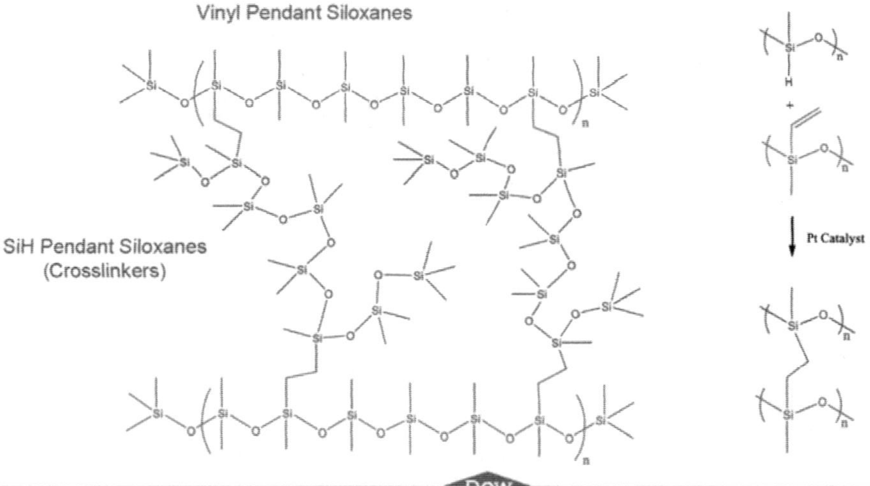

Development of Intermediate for Silicone Elastomers

- Silicone Elastomer Key Performance Properties
  - Elongation
  - Tensile
  - Hardness
  - Viscosity
  - Cure Rate
  - Adhesion
- Synthetic Approach
  - Siloxane Intermediate Synthesis
    - ✓ Copolymerization - Dimethyl / Methyl Hydrogen Siloxanes
    - ✓ Endblocker – Molecular Weight Control
    - ✓ Acid Catalyst
  - Elastomer Synthesis
    - ✓ Pt Catalyzed Hydrosilylation
    - ✓ Mix Part A and Part B
    - ✓ High Temperature Vulcanization (HTV)

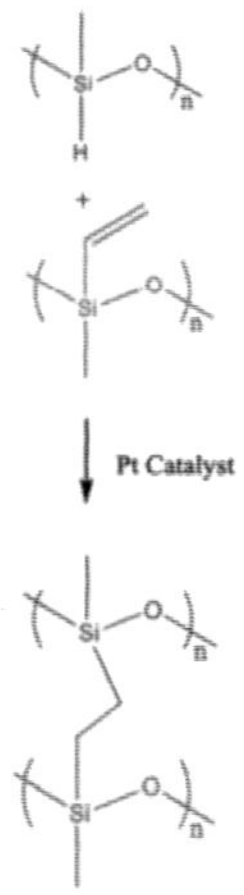

Full Factorial DOE

- Objectives:
  - Determine effect of SiH content and dP
  - Define general ranges required
- Key Responses:
  - Elongation, Tensile, Hardness

Results Summary

Sample #	2x2-1	2x2-2	2x2-3	2x2-4	2x2-5	2x2-6	2x2-7
Response	+SiH, +dP	-SiH, -dP	+SiH, -dP	-SiH, +dP	Center Pt	Center Pt	Center Pt
Elongation							
Hardness							
Tensile							
Viscosity							
Cure Rate							
Adhesion							

Screening Design (2x2)

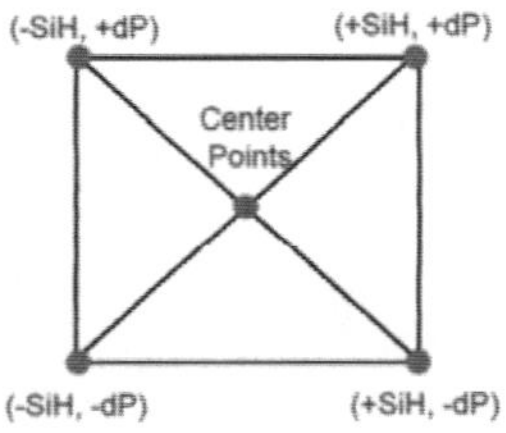

Interactions

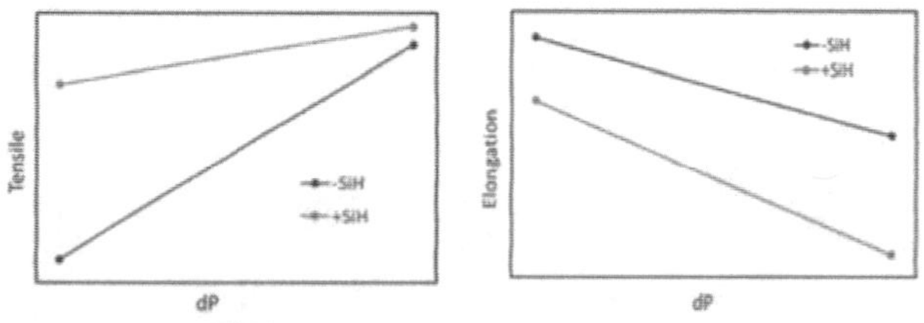

ANALYTICAL METHODS – *IN SITU* VS OFFLINE

Offline Testing Workflow

Pull Sample → Quench/ Prepare Sample → Run Method

45 minutes to 1 hour per sample

Online Testing Workflow

Continuous *In Situ* Monitoring → ~~Quench/ Prepare Sample~~ → Instant Conversion to Concentration & Other Parameters

Data point every ms

- More data points, better for accurately measuring fast kinetics
- No perturbations to reaction due to opening the system or change in volume
- Less handling of chemicals, Improved safety

PARALLEL AUTOMATED REACTORS

- 4 independent reactions
- 8 independent pumps
- (-)40 to 200 ºC
- PTFE and Glass MOC
  - Optimized mixing
- O/H and Magnetic agitation
- Working volume 450 mL
- Purge or blanket nitrogen
- Computer controlled
- Data logging
  - Temperatures
  - Agitation rates
  - Agitation torque
  - Customizable

RESPONSE SURFACE DOE

- Objectives:
  - Study factors (SiH and dP) in depth
  - Define the region of Interest
- Key Responses:
  - Elongation, Tensile, Hardness
- Central Composite Design (CCD)
  - 2 Factors
  - Each factor has 5 levels
  - 8 Center Points
  - 16 Experiments

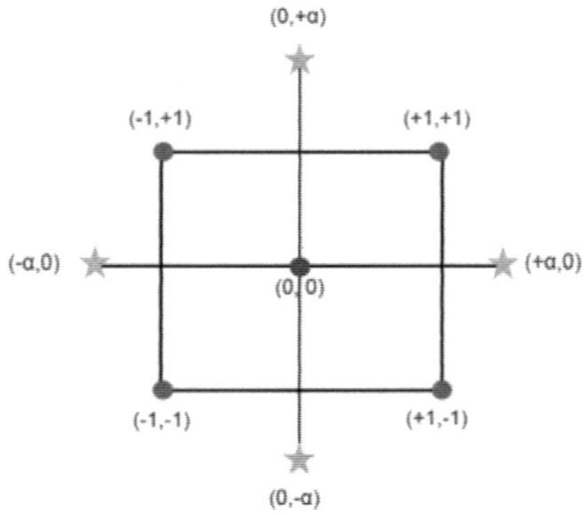

RESPONSE SURFACE DOE RESULTS – ELONGATION

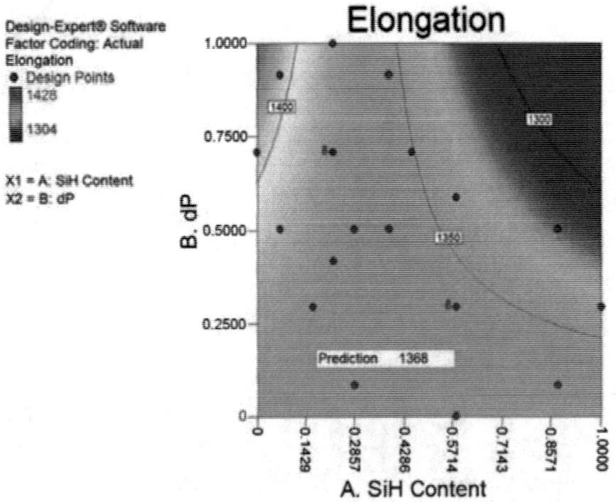

Generally

SiH ↑ Elongation ↓

@ Lower SiH

dP ↑ Elongation ↑

@ Higher SiH

dP ↑ Elongation ↓

Response Surface DOE Results - Tensile

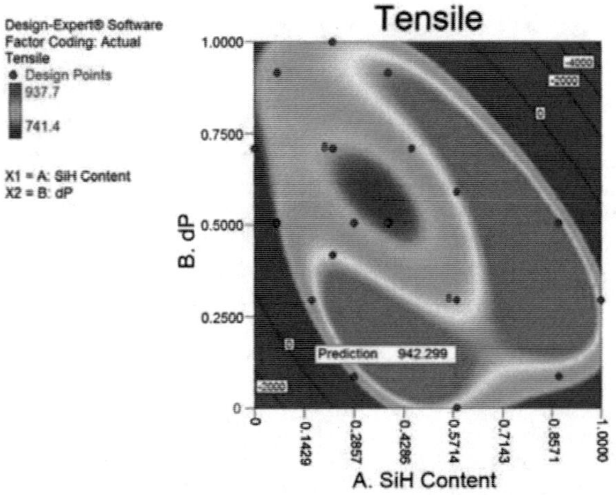

Generally

SiH ↑ Tensile ↑

Response Surface DOE Results - Hardness

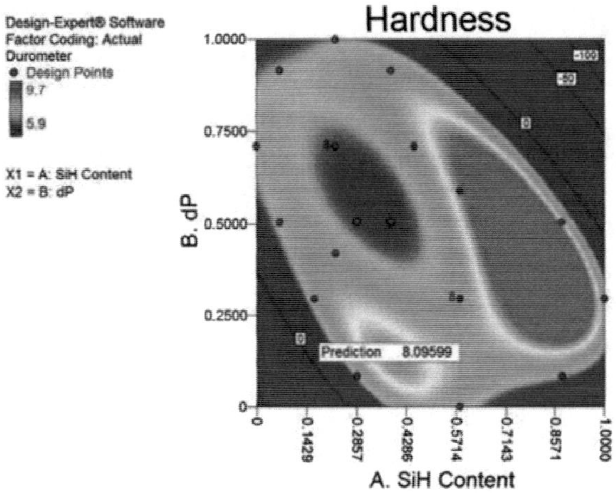

Generally

SiH ↑ Hardness ↑

CROSSLINKING VIA HYDROSILYLATION OF FUNCTIONALIZED SILICONES

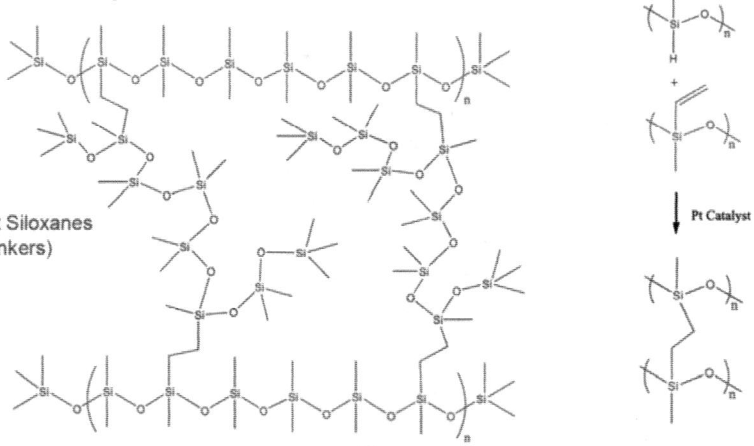

Brook, M. (2000). *Silicon in Organic, Organometallic, and Polymer Chemistry*. New York: John Wiley & Sons, Inc.

RESPONSE SURFACE DOE RESULTS

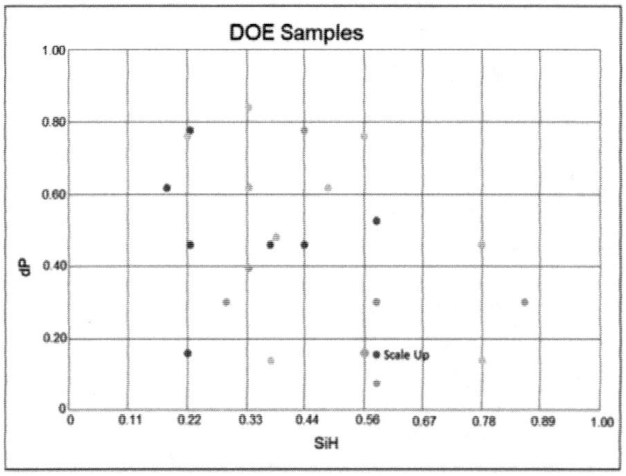

- Numerical system used to rate overall performance
- Weighting of each performance property based on importance in final applications
- Green = Best performance
- Yellow = Intermediate
- Red = Least acceptable
- Blue = Composition that was scaled up and met all performance requirements

SUMMARY

- Design of Experiments (DOE) methods have been useful in the development of organosiloxanes.
- Products with challenging performance requirements have been developed in short cycle times.
- Parallel automated reactor systems sped up development and reduced variability.
- This approach led to a greater understanding of structure / property relationships and improved predictive capability.
- Reduced risk and shortened timeline for scaling up new chemistry.

Dow November 9, 2021 17

ACKNOWLEDGMENTS

Dow Colleagues

Kara Milbrandt	Elizabeth Santos
Michael Hausinger	Mike Molnar
Nate Stelzer	Brandon Taitt
Karen Vo	Phil Irish
Joel Kerbleski	Tim Lueder
Steven Rausch	William Herron
Michael Telgenhoff	Andre Yvon Bessette
Don Eldred	Dan Roscioli
Eric Pearce	Shawn Chen
Dave Wassmann	Alix Schmidt
Jake Remacle	Andy Kenney

Dow November 9, 2021 18

Questions?

624459 Biorefinery Synthesis-a Route Towards Engineered Advanced Fuels

Abstract:

The consumption of liquid transportation fuels in the U.S. is close to 16 billion gallons per year, accounting for 28% of the country's greenhouse gas emissions. The development of sustainable biofuels, particularly those derived from lignocellulosic materials has been identified as an alternative for the mitigation of the negative impacts of uncontrolled fossil fuel consumption. However, most available biofuels have limited fungibility. This limitation has prompted the search for alternative biofuels with better fungibility. A promising platform to achieve this goal relies on the upgrading of ethanol. This platform offers three main advantages: (1) it can use the existing infrastructure for ethanol manufacture; (2) it can produce fuels in the whole distillation spectrum, from gasoline to diesel; and (3) it offers the possibility of using the advances in ethanol chemistry to produce fuels with superior properties. However, the identification of optimal strategies for ethanol upgrading remains challenging. On the one hand, the search space, characterized by the chemistries and catalysts that are used in a biorefinery for ethanol upgrading, is very large. Additionally, the multidisciplinary nature of the problem calls for the simultaneous design of processes and products integrating information of different areas (catalysis, process engineering, and fuel property modeling).

Accordingly, we present optimization-based approaches to address this problem. We give special emphasis to a recently developed superstructure-based framework that enables the early design of biorefineries for ethanol upgrading [1,2]. This framework relies on correlations and targeting methods to estimate capital and operating costs, enabling the identification of optimal upgrading strategies and the exploration of a large number of alternatives, without requiring extensive simulations. The design space considered is representative of the state-of-the-art for ethanol upgrading including twenty-two different chemistries and 112 different catalysts. The proposed framework is used to design optimal ethanol upgrading biorefineries for the production of gasoline, diesel, and jet fuel. We identify trade-offs between biorefinery complexity (measured as the number of chemical transformations) and profit. In general, we observe that increasing the number of chemical transformations improves the economic viability of the process due to an increase in the fuel yield. The possibility and the economic impact of producing fuels of superior quality (e.g. diesel with high cetane number, and gasoline with high octane number) is also studied. Specifically, we discuss how changing product specifications impacts fuel composition, process characteristics, and profit. The use of a system-level analysis allows us to identify non-intuitive processes for ethanol upgrading, both toward drop-in biofuels with properties similar to their fossil counterparts as well as toward biofuels with superior properties.

1. Restrepo-Flórez, J. M. & Maravelias, C. T. Advanced fuels from ethanol-a superstructure optimization approach. *Energy Environ. Sci.* **14**, 493–506 (2021).
2. Ryu, J. & Maravelias, C. T. Efficient generalized shortcut distillation model with improved accuracy for superstructure-based process synthesis. *AIChE J.* (2020)

626391 Molecular Design Targets and Optimization of Low-Temperature Thermal Desalination Systems

Abstract:

Water consumption is critical to modern society; the average American family uses approximation 1140 l of municipal water per day [1], and 130 million Americans face severe water scarcity at least part of the year [2]. Globally, 3.8 billion people currently experience water scarcity [3], and it is estimated that 66% of the world's population could be living under water-stressed conditions by 2025 [4]. Although the costs of water obtained from desalination have fallen in the last decade, these costs are higher than obtaining freshwater from rivers, groundwater, or water recycling. In 2019, less than 1% of the water consumed globally was produced by desalination [5], [6], [7]. The expansion of oil and gas extraction in the US has created new water and environmental challenges [1]. Around 0.50 liters of produced water may be generated to supply energy for one hour in an average American household. While evaporative and reverse osmosis desalination technologies are commonly deployed, they remain energy-intensive and unable to treat high salinity water sources. There is no one-size-fits-all technology for water treatment, including desalination. Instead, there is a growing emphasis on fit-for-purpose treatment in decentralized networks [8]. In this paradigm, water is treated to only the specifications needed for specific end uses. In this talk, we demonstrate, through technoeconomic optimization, ionic liquids are extremely promising for a new desalination process.

We present a mathematical optimization framework to rapidly design desalination processes. Directional solvent extraction (DSE) uses a thermoresponsive solvent to facilitate treatment over a wide salinity range [9]. DSE does not require membranes, which often foul at high salinities, and can utilize low-grade heat, including waste or renewable (solar) sources. Prior work in DSE includes characterization of molecular phenomena, bench-scale demonstrations, and limited process analysis [9], [10]. The framework has two new capabilities [11]: first, we perform simultaneous process optimization and heat integration to rapidly screen directional solvent candidates in seconds. Second, we perform a sensitivity analysis to identify the necessary solvent properties to enable cost-effective DSE processes for treating high salinity water, which is difficult/impossible to desalinate with other technologies. We emphasize these advances in process-scale models can rapidly accelerate DSE development by reducing the need for expensive experiments and guiding (computational) molecular design.

Our framework allows for rapidly screening and identification of quantitative solvent thermophysical property targets. We find the Levelized Cost Of Water (LCOW) is most sensitive to three thermophysical properties: i) the change in water solubility for a fixed temperature change (thermoresponsiveness); ii) the solubility of water in the solvent; iii) the solubility of the directional solvent in freshwater. [11]. Technoeconomic optimization was performed for five candidate fatty directional solvents ranging, giving LCOW predictions between $1.3/m^3 and $109/m^3. Sensitivity analysis shows significant improvements in three solubility properties (thermoresponsiveness of solvent, the solubility of water in the solvent, and solubility of the solvent in freshwater) are needed for the hypothetical fatty acid-like DS to achieve less than $0.5/m^3. In contrast, ILs show much greater promise as directional solvents. Using newly published data from [emim][Tf_2N] and assuming a moderate solvent price of $100/kg, we predict a modest $2.65/m^3 LCOW. Sensitivity analysis shows the required combination of thermophysical properties necessary to achieve LCOW to below $0.5/m^3. These results emphasize the potential of IL directional solvents to desalinate high salinity water, which is currently challenging with existing technologies. As ongoing work, we are considering extensions to hypersaline conditions.

References

[1] EPA, How we use water (DoA: 2020), https://www.epa.gov/watersense/how-we-use-water#Daily%20Life

[2] M.M. Mekonnen, A.Y. Hoekstra, Sustainability: four billion people facing severe water scarcity, Sci. Adv., 2 (February) (2016), pp. 1-7, 10.1126/sciadv.1500323

[3] M. Kummu, J.H. Guillaume, H. De Moel, S. Eisner, M. Flörke, M. Porkka, S. Siebert, T.I. Veldkamp, P.J. Ward, The world's road to water scarcity: shortage and stress in the 20th century and pathways towards sustainability, Sci. Rep., 6 (December) (2016), pp. 1-16, 10.1038/srep38495

[4] UNDESA, International decade for action "water for life" 2005–2015 (DoA: 2019), https://www.un.org/waterforlifedecade/scarcity.shtml

[5] N. Voutchkov, Energy use for membrane seawater desalination – current status and trends, Desalination, 431 (2018), pp. 2-14

[6] E. Jones, M. Qadir, M.T. van Vliet, V. Smakhtin, S. mu Kang, The state of desalination and brine production: a global outlook,, Science of The Total Environment, 657 (2019), pp. 1343-1356, 10.1016/j.scitotenv.2018.12.076

[7] H. Ritchie, Water use and stress, our world in data (DoA: 2020). URL https://ourworldindata.org/water-use-stress.

[8] K.R. Zodrow, Q. Li, R.M. Buono, W. Chen, G. Daigger, L. Dueñas-Osorio, M. Elimelech, X. Huang, G. Jiang, J.-H. Kim, B.E. Logan, D.L. Sedlak, P. Westerhoff, P.J.J. Alvarez, Advanced materials, technologies, and complex systems analyses: emerging opportunities to enhance urban water security, Environmental Science & Technology, 51 (18) (2017), pp. 10274-10281, 10.1021/acs.est.7b01679

[9] T. Luo, A. Bajpayee, G. Chen, Directional solvent for membrane-free water desalination—a molecular level study, Journal of Applied Physics, 110 (5) (2011), Article 054905, 10.1063/1.3627239

[10] S. Alotaibi, O. Ibrahim, S. Luo, T. Luo, Modeling of a continuous water desalination process using directional solvent extraction, Desalination, 420 (2017), pp. 114-124, 10.1016/j.desal.2017.07.004

[11] Alejandro Garciadiego, Tengfei Luo, Alexander W. Dowling, Molecular design targets and optimization of low-temperature thermal desalination systems, Desalination, 504 (2021), 10.1016/j.desal.2021.114941.

629691 Energy Integration through Retrofitting of Heat Exchanger Network at Equinor Kalundborg Oil Refinery

Abstract:

Heat integration studies are commonly performed in the wider chemical industry to map current energy utilization and identify potential improvements with respect to energy efficiency. In this regard there are several established methodologies such as: Pinch analysis, mathematical programming and hybrid methods [1][3][4][5]. One of the chief challenges in the practical application of retrofitting in HEN, remains the difficulty associated with the data collection and their validation. In this contribution, we report a heat integration studies to a full-scale refinery plant. We present a methodology that involves iterative application of a process simulator with plant data as well as Monte Carlo simulation to analyze the potential sources of uncertainty in the data collection part.

In 2019 Equinor oil refinery in Kalundborg Denmark used 2908103 MWh, in which 88% was fuel-gas and oil, 6.2% electricity, 2.7% steam and 3% natural-gas [2]. Combined with a Solomon index of 80.3, the potential for energy optimization and therefore also economical and environmentally prospects are substantial. An energy optimization was conducted at Equinor oil refinery in Kalundborg, based on the methodology of pinch analysis and HEN retrofitting. The optimization was constrained to the Naphtha hydrotreating section, Catalytic reforming section, VBGO/LVGO/LGO hydrotreating section and finally the hydrotreating section of KERO/LKERO and LGO.

Design duties of heat exchangers could not be used, as process parameters and composition changes from day to day. All unit operations in the 4 sections were simulated in PRO/II, whereas out the 36 heat exchangers, 17 are shell and tube, 15 are air coolers and 4 are red heaters. Average temperature, pressure, volume flow rate and assay of feed and residue was taken over a month. The date of month was chosen, when the refinery had been cleaned and fluctuations was relatively low, in order to easier balance mass and energy.

Based on the SRK EOS, enthalpies were segmented and exported into UniSim ExchangerNet, to evaluate the energy targets and network performance. Considering a base case of ΔT_{min} = 20°C retrofit solutions were proposed by locating pinch violations and inefficiencies.

The retrofit solutions were economically evaluated by considering the investment cost of new heat exchangers and the energy savings from each retrofit option. A sensitivity analysis was conducted by changing ΔT_{min} and investment cost.

Finally, a Monte Carlo simulation was conducted to evaluate the uncertainty of the pinch temperature, external minimum utility heat and external minimum cooling utility, by considering the fluctuation of temperature and massflow.

References

[1] Mary O. Akpomiemie and Robin Smith. \Cost-eective strategy for heat ex-changer network retrot". In: Energy 146 (Mar. 2018), pp. 82{97. issn: 03605442. doi: 10.1016/j.energy.2017.09.005.

[2] Green account reports - equinor.com. url: https ://www.equinor.com/no/ where-we-are/denmark/green-account-reports.html.

[3] Ning Jiang et al. \Heat exchanger network retrot with a xed network structure". In: Applied Energy 127 (Aug. 2014), pp. 25{33. issn: 03062619. doi: 10.1016/j.apenergy.2014.04.028. url: http://dx.doi.org/10.1016/j.apenergy.2014.04.028.

[4] Kitipat Siemanond and Supachai Kosol. \Heat exchanger network retrot by pinch design method using stage-model mathematical programming". In: Chemical Engineering Transactions. Vol. 29. Italian Association of Chemical Engineering - AIDIC, 2012, pp. 367{372. doi: 10.3303/CET1229062.

[5] TN TJOE and B LINNHOFF. \USING PINCH TECHNOLOGY FOR PROCESS RETROFIT". In: (1986), pp. 47{60. issn: 00092460.

630586 Simultaneous Synthesis of Metabolic and Process Engineering for the Production of Muconic Acid

Abstract:

Recent advances in metabolic pathway reconstruction and in silico modelling have enabled the development of microbial strains suitable for efficient cell-factories. Emerging infrastructures of Design-Build-Test-Learn paradigms hold a tremendous promise to innovate with new products, also novel integrated processes that can be intensified for the highest efficiency[1]. Still, the current Design-Build-Test-Learn (DBTL) cycle approach in biofoundries and biorefineries remains a disconnected sequence of tasks with an undisclosed potential to integrate. Specifically, the genetic modifications of strains are usually driven by the overproduction of target metabolites without considering downstream processing which can account for up to 80% of the overall bio-production cost[2]. Therefore, there is a need for an inclusive methodical approach that incorporates separation processes in metabolic engineering upstream design.

The paper presents a systems integration approach for the computational strain design workflow for the identification of reaction eliminations that reshape network connectivity in way that both biomass production and revenue are simultaneously maximized by utilizing a bilevel optimization framework. The research expands the outer problem of the bilevel approach that is established in the literature with the use of a superstructure scheme that addresses several design options simultaneously with the several options to select the pathways. The superstructure scheme is laid out to reduce model and optimization complexities. The existence of each process is denoted by binary variables. Our method entails the identification and the categorization of these technologies as well as the insertion of the model equations and their economic parameters into the optimization problem. Separation steps include pretreatment, cell removal, product isolation, concentration, purification and refinement. The superstructure's input stream is a variable determined by the optimization problem and is separated into five components: product, liquid by-product, water, cells, and solid by-product. Each process redistributes the stream's components through linearized model equations. The problem can be constrained based on the specific bio-process characteristics and objectives. The model has been developed in GAMS environment and solved using the BARON global optimization solver.

We assess the effect of the new optimization goal, for varying number of metabolic interventions to the downstream separation network and the bioprocess revenue. We then compare the enriched model to the previous analysis, which only aims to maximize the production of a target metabolite. To showcase the functionality and effectiveness of the developed model we applied the workflow to a muconic acid producing strain of *S.cerevisiae* (iMM904 GEM) that includes the necessary heterologous pathways. Additionally, by utilizing different metabolic engineering tools, it is possible to acquire a variety of promising metabolic interventions that could lead to profitable biorefineries. Overall, this computational framework could be an important step to bridge the gap between strain design and process engineering.

References:

[1] IBISBA, www.ibisba.eu

[2] Kiss, A. A., & Rito-palomares, M. (2014). *A systems engineering perspective on process integration in industrial biotechnology*. *October*. https://doi.org/10.1002/jctb.4584

629275 Process Design for the Production of Xylitol in a Multi-Product Biorefinery

Abstract:

Xylitol is a sugar substitute with various beneficial health properties and a potential platform chemical that gained significant attraction in the last decade [1]. There exists a great potential for xylitol to be produced biotechnologically with engineered cell factories instead of the current chemical production process [2]. Moreover, the use of lignocellulosic biomass as feedstock for the biotechnological process adds further benefits in terms of sustainability. However, using lignocellulosic biomass as feedstock introduces several challenges for the conceptual design of the biotechnological processes, as considerations about potential value-added co-products, pretreatment technologies, and possibilities for process integration, amongst others [3].

In the presented study for a base-case process design, a synergistic optimization-based process design framework (S3O) is used to overcome the named hurdles and to design this process conceptually [4]. In this base case design, succinic acid and sustainable aromatic kerosene are chosen as value-added co-products and wheat straw as feedstock. Besides, the generation of heat as a product for possibly integrating it with the other products' downstream processes is considered. For all unit operations, mechanistic models are developed, validated, and assessed regarding their robustness via an uncertainty and sensitivity analysis. A design space exploration is performed to determine the most sensitive variables with respect to the objective function, and flowsheet sampling simulations are performed to fit different surrogate models for each flowsheet. As the combination of the surrogate model and optimization technique in the following superstructure optimization step has a crucial influence on the performance in the optimization problem, a benchmark of different options is performed and compared with previously obtained results [4,5]. The superstructure optimization step is a set of candidate process topologies, which are subsequently optimized in a simulation-based optimization approach to consolidate the process design under uncertainty [6].

As objective function in the framework, key performance indicators are selected, e.g., net present value or minimum selling price. The resulting base case process itself is evaluated against both the criteria of being economically viable and its sustainability. Potential options for process integration regarding the reuse of heat and CO_2 are assessed and compared to the base case design. Ultimately, the presented process design is evaluated regarding its feasibility compared to the established chemical production process, and potentials for further improvements considering the whole value chain and other potential value-added co-products and feedstocks are given.

Acknowledgments

The project is part of the Fermentation-Based Biomanufacturing Initiative at DTU and is funded by the Novo Nordisk Foundation under the grant NNF17SA0031362.

References

[1] A.F. Hernández-Pérez et al. 'Xylitol bioproduction: state-of-the-art, industrial paradigm shift, and opportunities for integrated biorefineries', *Critical Reviews in Biotechnology*, 2020, Vol. 39(7), pp. 924-943.

[2] Y.D. Arcaño et al.' Xylitol: A review on the progress and challenges of its production by chemical route', *Catalysis Today*, Vol. 344, 2020, pp. 2-14.

[3] T. Chaturvedi et al. 'Developing Process Designs for Biorefineries – Definitions, Categories and Unit Operations', *Energies*, 2020, Vol. 13(6), 1493.

[4] N.I. Vollmer et al.' Synergistic Optimization Framework for the Process Synthesis and Design of Biorefineries', *Frontiers in Chemical Science and Engineering*, under review.

[5] N.I. Vollmer et al.' Benchmarking of Surrogate Models for the Conceptual Process Design of Biorefineries', *Computer-Aided Chemical Engineering*, accepted.

[6] R. Al et al. 'Stochastic simulation-based superstructure optimization framework for process synthesis and design under uncertainty', *Computers & Chemical Engineering*, 2020, Vol. 143, 107118.

632749 Digital Design of a Lomustine Manufacturing Process Using Pharmapy

Abstract:

Model-based process design is being increasingly adopted by the pharmaceutical industry as it seeks to modernize process development methods and embrace Industry 4.0 technology and tools. This encompasses modeling, computational and numerical methods development, and experimental planning, leading to the construction of digital twins/virtual plants of the processes under study. In this way, it is possible to explore a variety of process conditions and configurations, aiming for higher productivity, lower processing times, lower waste formation and higher product (consumer) and process (operational) safety. Also, model-based process design supports industry and government decision-making to ensure a safe and affordable supply of critical medicines to the public.

Recently, lomustine, a high-value, low-volume cancer drug has seen a dramatic increase in price as a consequence of limited supply [1]. This unfortunate situation has spurred research into discovery of novel routes for its manufacture. One such effort is recent experimental work which identified a path for continuous lomustine synthesis through a series of telescoped reaction steps [2]. This innovation in lomustine synthesis have motivated the next steps towards optimal production of the drug substance lomustine, specifically i) the analysis of extended processes for the formation and separation of the crystalline active pharmaceutical ingredient (API), and ii) the evaluation and comparison of different process configurations, viz. end-to-end batch or continuous processes, and hybrid schemes involving batch, semi-batch and continuous operations. This extended analysis supports informed decision-making aimed at the successful commercial production of this important API.

The process alternatives considered in this work differ in their operating mode, i.e. batch/continuous/hybrid manufacturing, where unique benefits and concerns exist for each case. In the context of pharmaceutical manufacturing, where small product campaigns are common, the start-up of processes involving continuous operation can be a major issue, greatly impacting product yield and equipment size [3]. On the other hand, batch and semi-batch operation offer operational flexibility and are not limited by residence time; however, some advantages of continuous processing such as consistency of product critical quality attributes (CQAs), better controllability, and reduced equipment size [4] may be lost. For this reason, a framework that enables the systematic evaluation of process alternatives is necessary, to make valid comparisons based on technical and/or economic criteria.

In this work, we present the evaluation and comparison of processing alternatives for lomustine production via digital twin construction and analysis. This is made possible by the use of PharmaPy, a Python-based, object-oriented platform for the development of pharmaceutical processes [5]. PharmaPy offers tools for dynamic model construction and execution under alternative operating modes, and also permits the incorporation of experimental data for model calibration purposes. As a framework for pharmaceutical process development, PharmaPy is designed to facilitate generating digital twins via robust parameter estimation, in this case, using experimental data from ongoing experimental efforts on lomustine [6]. Then, different process alternatives are constructed via object creation and manipulation using Python syntax, as offered by the PharmaPy platform. Each processing alternative includes API synthesis, crystallization and vaporization (solvent switch), and a series of solid-liquid separation units, viz. filtration, deliquoring, washing and drying of the resulting solid cake [7]. Using the set of unit operations included in PharmaPy, a complete representation of the drug substance manufacturing process of lomustine can be conveniently constructed.

Comparison of the process alternatives is made on the basis of economic criteria (capital and operational expenses) for a desired product throughput, while meeting operational constraints (maximum concentrations in relation to solubility constraints, maximum operating temperature/pressure), and product CQAs, e.g. crystal size distribution of the solid cake, residual impurities/solvent(s) of the dried product. To analyze uncertainty in generated digital twin predictions, global sensitivity analysis using random sampling (viz. MonteCarlo and Sobol) is used to evaluate the impact of model parameters (kinetic constants, transport coefficients) and operating conditions (input flows, operating temperature/pressure) on the mentioned CQAs and on capital operational costs. Sensitivity analysis serves as a valuable input for further optimization efforts, identifying critical process parameters (CPPs) that have most impact on the outputs of interest, and also providing insights on of how parametric uncertainty propagates into the model predictions, which can support the determination of the operational design space [8], [9].

References

[1] J. Funaro, H. Friedman, and M. Weant, "A costly 'rebranding' of an old drug comes with a 700% price increase," *The Cancer Letter*, 27-Sep-2017.

[2] Z. Jaman, T. J. P. Sobreira, A. Mufti, C. R. Ferreira, R. G. Cooks, and D. H. Thompson, "Rapid On-Demand Synthesis of Lomustine under Continuous Flow Conditions," *Org. Process Res. Dev.*, vol. 23, no. 3, pp. 334–341, 2019.

[3] D. Casas-Orozco *et al.*, "Application of PharmaPy in the digital design of the manufacturing process of an active pharmaceutical ingredient," in *31st European Symposium on Computer Aided Process Engineering*, 2021.

[4] E. Içten *et al.*, "A Virtual Plant for Integrated Continuous Manufacturing of a Carfilzomib Drug Substance Intermediate, Part 1: CDI-Promoted Amide Bond Formation," *Org. Process Res. Dev.*, 2020.

[5] D. Casas-Orozco *et al.*, "PharmaPy: an object-oriented tool for the development of hybrid pharmaceutical flowsheets," *Comput. Chem. Eng.*, 2021 (submitted).

[6] J. Mackey *et al.*, "End-to-End Reconfigurable Process Development of a Hybrid Manufacturing System for the Cancer Drug Lomustine," in *AIChE Annual Meeting*, 2021.

[7] F. Destro *et al.*, "Digital design of an intensified filtration-drying unit for pharmaceutical upstream manufacturing," in *AIChE Annual Meeting*, 2020.

[8] D. Laky, S. Xu, J. S. Rodriguez, S. Vaidyaraman, S. G. Muñoz, and C. Laird, "An optimization-based framework to define the probabilistic design space of pharmaceutical processes with model uncertainty," *Processes*, vol. 7, no. 2, 2019.

[9] L. X. Yu *et al.*, "Understanding pharmaceutical quality by design," *AAPS J.*, vol. 16, no. 4, pp. 771–783, 2014.

630699 SPICE_ED: A Framework for Simultaneous Materials Screening and Process Synthesis for Extractive Distillation

Abstract:

Distillation is the most widely used technology for separating mixtures into pure components in the chemical process industry [1]. However, conventional distillation is energy-intensive and often fails to meet the purity requirement for azeotropic mixtures. Extractive distillation (ED) is an intensified process that uses a low-volatile solvent as entrainer and can separate azeotropic mixtures into pure components [2]. ED has additional degrees of freedom (DOF) because of the choices for solvent selection and flow rates [3]. While many works (e.g., [3-5]) have addressed ED design, a systematic framework is still required for optimal synthesis, optimization, and further integration of ED configurations in an end-to-end processing network *aka* process flow diagram. Furthermore, recent emphasis on inherently sustainable, energy-efficient, and environmentally benign process design requires in-tandem selection of candidate solvents at the conceptual design stage.

Building block-based representation [6] of chemical process flowsheets was first used in developing the prototype design tool SPICE (Synthesis and Process Intensification of Chemical Enterprises) and was further modularized to specifically incorporate reactive distillation [7], membrane separation [8], and membrane reactors [9]. In this work, we introduce a new module SPICE_ED (SPICE involving Extractive Distillation). SPICE_ED specializes in systematic process synthesis, intensification, and optimization of ED columns and associated process configurations. SPICE_ED employs a mixed-integer nonlinear model (MINLP)-based equation-oriented design approach, where the conceptual decisions are represented using design building blocks. Based on user-defined specifications on feed (flow rates, composition, thermo-physical properties) and product purity and demand constraints, SPICE_ED performs automatic synthesis of optimal ED flowsheets. A key feature is the efficient prediction of fluid properties, phase equilibrium, and solvent-extraction phenomena using both rigorous and surrogate thermodynamic packages and detailed mathematical models e.g., equation-of-states (Peng Robinson or van der Waals) and Gamma-Phi method with NRTL model. Computational efficiencies are obtained by using reliable surrogate models ($R^2 > 0.99$) for thermodynamic and physical properties. This allows us to generate detailed configurations of ED columns considering appropriate feed inlet trays, boil-up, and reflux ratios, and side product stream trays. Furthermore, we can perform large-scale computational screening and selection of solvents, such as ionic liquids (ILs), for designing ED-based separation process involving azeotropic mixtures. In this context, SPICE_ED can be also used to generate structure-property-performance-process relationships mapping the entire range of solvent design space. We illustrate these capabilities by solving a simultaneous material selection and process design problem involving IL-based extractive distillation for hydrofluorocarbons (HFC) separation. Specifically, for the separation of R-410A, which is a refrigerant mixture of two HFCs (R-32 and R-125), SPICE_ED suggests an optimized process flowsheet that reduces the energy consumption by 65% from the base case flowsheet [11]. Additionally, the optimized process is environmentally sustainable as it emits 72% less indirect CO_2.

References:

[1] Humphrey, J.L., 1995. Separation processes: playing a critical role. *Chemical Engineering Progress, 91*(10).

[2] Lei, Z., Dai, C., Zhu, J. and Chen, B., 2014. Extractive distillation with ionic liquids: a review.

AIChE Journal, *60*(9), pp.3312-3329.

[3] Kossack, S., Kraemer, K., Gani, R. and Marquardt, W., 2008. A systematic synthesis framework for extractive distillation processes. *Chemical Engineering Research and Design*, *86*(7), pp.781-792.

[4] Emhamed, A.M., Czuczai, B., Rev, E. and Lelkes, Z., 2008. Analysis of extractive distillation with mathematical programming. *Industrial & engineering chemistry research, 47*(24), pp.9983-9995.

[5] Valencia-Marquez, D., Flores-Tlacuahuac, A. and Vasquez-Medrano, R., 2012. Simultaneous optimal design of an extractive column and ionic liquid for the separation of bioethanol–water mixtures. *Industrial & engineering chemistry research, 51*(17), pp.5866-5880.

[6] Demirel, S.E., Li, J. and Hasan, M.M.F., 2017. Systematic process intensification using building blocks. *Computers & Chemical Engineering*, *105*, pp.2-38.

[7] Demirel, S.E., Li, J., El-Halwagi, M.M. and Hasan, M.M.F., 2020. Sustainable Process Intensification Using Building Blocks. *ACS Sustainable Chemistry & Engineering*, *8*(48), pp.17664-17679.

[8] Demirel, S.E., Li, J. and Hasan, M.M.F., 2021. Membrane Separation Process Design and Intensification. *Industrial & Engineering Chemistry Research*, DOI: 10.1021/acs.iecr.0c05072.

[9] Monjur, M.S., Demirel, S. E., Li, J. and Hasan, M.M.F., 2021. SPICE_MARS: A Process Synthesis Framework for Membrane-Assisted Reactive Separations. *Industrial & Engineering Chemistry Research*, Under Review.

[10] Shiflett, M.B. and Yokozeki, A., 2006. Separation of difluoromethane and pentafluoroethane by extractive distillation using ionic liquid. *Chimica oggi*, *24*(2), pp.28-30.

621961 Electromagnetic Simulations Aided Stable Microwave Reactor Configuration

Abstract:

The application of microwaves to enable sustainable manufacturing is rapidly gaining importance as the pathway to process intensification. Application of electromagnetic radiation to traditional reactor configurations, such as fixed beds, is hindered by the development of hotspots owing to spatial focusing of energy. Structured reactors, such as catalyst coated monoliths, offer a solution but are limited by the small amount of catalyst loading.

In this talk, we present a novel strategy to mitigate these effects. By using electromagnetic simulations, we predict a new reactor configuration, called packed monoliths, consisting of catalytic particles packed in a SiC monolith. We show that the SiC monolith acts as a preferential absorber of microwaves, suppressing the hotspots arising inside the catalytic bed. We also highlight the role of wall thickness and material properties in suppressing hotspots. Experiments conducted with the predicted configuration show an order of magnitude improvement in productivity and the suppression of hotspot formation.

624840 Decoupling MW Sensitivity and Reactivity: Towards Understanding Fe-ZSM-5@SiC As Effective Microwave Catalyst for Methane Dehydro-Aromatization

Abstract:

Due to the limitations in transportation of abundantly available natural gas to point-of-use locations, large amounts of natural gas are flared causing significant environmental pollution. Decentralized conversion of natural gas to value-added chemicals could help address this issue. One such conversion route is the direct conversion of methane to benzene via methane dehydro-aromatization (MDA). Typically, metal/zeolite catalysts are used for this reaction, where methane is activated on a metal site and then oligomerized on the Bronstedt acid sites of the zeolite where the zeolite provides shape selectivity towards the desired product benzene. Thermodynamically, MDA requires high temperatures to attain sufficient conversion. However, at these high temperatures, carbon deposition becomes favorable and results in rapid deactivation of the catalyst, rendering the process uneconomic. This motivates the search for novel ways of conducting this reaction. One such approach is based on microwave-assisted heterogeneous catalysis.

In microwave-assisted heterogeneous catalysis, microwave radiation is used to selectively heat the (solid) catalyst. By delivering the energy directly to the active site, this not only promises higher energy efficiency than conventional thermal heating, but also might mitigate coke formation by keeping the gas phase relatively cold and hence suppressing undesired side-reactions that result in coke forming. However, due to their complexity, microwave-assisted heterogeneous reactions are poorly understood to-date and systematic studies on how microwaves interact with catalytic materials are largely lacking. We had previously identified Fe-ZSM-5 as a promising catalyst for microwave-assisted MDA and had systematically explored the effect of catalyst properties, including acidity, metal loading, and particle size, on conventional reactivity, microwave absorption, and microwave reactivity.

Building on this work, we are now exploring Fe-ZSM-5@SiC core@shell catalysts which combine the activity of the metal-exchanged zeolite with the well-known high MW absorption of SiC. By designing these core@shell structures, we aim decouple MW heating and catalytic activity via separation into different material components while keeping the two components in intimate contact and hence maximizing heat transfer. We developed a synthetic route for these core@shell catalysts and systematically explored key synthesis parameters, such as solvent amount and ZSM-5/SiC ratio, to obtain homogeneous, well-defined Fe-ZSM-5@SiC structures while minimizing undesired homogeneous nucleation of ZSM-5 in the synthesis mixture. The Fe-ZSM-5@SiC samples were carefully characterized via SEM, EDX, XRD, XPS and BET to confirm the synthesis of desired catalyst structure and composition. Comparative reactive evaluation at conventional thermal heating and microwave-assisted conditions are currently on-going in comparison between various catalyst configurations, including core@shell structures and physical mixtures. Overall, we expect that such intentional design of catalysts for use in microwave-assisted heterogeneous catalysis will enable taking full advantage of this promising reaction concept.

627080 Controlling Homogeneous/Heterogeneous Reactions in Alkane Dehydrogenation

Abstract:

Heterogeneous reactions at sufficiently high temperatures, such as in the hydrocarbon conversion, are commonly accompanied by homogeneous reactions. The homogeneous reactions follow a radical mechanism, which typically has a low selectivity and results in low yield of products. One way to suppress the low selectivity homogeneous reactions is through localized heating, for example using a microwave reactor. Development of a strategy to control the homogeneous/heterogeneous reactions can guide improvement of the process.

In this talk, we use propane dehydrogenation over Ga-alumina based catalyst as a model reaction to compare the conversion and selectivity of localized heated reactor vs. a conventional reactor. We use a coprecipitation method to prepare the Ga–doped alumina catalyst, which shows great reactivity and stability thanks to the synergy between Ga and Al atoms. XRD and Al27 NMR characterization of the catalyst shows incorporation of Ga atoms into the bulk of Al2O3 and occupied the tetrahedral Al sites, which shows a strong correlation with Ga-alumina activity. We also present a combined parametric experimental study and computational fluidic dynamics simulation to discuss the influence of temperature, flow rate, packing density, and reactor size on homogeneous/heterogeneous reactions. We extend our validated computational model for scale-up and provide insights into optimized operation.

631550 Microwave Heating of Liquid-Liquid Biphasic Systems

Abstract:

Microwave (MW) technology can be a powerful tool for the intensification of chemical processes and is often recognized for the benefits of high-efficiency and fast and volumetric heat delivery. In multiphase systems, MWs offer the additional advantage of selective heating, which may allow for further improvement of processes such as reactive extractions beyond what is possible in conventional systems. However, limited fundamental understanding of design principles has hindered the effective use of MWs. Further complicating the issue – inadequate temperature measurement strategies often yield misleading results or leave the analysis of topics like selective heating wholly unapproachable. In this work, a commercial single-mode microwave applicator is modified to allow for spatial resolution in temperature measurements, both a continuous-flow microreactor and in batch mode. Simultaneously, corresponding fluid dynamics models are built to simulate temperature profiles. This complementary experimental and computational approach allows us to explore selective MW heating of liquid-liquid biphasic systems. An analytical expression for temperature difference between phases is developed and compared to experimental measurements for various organic solvents in contact with water and various key variables in both batch and continuous flow systems. The agreement between simulations and experiments is excellent. We present design guidelines to tune the selective heating in microwave reactors.

631943 Microwave Fluidized Bed and Microwave Hybrid Fixed Bed Reactor for the Ethane Dehydroaromatization Reaction

Abstract:

Ethane dehydroaromatization under microwave heating has been demonstrated as an efficient process for the conversion of ethane to ethylene over a metal loaded zeolites. However, microwave heating has displayed a deficit in the capability to efficiently aromatize the abundant amount of ethylene produced into the target products benzene and toluene. The low aromatic selectivity is attributed to the underutilization of the catalyst bed due to hot spot formation that drives the high conversion at low temperatures. The objective of this research was to take a systems approach by changing the microwave reactor configurations so that catalyst bed utilization can be increased. In this research a 6%Mo/ZSM-5 catalyst was tested in thermal, microwave, and microwave hybrid heated fixed bed setups. Thermal or microwave heated non-optimized fluidized bed configurations for the ethane dehydroaromatization reactions were also tested. The effects of each reactor configuration were concluded by comparing similar ethane conversions by varying the temperature and power needed in each reactor setup while holding all other variables conditions constant. The temperature distributions on the different microwave heating was examined by a thermal imager and the reaction products were tracked by a micro-GC. It was determined that the addition of the thermal heating to the microwave fixed bed was able to increase aromatic selectivity in comparison to only microwave heating. The fluidization of the catalyst bed resulted in a three times higher aromatic selectivity than when compared to the microwave fixed bed. The microwave fluidization prevented localized hot spot formation thus preventing the detrimental degradation of the zeolite structure. However, the microwave fluidized bed required three times the amount of power to heat the catalyst in comparison to the microwave fixed bed.

626817 Modeling, Engineering, and Integration of a Smart Moisture Absorbing Foam (SMAF) into a Man-Portable Atmospheric Water Extraction Device

Abstract:

This poster details the design and development of a new Atmospheric Water Extraction (AWE) device capable of >5.5 L/day of water production. The device leverages the unique properties of a Smart Moisture Absorbing Foam (SMAF): high water uptake and compressive water release. Operation of the AWE system is simple and efficient. The unit pushes air through a pre-filter and manifold using a battery-powered fan. Air is distributed through a thin macroporous sorbent matrix, resulting in maximum contact between atmospheric water vapors and the embedded hydrophilic nanoparticles. Water release occurs when the operator turns a crank-shaft, compressing the SMAF bed and releasing the potable water into a catchment container for consumption.

The use of compression to switch between hydrophilicity and hydrophobicity enables the AWE system to produce 5.5 – 7.5 L/day of water within a strict weight envelope <2.5 kg. The device operates 24 hrs/day over a wide range of temperature (40-110 °F) and humidity (RH = 10-100%) with minimal interaction. This poster describes modelling of such a system from first-principles through a detailed computational fluid dynamic (CFD) simulation. These results are used to develop a computer aided design of a prototype system. The output of this research is an AWE system capable of Defense Advanced Research Projects Agency (DARPA) water production requirements within the size, weight and power envelope. Such technology has implications as a technology disruptor for both military and humanitarian purposes.

Acknowledgement of Support and Disclaimer: This material is based upon work supported by Defense Advanced Projects Agency (DARPA) under Contract No. HR001121C0032. Any opinions, findings and conclusions or recommendations expressed in this material are those of the author(s) and do not necessarily reflect the views of Defense Advanced Projects Agency (DARPA).

628210 Multi-Level Modelling of Low-Carbon Heating Systems: Integrating Household-Level Cost-Benefit Analysis with National-Level Value Chain Optimisation

Abstract:

Achieving net zero greenhouse gas emissions, as a number of countries have pledged to do, requires decarbonising all sectors of the energy system. In the UK, the decarbonisation of residential space and water heating presents a unique challenge because the current natural gas-based heating systems, present in around 85% of the housing stock, are often lower cost than the low-carbon alternatives being considered by the UK government, such as heat pumps, electric storage heaters, and a variety hydrogen heating technologies. In addition, there is a considerable variability in the characteristics, thermal efficiency, occupancy, and heating patterns in the current housing stock that is not represented in whole-system level modelling studies. This study aims to combine modelling at the household and whole-system level to answer the question *"How do the costs and benefits of low-carbon heating, at the household level, support or undermine a whole-system optimisation approach"*.

A cost-benefit model is being developed based on existing data from housing stock surveys, low-carbon heating field trials, and cost-benefit literature, to estimate the capital and operating costs of low-carbon heating technologies. This study will consider a wide range of building types, representative of the UK housing stock, taking account of building characteristics such as floor area and thermal efficiency. In addition, the model will take account of uncertainties such as occupancy levels and heating patterns. The results from the household level analysis will be used to identify relationships between the building characteristics and the levelised cost of heat, and highlight the types of households where the levelised cost of heat is most sensitive to personal preferences such as heating patterns and uncertainties such as fuel costs.

The Value Web Model, a mixed-integer programming optimisation model for integrated energy value chains, will be used to optimise the design and operation of the UKs heat and electricity system at the whole-systems level and the capital cost distribution from the cost-benefit model will be used to provide a detailed spatial representation of the UK's housing stock by taking account of how capital costs vary between regions of the UK. These results will identify the mix of low-carbon heating technologies at the regional level that leads to the lowest whole-system costs and the energy generation and transportation technologies required to support heat decarbonisation. This will be compared to the results of the household level cost-benefit analysis to highlight the extent to which these approaches align, and consider the regional impacts of decarbonisation and suggest policy instruments that might be required to promote a low-carbon technology mix.

ENERGY INTEGRATION FOR WASTE HEAT IN INDUSTRIAL PLANTS THROUGH A METAHEURISTIC-DETERMINISTIC APPROACH

Francisco J. López-Flores[a], Luis G. Hernández-Pérez[a], Luis F. Lira-Barragán[a], Eusiel Rubio-Castro[b] and José M. Ponce-Ortega[a]
[a]Universidad Michoacana de San Nicolás de Hidalgo, Morelia, Mich., México.
[b]Universidad Autónoma de Sinaloa, Culiacán, Sin., México.

Abstract

This work presents a novel approach to recover industrial waste heat and to integrate it into utilities, refrigeration, and electricity production through the incorporation of heat exchanger networks and thermal engines (steam Rankine cycle, organic Rankine cycle, and absorption refrigeration cycle). The solution approach is based on the iteration between metaheuristic–deterministic optimization strategies. Metaheuristic optimization is established through the MS Excel-VBA-Aspen Plus link to obtain accurate modeling results. Deterministic optimization is implemented in the GAMS platform, where the mathematical formulation is based on a superstructure that considers all of the energy interconnections between the heat exchanger network, utilities, and thermal engines. Furthermore, economic, environmental, and social targets are evaluated. A case study is presented to show the applicability of the proposed methodology. The operating conditions obtained are presented (working fluid flow rate, temperatures, pressures, and efficiencies) for each thermal engine. Furthermore, the results show an increase in the total annual profit by 148%, and a decrease in greenhouse gas emissions equivalent to 18.7%.

Introduction

Several industrial processes produce great amounts of energy that has no practical use and is discharged into the environment. Wasted energy is an environmental problem, which produces considerable amounts of greenhouse gas emissions, while the reuse of the produced waste heat can increase the thermal efficiency and minimize the consumption of primary energy sources (reducing the environmental impact). Waste heat is classified according to temperature ranges as high temperature (waste heat > 400 °C), medium temperature (100 °C < waste heat < 400 °C), and low temperature (waste heat < 100 °C) [1]. It has been reported that, annually in the US, a total of 18.9 exajoules of waste heat is discharged by thermal power plants, where 4% of this energy corresponds to waste heat at temperatures above 90 °C [2]. In China, it is estimated that around 7.6 exajoules of low-temperature waste heat is lost per year [3]. Additionally, the increment in fuel prices, the progressive growth in world energy demand, and the concern related to global warming establish the efficient use of energy as a very important challenge to solve in the industrial sector, which represents the highest energy consumption with more than 20% of the total [4]. In this context, there is a set of systems available for waste heat recovery (WHR), with the purpose of reusing the residual energy and, as a consequence, to improve the overall energy efficiency. The organic Rankine cycle (ORC) is a thermodynamic cycle for the generation of electricity, which uses an organic liquid with a high molecular weight and a low liquid−vapor phase change as a working fluid; the main application is the WHR [5] The absorption refrigeration (AR) cycle represents an efficient alternative to satisfy the refrigeration requirements in multiple processes The AR cycle is driven by a low-temperature heat source.

Heat exchanger networks (HENs) deal with the heat recovery between process streams; with this approach, the consumption of utilities (i.e., steam, fuels, cooling water, refrigeration) is reduced, and consequently, economic and environmental benefits are obtained. Since the pinch analysis was developed, several methods and solution approaches have been proposed [6]. In this sense, different optimization techniques have been developed for WHR that simultaneously involve HENs and thermal engines.

The main disadvantage of works where HENs are integrated with thermal engines for waste heat recovery (i.e., SRC, ORC, or AR) is the significant number of assumptions related to the operation of the thermal engines because to simulate them through deterministic optimization approaches is extremely complicated due to the use of rigorous thermodynamic models; this is associated with the highly nonconvex mathematical terms, which lead to difficult convergence problems. Then, the previous attempts made significant assumptions through fixed efficiency factors as well as the internal variables such as operating temperatures, flow rates, film heat transfer coefficients, data for costs, thermodynamic properties, etc. Therefore, in order to overcome previous limitations, in this work, we propose a hybrid optimization approach, which combines metaheuristic and deterministic techniques together with process simulators; this way, thermal engines are properly simulated, and the optimization of an entire integrated system between HENs and thermal engines can obtain more accurate results.

Methodology

The methodology used in this work consists of a set of tools that allows solving a specific part of the problem. The proposed general solution approach (see **Figure 1**) consists of a metaheuristic algorithm (programmed in MS Excel), a process (developed in GAMS). Initially, the models for the used thermal engines are simulated in Aspen Plus. Additionally, the data are collected (temperature, pressure, etc.), then the thermal engines proposed in Aspen Plus are simulated with the collected data. Subsequently, the iteration between deterministic optimization (GAMS) and metaheuristic optimization (the link between MS Excel-VBA-Aspen Plus) is established; GAMS solves the mathematical model that describes the process to be optimized, whereas metaheuristic optimization solves the highly nonconvex part found in thermodynamic models. Finally, after several iterations and with a minimal difference between successive iteration, an optimal solution is obtained.
The thermodynamic models selected in Aspen Plus are the following: NBS steam tables (STEAMNBS) for the SRC [7]. Peng−Robinson equation of state property method for ORC [7].

The PR-BM property method for the AR cycle (Peng– Robinson Equation of State modified by Boston- Mathia) [8].

Furthermore, the efficiencies used for optimization are: $\eta_{t,isentropic} = 0.9$ for the turbines and $\eta_p = 0.8$ for the pumps.

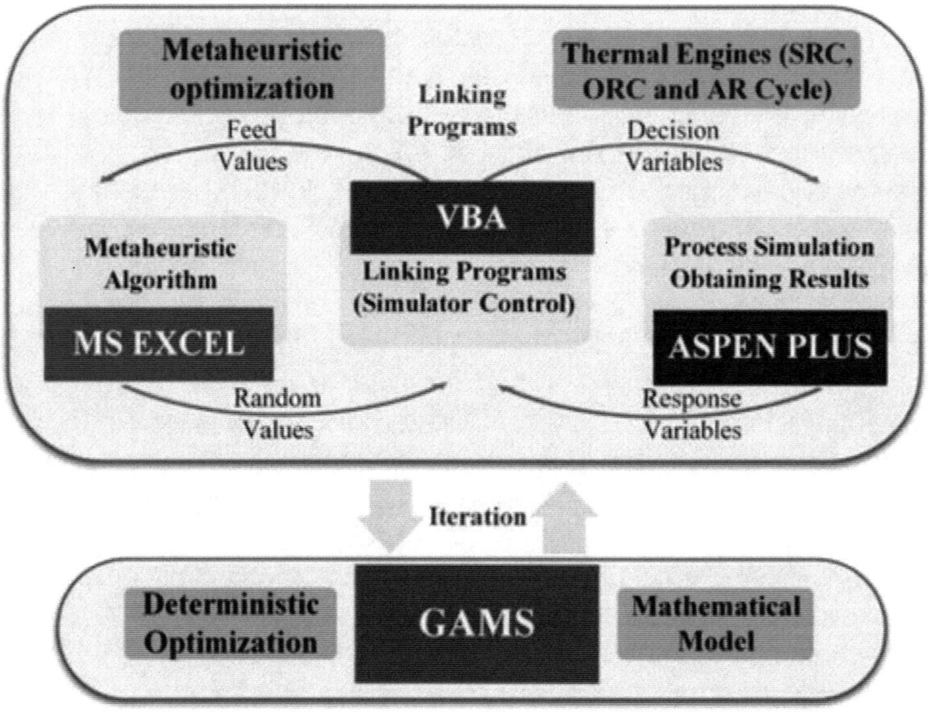

Figure 1. Proposed solution approach.

The deterministic approach is based on the superstructure for the integrated energy system, and the mathematical formulation is defined by the following sections: the model for the heat exchanger network, optimal selection of working fluid (ORC and AR cycle), thermal engines and their interactions, optimal size for the solar collector, maximum availability for biofuels and finally, the objective functions (economic, environmental, and social). The economic objective function considers revenues from the selling of power plus tax credits reduction minus capital costs, fixed costs, operating costs, and energy source costs. The environmental objective function is determined by the overall NGHGE, which is calculated using the emissions generated by fossil fuels and biofuels implemented in the boiler of the SRC. The social objective function considers the number of generated jobs, which is based on an input–output analysis; therefore, the number of jobs is the sum of those created for the production of fossil fuels, biofuels, and the solar collector.

Economic objective function

The economic objective function aims to maximize profit, which includes the *RSP* (Revenues from the selling of power) and *TCR* (Tax credits reduction), minus the *CaC* (Capital costs), *FiC* (fixed cost), *OC* (operating cost) and *ESC* (Energy sources cost):

$$Max\ profit = RPS + TCR - CaC - FiC - OC - ESC \tag{1}$$

General quantification of *GHG*

The environmental impact assessment is carried out through the general quantification of *GHG* because fossil fuels and biofuels release carbon dioxide when burned:

$$Min\ NGHGE^{Overall} = \sum_{t\in T}\sum_{f\in F}\left[GHGE_f^{Fossil} Q_{f,t}^{Fossil} D_t\right] + \sum_{t\in T}\sum_{b\in B}\left[GHGE_b^{Biofuel} Q_{b,t}^{Biofuel} D_t\right] \tag{2}$$

where $Min\ \ NGHGE^{Overall}$ are generally global *GHGs* released to the environment, while $GHGE_f^{Fossil}$ and $GHGE_b^{Biofuel}$ are the individual *GHG* for fossil fuels *f* and biofuels *b*. It is important to note that individual GHGs are determined through life cycle analysis given in units of tonnes of CO_{2eq} per *kJ* generated.

Number of jobs created

The number of jobs created by the project to produce fossil fuels, biofuels and the solar collector:

$$Max\ NJOBS^{Overall} = \begin{pmatrix} \sum_{t\in T}\sum_{f\in F}\left[NJOB_f^{Fossil} Q_{f,t}^{Fossil} D_t\right] + \sum_{t\in T}\sum_{b\in B}\left[NJOB_b^{Biofuel} Q_{b,t}^{Biofuel} D_t\right] + \\ \sum_{t\in T}\left[NJOB^{Solar} Q_t^{Solar} D_t\right] \end{pmatrix} \tag{3}$$

where $NJOB_f^{Fossil}$, $NJOB_b^{Biofuel}$ and $NJOB^{Solar}$ represent the number of jobs generated by *kJ* provided by fossil fuels, biofuels, and solar collectors, respectively.

Objective Function for the Metaheuristic Optimization

The objective function consists in maximizing the revenue from electricity generation and minimizing the total annual cost, $OF = \{M\text{ax}\, REVENUE, MinTAC\}$.

The total profit from the sale of the generated electricity is given by:

$$REVENUE = (Power^{src} + Power^{orc})SP^{elec} t_{OP} \tag{4}$$

where SP^{elec} is the sale price of electricity ($/kWhr), t_{OP} represents the hours of operation per year of the plant (hr), $Power^{src}$ and $Power^{orc}$ denote the electricity produced by the turbines in the SRC and ORC (kW), respectively.

Solution procedure

The proposed approach involves a hybrid metaheuristic-deterministic optimization strategy, which is described as follows: The metaheuristic optimization approach implemented for each thermal engine uses the I-MODE (Improved Multi-Objective Differential Evolution) algorithm [9] and the component object module (COM) technology is implemented to establish a linking between Aspen Plus and Microsoft Excel [10] With COM technology, it is possible to add code so that applications behave like an object link and embedding (OLE) automation server where Visual Basic for Applications (VBA) is used to interface with other applications. Regarding the optimization stage, a decision vector of design variables is sent from MS Excel to Aspen Plus (for example, temperatures, pressures, mass flow rates, etc.) where rigorous calculations are obtained for the design data of each thermal engine through the resolution of mass and energy balances and considering rigorous methods for thermodynamic properties. After running the simulations, the data is returned from Aspen Plus to Excel in order to calculate the objective functions, which are evaluated, and new vectors of design variables are generated. After Maximum Number of Generations (MNG), the results for each thermal engine (efficiencies, performance, operating temperatures, operating pressure, etc.) were selected and sent to the deterministic optimization platform for evaluation.

After several iterations, the convergence or the last iteration of the general solution procedure is obtained when the difference between the decision variables (efficiencies, temperatures, performance coefficients, etc.) is minimal with respect to the previous iteration.

Results

Case Study

In order to apply the proposed methodology, a case study located in the city of Morelia in Mexico is addressed, which has the coordinates N 19° 42´ 08´´ and W 101° 11´ 08´´. Six *HPS* (hot process streams) and five *CPS* (cold process streams) are considered for energy integration

Solution Approach

The selection of the I-MODE parameters includes a sensitivity analysis, which consists in broad strokes of the following: a number is proposed for each parameter (number of individuals, number of generations, etc.) separately, then the result obtained after running the optimization is analyzed, it is proposed to double that value for the same parameter, if it improves significantly it means that the proposed value is not enough, then the procedure is repeated by doubling the value again until the solution does not change in a considered way. Once an appropriate value for a parameter has been found, the same procedure is followed to find a suitable value for another parameter. In this case study, the values for the parameters associated with metaheuristic optimization (I-MODE) are the following: population size (NP) of 10 individuals, generation number (GenMax) of 50, taboo list size (TLxS) of 5 individuals, taboo radius (TR) of 0.01, crossover fractions (Cr) of 0.8, and mutation fractions (F) of 0.5. The computing time required to obtain the Pareto optimal solutions varied from 40 to 70 min for each iteration. Moreover, the solvers DICOPT, CONOPT and CPLEX were implemented in GAMS for solving the deterministic optimization model. This model consists of 1,001 equations, 1,015 continuous variables and 270 binary variables. The CPU time was 0.22 s for each iteration only for the solution of the deterministic model.

Result of Deterministic Optimization

Table 1 shows the detailed results for the last iteration. These results are compared favorably with the results found by Lira-Barragan et al [11] The produced power is lower 19.5%, the capital cost for SRC and ORC is considerably lower 44.4%, this because the power production unit costs ($PP\text{Cos}\,t^{src}$ and $PP\text{Cos}\,t^{orc}$) determined by the metaheuristic optimization have lower values, between 0.078 and 0.084 $/kWh for the SRC, and for the ORC between 0.0122 and 0.009 $/kWh, comparable with 0.1 $/kWh for the SRC and 0.12 $/kWh for the ORC [11] The cost of heat exchangers is lower 37.4%, the capital cost of AR cycle is lower 50.8%, for all iterations there are no operating costs (cooling water), the energy source cost is lower 17.2%. Based on the above, the following most important results were obtained, the total annual profit is notably higher by 148%, the overall NGHGE are 18.7% lower, and the number of jobs generated are practically the same. Therefore, it is concluded that better economic, environmental, and social solutions were obtained.

Table 1. Comparison of Results

Concept	Ref [11]	This work
Power produced, kW	10,000	8051.9
Selling power by the SRC and ORC, $/y	12,263,999.9	9,874,898.2
Total cost for SRC and ORC, $/y	8,934,793.5	4,969,298.2
Total revenue from SRC and ORC, $/y	3,329,205.7	4,905,600.0
Tax credits reduction, $/y	67,742.6	67,742.6
Capital cost for exchangers, $/y	79,576.6	49,818.0
Capital cost for AR cycle, $/y	283,172.2	139,214.2
Cooling water cost, $/y	11,160.6	0
Profit, $/y	1,366,799.1	3,393,014.2
Overall NGHGE, ton CO_{2eq}/y	210,889.3	171,409.2
Jobs	4617	4615.0

Figure 2 shows the optimal configuration for the last iteration with a power production of 6597.4 kW for the SRC and 1454.5 kW for the ORC, there are no cooling water requirements, the refrigeration is 5,214.5 kW, the ORC only receives heat from the SRC, only cold streams 1 and 2 require *LPS* heating, the cold streams 3 and 4 receive heating from the ORC. The cold streams 2, 3, 4 and 5 have two heat exchangers in the same stage (in parallel). Furthermore, the solar collector is not considered as a heat source.

Notice that the heat transfer areas of the *LPS, C1* and *LPS, C4* exchangers are small. This is due to temperature differences and film heat transfer coefficients between process streams. Furthermore, in the industry there is a large versatility of heat exchangers that range in size from small (0.1 m^2) to supergiant (over 10^5 m^2).

The ORC and AR cycle can receive heat from both hot process streams and *LPS*, but in the optimal configuration obtained from the case study this is not present. The main problem is that the film heat transfer coefficients determined for *ORC1* and *AR1* are exceedingly small (values less than 1 kW/m^2 K) compared to the values of the rest of the process streams, whereby the existence of heat exchangers to supply heat to the ORC and AR is not considered a benefit for the optimal solution (**Figure 2**). However, it could be considered a case study with a greater number of hot process streams, a greater diversity of temperatures, different working fluids that provide different film heat transfer coefficients and constraints for the existence of at least one heat exchanger between HPS and thermal engines (ORC and AR), so it is expected to obtain an optimal solution where the hot streams supply the heat to the cycles.

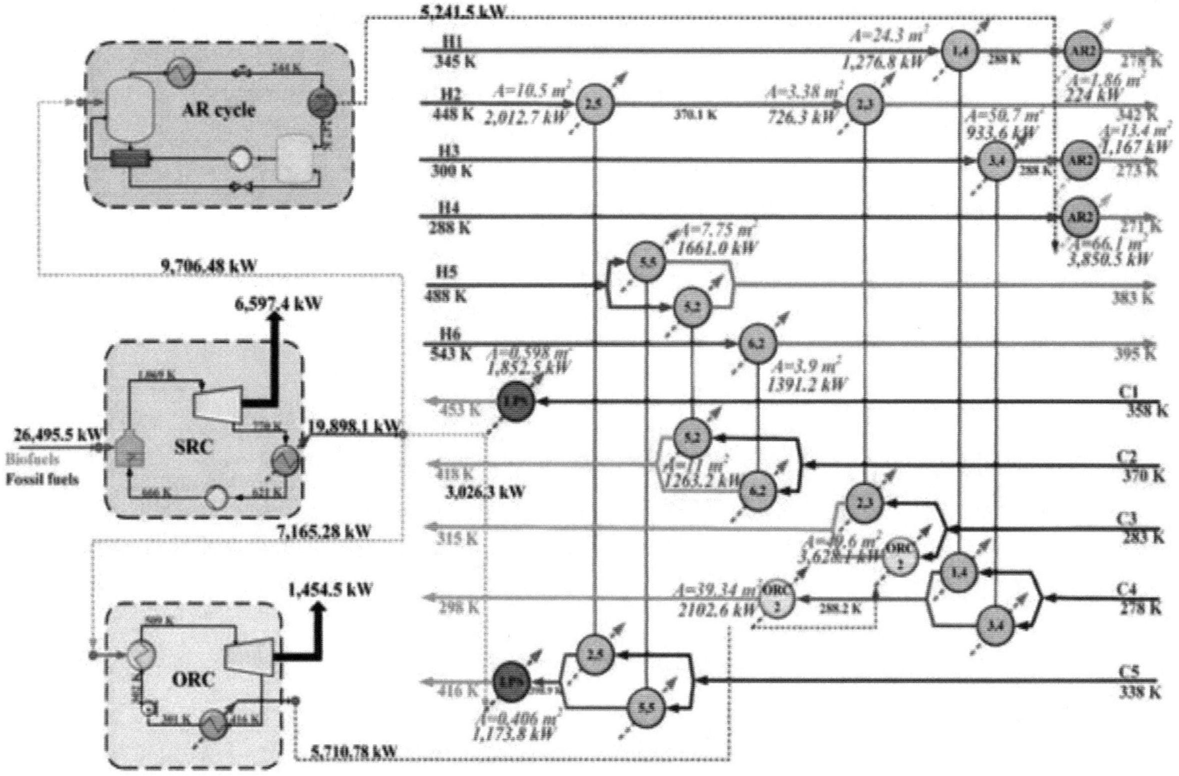

Figure 2. Optimal configuration.

Conclusions

In this work, a new approach based on metaheuristic-deterministic optimization strategies has been proposed for the WHR in industrial plants that simultaneously involves the integration of heat exchanger networks (HEN), and thermal engines. In the proposed approach, variables are selected and sent (efficiencies, temperatures, supplied heat, unit costs, etc.) between the iteration of the metaheuristic and deterministic optimization platforms.

A case study was used to illustrate the application of the proposed approach. The results for the thermal engines are shown in figures that show the solutions after MNG, where several interesting solutions are identified through the beneficial trade-offs in terms of economics and refrigeration heat for the AR cycle. In addition, the operating conditions of the best solution are presented for each heat engine that are different from the evaluated example problem, but they are favorable compared to results previously reported in the literature.

The results of the deterministic optimization show interesting aspects such as the decrease in energy production, thus reducing the costs of energy sources, and consequently the diminishing in sale of energy produced by the SRC and ORC. However, the most interesting result is the reduction in the total annual costs of the thermal engines; therefore, higher total profits, lower overall NGHGE and similar produced jobs are generated.

Finally, the proposed solution approach is general, and it can be implemented in any energy integration system that simultaneously considers the use of thermal engines.

References

1. Brückner, S., Liu, S., Miró, L., Radspieler, M., Cabeza, L. F., and Lävemann, E. (2015), Industrial waste heat recovery technologies: An economic analysis of heat transformation technologies. Appl. Energy, 157-167.
2. Gingerich, D. B. and Mauter, M. S. (2015), Quantity, quality, and availability of waste heat from United States thermal power generation. Environ. Sci. Technol., 8297-8306.
3. Fang, H., Xia, J., Zhu, K., Su, Y. and Jiang, Y. (2013), Industrial waste heat utilization for low temperature district heating. Energy Policy, 236-246.
4. Gharaie, M., Zhang, N., Jobson, M., Smith, R. and Panjeshahi, M.H. (2013), Simultaneous optimization of CO_2 emissions reduction strategies for effective carbon control in the process industries. Chem. Eng. Res. Des., 1483-1498.
5. Mahmoudi, A., Fazli, M. and Morad, M.R. (2018), A recent review of waste heat recovery by Organic Rankine Cycle. Appl. Therm. Eng., 660-675.
6. Silva, A. P., Ravagnani, M.A.S.S., Biscaia, E.C., Caballero, J. A. (2010), Optimal heat exchanger network synthesis using particle swarm optimization. Optim. Eng., 459-470.
7. AspenTech. Aspen Physical Property System, 2013. https://www.aspentech.com/en/products/engineering/aspen-properties (Accessed December 2020)
8. Mansouri, R., Boukholda, I., Bourouis, M. and Bellagi, A. (2015), Modelling and testing the performance of a commercial ammonia/water absorption chiller using Aspen-Plus platform. Energy, 2374-2383.
9. Sharma, S. and Rangaiah, G.P. (2013), An improved multi-objective differential evolution with a termination criterion for optimizing chemical processes. Comput. Chem. Eng.
10. Ponce-Ortega, J.M. and Hernández-Pérez, L.G. (2019), Optimization of Process Flowsheets through Metaheuristic Techniques; Springer International Publishing, Gewerbestrasse, Cham, Switzerland.
11. Lira-Barragán, L.F., Ponce-Ortega, J.M., Serna-González, M. and El-Halwagi, M.M. (2014), Optimal design of process energy systems integrating sustainable considerations. Energy, 139-160.

A RECYCLE/REUSE NETWORK FOR THE OPTIMAL WATER MANAGEMENT IN HYDRAULIC FRACTURING OPERATIONS

Tania I. Serrano-Arevalo[1], Luis F. Lira-Barragán[1], José M. Ponce-Ortega[1] and Mahmoud El-Halwagi[2]
(1) Universidad Michoacana de San Nicolás de Hidalgo, Morelia, Mich, México,
(2) Texas A&M University, College Station, TX, USA.

Abstract

Recently, there have been substantial discoveries of shale gas reserves in North America. Hydraulic fracturing technologies have facilitated the production of shale gas trapped in tight formations. According to the Energy Information Administration (EIA), by the year 2035 shale gas is expected to provide about half of the total natural gas supply in the U.S. The EIA also estimates that the U.S. has enough natural and shale gas to meet domestic electricity demands for 575 years at current electricity generation levels. Certainly, this information highlights the importance and the enormous potential of shale gas in the future. However, this industry deals with several challenges, where the most important concern is related to the water owing to huge amounts of water demanded by hydraulic fracturing phase to be successfully implemented and once the completion phase has finished a portion of the injected water during the fracking returns to the surface and according to the time (and even the quality water is different) it can be classified in flowback, transition and produced water. Then, a key process consists in treat and storage these streams with the purpose to recycle or reuse them and reduce the fresh water consumption. This aspect is essential for regions with a strong water scarcity. In developing a water recycle/reuse strategy, it is important to consider the segregation and avoid the mixing of flowback, transition and produced water to be treated and storage. The recycled water is able to be sold and used in other activities. Thus, this work proposes a formal optimization formulation for the recycle/reuse network implemented in shale gas operations considering the economic target, which is composed by the costs of fresh water, treatment, storage, disposals (including operating and capital costs) and transportation as well as the sale of the recycled water. The environmental objective aims to minimize the fresh water requirements. The methodology considers a fixed scheduling for the completion operations in order to implement the approach. Finally, an example problem is presented to show the applicability of the proposed methodology. The proposed multi-objective model is solved by mixed integer linear programming (MILP). Fracking in the US is presented as a case study, where it is possible to reduce up to 32.43% of fresh water consumed and reuse up to 12.26% of total water. well waste for fracking needs, which has a significant environmental impact.

Introduction

Shale gas reservoirs have recently been discovered that can contribute and positively impact energy and economic development. Reserves in North America in particular have been discovered, shale gas is expected to provide approximately half of the total natural gas supply in the United States by 2035 [1]. Hydraulic fracturing results in the discharge of large amounts of fresh water, due to the amount of water that is injected during the process. To achieve this task, advances in hydraulic fracturing technology are necessary, [2] such as infrastructure, sustainable technologies, as well as determining the optimal selection and technologies that incorporate economic aspects to meet a certain quality of gas. [3] The characteristics of shale gas wastewater influence the selection of treatment options to minimize its environmental impact, [4] because it is contaminated by total dissolved solids (TDS), thorium, metals, organic compounds such as kerosene, benzene and toluene and radioactive substances such as uranium, radium 226 and 228 [5,6].

Methodology

The multi-objective mathematical model seeks an economic-environmental balance. A set of wells *n* will be fractured according to a fixed schedule where one or more fracking crews can participate simultaneously. During hydraulic fracturing, large amounts of water need to be injected, when the stage ends, a fraction of said water (residual) returns to the surface that can be classified into flowback, transition and produced (according to the time *t* at which it returns to the surface), which needs treatment to clean the streams, which is why a set of treatment technologies is given for the flowback, transition and produced water (*f, m, p*). Another quantity of water is stored and recirculated in the hydraulic fracturing of simultaneous wells according to the flowback, transition and produced classification (*c, d, g*), finally the treated water is sent to reuses *b, a, e,* for the different types of water (**Figure 1**). The total cost (*TC*) is composed of the operating costs (*TOC*) and capital costs (*TCC*), as well as the income generated by the sale of water streams to be reused after treatment technologies (*TVC*). The *TOC* includes the costs of fresh water, transportation costs in the trajectories and the operating costs of the treatment units, the *TCC* the costs associated with the acquisition of treatment and storage units, and the *TVC* represents the total sale of the water residual. It is proposed to determine the optimal configuration of water management in the extraction of shale gas in the hydraulic fracture considering economic objectives (minimizing the total cost) and environmental (minimizing the consumption of fresh water). Once the proposed approach is resolved, the number and capacities of treatment technologies (for each classification of wastewater that returns to the surface), storage units, treatment and required reuse will be determined.

Model Formulation

To model the above methodology, a set of mathematical relationships is needed, which evaluate the economic and environmental part in order to achieve the desired objective, which is, through optimization, to determine the appropriate selection of treatment technologies for the wastewater streams that leave the wells, storage units and necessary reuses. As well as the use of fresh water in hydraulic fracturing and the water that returns to the surface. A disjunctive optimization model is proposed to select the set of different treatment and storage technologies, considering as a variable the fracture schedule of wells that are operated simultaneously over a time horizon.

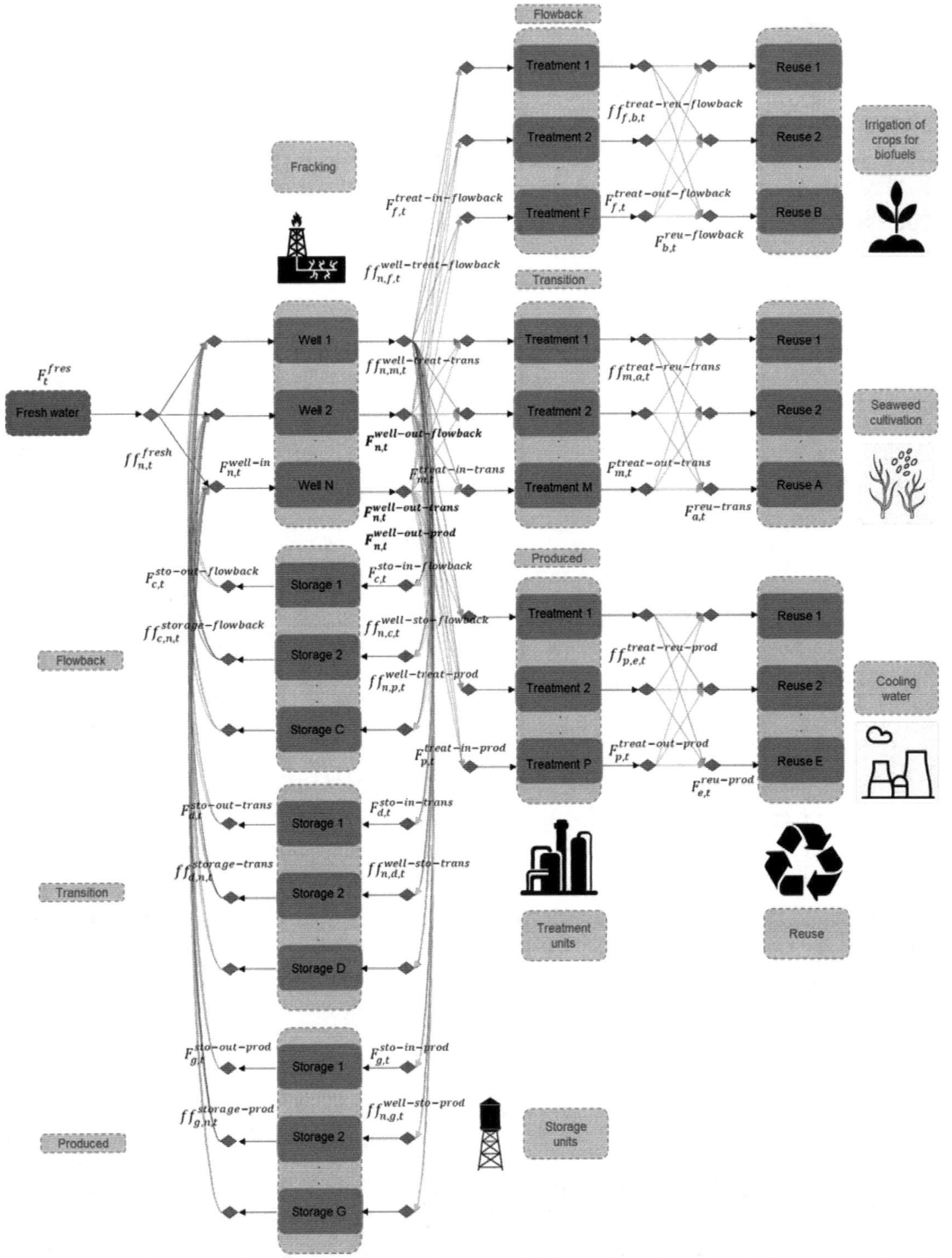

Figure 1. Superstructure for the optimization model for planning in shale gas extraction and return flow.

This mathematical model includes 76 equations in total, then some are shown that generally involve balances of matter (water) in hydraulic fracturing, treatment, storage and reuse units, as well as dimension restrictions, operation, efficiencies and associated costs. The objective function (minimization of *TC* and fresh water requirements *TWR*) is also shown.

$$F_t^{fresh} \leq \left(\frac{\text{AvailV}_t^{\text{fresh-max}}}{\text{H}^{\text{time}}} \right), \quad \forall t \tag{1}$$

$$F_t^{fresh} = \sum_n ff_{n,t}^{fresh}, \quad \forall t \tag{2}$$

$$\text{F}_{\text{n,t}}^{\text{well-in}} = ff_{n,t}^{fresh} + \sum_c ff_{c,n,t}^{storage-flowback} + \sum_d ff_{d,n,t}^{storage-trans} + \sum_g ff_{g,n,t}^{storage-prod}, \quad \forall n, \forall t \tag{3}$$

$$\text{F}_{\text{n,t}}^{\text{well-out-flowback}} = \beta_{n,t}^{flowback}\, \text{F}_{\text{n,t}}^{\text{well-in}}, \ \forall n, \forall t \tag{4}$$

$$\text{F}_{\text{n,t}}^{\text{well-out-}trans} = \beta_{n,t}^{trans}\, \text{F}_{\text{n,t}}^{\text{well-in}}, \ \forall n, \forall t \tag{5}$$

$$F_{f,t}^{treat-out-flowback} = \sum_b ff_{f,b,t}^{treat-reu-flowback}, \quad \forall f, \forall t \tag{6}$$

$$F_{m,t}^{treat-out-trans} = \sum_a ff_{m,a,t}^{treat-reu-trans}, \quad \forall m, \forall t \tag{7}$$

$$F_{p,t}^{treat-out-prod} = \sum_e ff_{p,e,t}^{treat-reu-prod}, \quad \forall p, \forall t \tag{8}$$

$$\text{V}_{\text{d}}^{\text{sto-initial-trans}} = V_{d,t}^{sto-trans}, \quad \forall t = t^{final} \tag{9}$$

$$\text{V}_{\text{c}}^{\text{sto-initial-flowback}} = V_{c,t}^{sto-flowback}, \quad \forall t = t^{final} \tag{10}$$

$$\text{V}_{\text{g}}^{\text{sto-initial-prod}} = V_{g,t}^{sto-prod}, \quad \forall t = t^{final} \tag{11}$$

$$\text{F}_{\text{n,t}}^{\text{well-out-flowback}} = \sum_f ff_{n,f,t}^{well-treat-flowback} + \sum_c ff_{n,c,t}^{well-sto-flowback}, \quad \forall n, \forall t \tag{12}$$

$$\text{F}_{\text{n,t}}^{\text{well-out-trans}} = \sum_m ff_{n,m,t}^{well-treat-trans} + \sum_d ff_{n,d,t}^{well-sto-trans}, \quad \forall n, \forall t \tag{13}$$

$$\text{F}_{\text{n,t}}^{\text{well-out-prod}} = \sum_p ff_{n,p,t}^{well-treat-prod} + \sum_g ff_{n,g,t}^{well-sto-prod}, \quad \forall n, \forall t \tag{14}$$

$$F_{f,t}^{treat-out-flowback} = \alpha_{\mathrm{f}}^{\text{treat-flowback}} F_{f,t}^{treat-in-flowback}, \quad \forall f, \forall t \tag{15}$$

$$F_{m,t}^{treat-out-trans} = \alpha_{\mathrm{m}}^{\text{treat-trans}} F_{m,t}^{treat-in-trans}, \quad \forall m, \forall t \tag{16}$$

$$F_{p,t}^{treat-out-prod} = \alpha_{\mathrm{p}}^{\text{treat-prod}} F_{p,t}^{treat-in-prod}, \quad \forall p, \forall t$$

$$\begin{aligned} TOC = {} & \mathrm{Cos}\, t^{fresh} + \mathrm{Cos}\, t^{op-treat-flowback} + \mathrm{Cos}\, t^{op-treat-trans} + \mathrm{Cos}\, t^{op-treat-prod} + \\ & \mathrm{Cos}\, t^{trans-fresh} + \mathrm{Cos}\, t^{trans-flowback} + \mathrm{Cos}\, t^{trans-trans} + \mathrm{Cos}\, t^{trans-prod} + \mathrm{Cos}\, t^{trans-treat-reu-flowback} + \\ & \mathrm{Cos}\, t^{trans-treat-reu-trans} + \mathrm{Cos}\, t^{trans-treat-reu-prod} + \mathrm{Cos}\, t^{trans-well-sto-flowback} + \mathrm{Cos}\, t^{trans-well-sto-trans} + \\ & \mathrm{Cos}\, t^{trans-well-sto-prod} + \mathrm{Cos}\, t^{trans-sto-well-flowback} + \mathrm{Cos}\, t^{trans-sto-well-trans} + \mathrm{Cos}\, t^{trans-sto-well-prod} \end{aligned} \tag{17}$$

$$\begin{aligned} TCC = {} & Cap\, \mathrm{Cos}\, t^{treat-flowback} + Cap\, \mathrm{Cos}\, t^{treat-trans} + Cap\, \mathrm{Cos}\, t^{treat-prod} + \\ & Cap\, \mathrm{Cos}\, t^{storage-flowback} + Cap\, \mathrm{Cos}\, t^{storage-trans} + Cap\, \mathrm{Cos}\, t^{storage-prod} \end{aligned} \tag{18}$$

$$TVC = \mathrm{H}^{\mathrm{time}} \left[\begin{array}{l} \sum_b \sum_t \mathrm{UVC}_{\mathrm{b}}^{\text{reu-flowback}} F_{b,t}^{reu-flowback} + \\ \sum_a \sum_t \mathrm{UVC}_a^{\text{reu-trans}} F_{a,t}^{reu-trans} + \\ \sum_e \sum_t \mathrm{UVC}_e^{\text{reu-prod}} F_{e,t}^{reu-prod} \end{array} \right] \tag{19}$$

Total cost

$$TC = TOC + TCC - TVC \tag{20}$$

$$O.F. = \min \{TC, TWR\}$$

Results and Discussion

The proposed optimization formulation was coded in the GAMS® software using the CPLEX solver, the problem includes 76 equations, 26,236 continuous variables, 9 discrete variables and 10,930 restrictions. The results are presented below.

Study case

Reuse in algae cultivation, irrigation water for biofuels and cooling water in power plants are considered; membrane distillation and reverse osmosis are considered for the treatments. However, other additional reuses can be included in this approach or treatment technologies without major changes in the mathematical formulation. To implement the proposed approach, the actual parameters of technical reports with data for the Marcellus and Barnett regions have been taken. For the macroscopic exploitation of shale gas wells in the US (Texas and Pennsylvania), the scheduling is considered as a variable in the simultaneous exploitation of these wells (divided into 3 crews). Flowback water returns the first 3 weeks after drilling is completed, transition water the next 12 weeks (after flowback water is completed), and produced water returns the next 17 weeks (after transition water is completed), with 20.59%, 12.13%, 6.58% with respect to the total water injected. The following scenarios are considered: in scenario A the crews of 1-3 are considered and wells

1-8, 9-15,16-20 operate, respectively, scenario B, wells 1-7, 8-14, 15-20, while stage C with 1-8, 9-16, 17-20 and finally D 1-8, 9-16, 17-20. For scenarios A and C, crews 1 and 2 work simultaneously and for B and C, crews 1, 2 and 3 work simultaneously. For all cases, the exploitation of 20 wells (*n*) is considered water scarcity (weeks 24-26), each one requiring 15,000 m³ of fresh water F_t^{fresh}.

Once the mathematical model is solved with the parameters described above, the following results will be obtained. **Table 1** shows the results obtained in terms of *TWR* and *TC* (including their composition) detailed for the scenarios (A, B, C and D). As shown, Scenario C implies the highest value for *TWR* and the highest *TC*, this because it implies a greater amount of reused water and once the least amount of recirculated water. Otherwise, Scenario D involves the lowest *TWR* and the highest amount of recirculated water with which the *TC* decreases. Scenarios A and B represent intermediate solutions between optimistic and pessimistic scenarios. When fracturing 20 wells, in all the study cases the outlet water is the same regardless of the schedule, because the injected water is the same. Scenario C uses only 212,983.61 m³ of fresh water, which saves 29.00% and scenario D uses the least amount of fresh water and saves 32.43% of water. According to the amount of water reused, Scenario C reuses 18.30% of the total water that comes out of the wells (it uses a greater amount of fresh water and in turn does not recirculate a large amount), while Scenario D reuses 12.26 % (because more water is recirculated than C). It should be considered that the original quantity required without recirculation in all scenarios is 300,000.00 m³ at the end of the exploitation of all wells and that it should be reduced by using the recirculation system.

Table 1. Configuration for each scenario with a minimum *TC*.

CONCEPT	**SCENARIO A**	**SCENARIO B**	**SCENARIO C**	**SCENARIO D**
Fresh water used without recirculation (m³)	300,000.00	300,000.00	300,000.00	300,000.00
Fresh water used with recirculation (*TWR*) (m³)	204,686.65	206,707.41	212,983.61	202,700.96
Percentage of freshwater saved (%)	31.77	31.09	29.00	32.43
TOC ($)	2.69E+06	2.67E+06	2.82E+06	2.66E+06
TCC ($)	41,833.91	49,235.91	48,177.73	49,216.53
TVC ($)	11,938.61	13,424.50	17,731.93	11,840.83
TC ($)	2.720E+06	2.708E+06	2.850E+06	2.697E+06
Flowback reused water (m³)	3,847.23	5,455.46	8,288.18	5,455.46
Transition reused water (m³)	6,583.25	6,065.98	7,375.54	4,917.02
Produced reused water (m³)	4,707.98	5,193.55	5,905.54	4,075.68
Total reused water (m³)	15,138.45	16,714.99	21,569.25	14,448.16
Total output water from the *n* wells (m³)	117,881.40	117,881.40	117,881.40	117,881.40
Percentage of reused water concerning the outlet (%)	12.84	14.18	18.30	12.26

Figure 2 shows a comparison of the amount of water required when the proposed superstructure (storage, treatment and reuse systems) is implemented. As can be seen in scenario A, the required quantity with recirculation is 204,686.65 m^3 (this is obtained by adding all the points shown in (a)), with a saving of unused fresh water of 95,312.35 m^3 (compared to the original quantity of required water that would be used without recirculation), in weeks 49-90 the fresh water requirements are null because these weeks in this period no longer complete wells according to the schedule, the week with the highest amount of water requirements It is week 5 in which the 3 crews operate simultaneously with 8,999.97 m^3. For scenario B, the fresh water requirements with recirculation are 206,707.41m3, achieving a saving of 93,291.59 m^3 and in week 43 the last well is completed. Finally, scenario C has a fresh water saving of 87,015.39 m^3 and scenario D has fresh water requirements of 202,700.96 m^3 with a 32.43% savings.

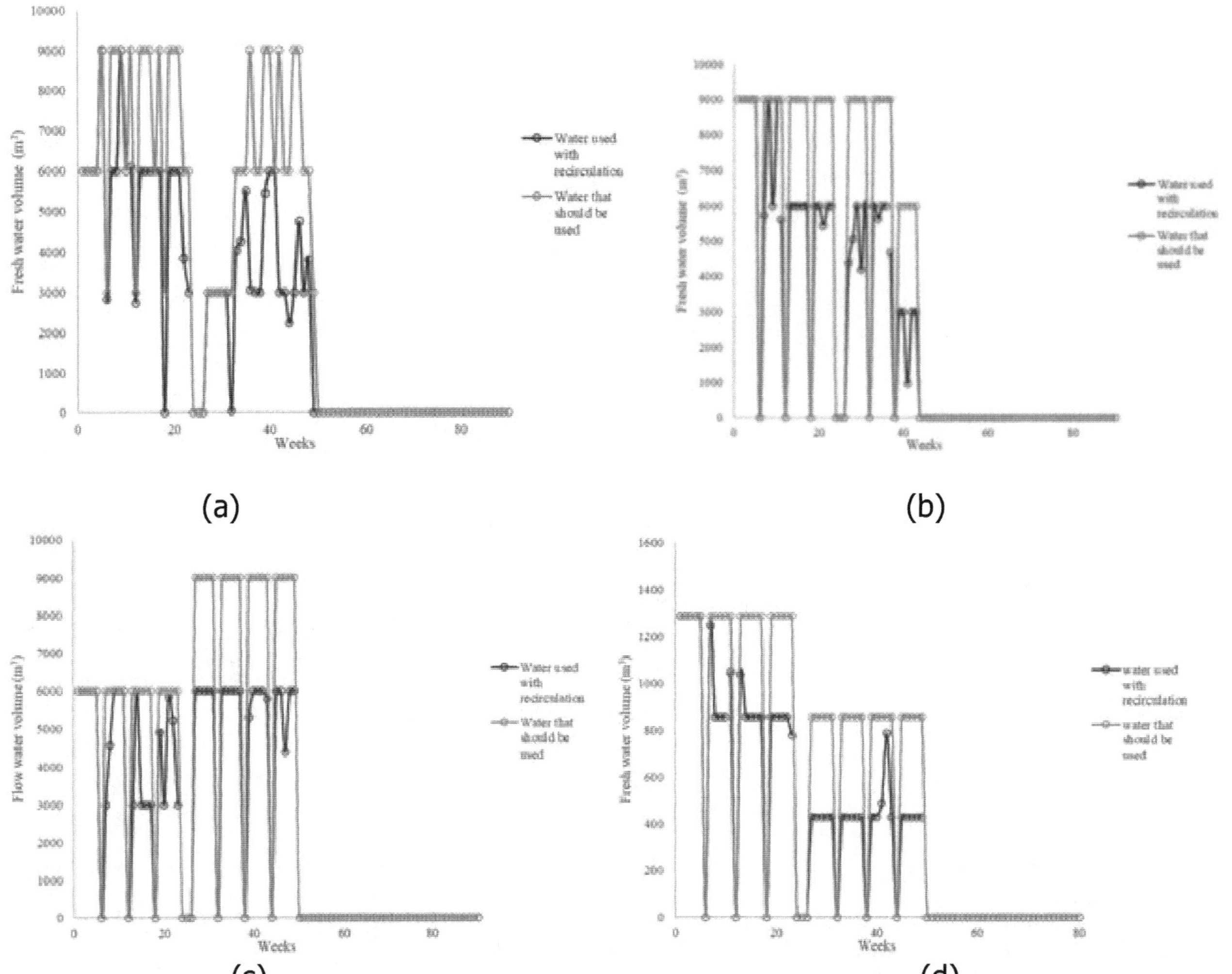

Figure 2. Fresh water consumption. (a) scenario A, (b) scenario B, (c) scenario C, (d) scenario D.

Conclusions

This shows the mathematical formulation and its implementation in shale fracking gas extraction in different case studies in order to make a comparison of the operation of the proposed methodology. Making a comparison between the results for the proposed scenarios (A, B, C and D), it is more convenient that the three crews work simultaneously without waiting for the drought to pass (weeks 24-26) and immediately complete the wells (scenario D). The proposed approach achieves important benefits in terms of reductions in total cost (*TC*) and total freshwater needs (*TWR*), scenario D reduces freshwater requirements by 32.43% where 12.26% of freshwater is reused. the total production of water in the wells. Additionally, it is possible to achieve a savings of $ 160,000 in total cost compared to worst case scenario (C). Furthermore, it is shown that it is possible to obtain an economic benefit - at the same time, which makes it an attractive solution for environmental implementation in the future.

References

1. U.S. Energy Information Administration (EIA). U.S. 2012 Annual Energy Outlook with Projects to 2035. Washington, DC: US Department of Energy, 2012. Available at: https://www.eia.gov/outlooks/aeo/er/pdf/0383er(2012).pdf (Accessed March 2021)
2. Xu, J., Zhai, C., Qin, L., Liu, S (2018), "Pulse hydraulic fracturing technology and its application in coalbed methane extraction", *Int. J. Oil Gas Coal Technol.* 19(1), 115–133, DOI: 10.1504/IJOGCT.2018.093962.
3. Martinez-Gomez, J., Napoles-Rivera, F., Ponce-Ortega, J. M., El-Halwagi, M. M (2017), "Optimization of the production of syngas from shale gas with economic and safety considerations", *Applied Thermal Engineering.* 110, 678-685, DOI: 10.1016/j.applthermaleng.2016.08.201.
4. Sun, Y., Wang, D., Tsang, D. C. W., Wang, L., Ok, Y. S., Feng, Y (2019), "A critical review of risks, characteristics, and treatment strategies for potentially toxic elements in wastewater from shale gas extraction", *Environ. Inter.* 125, 452-469, DOI: 10.1016/j.envint.2019.02.019.
5. Manzanares-Rivera, J. L (2014), "Uso de agua en la extracción de gas de lutitas en el noreste de México: Retos de regulación ambiental", *Estudios sociales (Hermosillo, Son.).* 22(44), 172-197, DOI: http://www.scielo.org.mx/scielo.php?script=sci_arttext&pid=S0188-45572014000200007.
6. Haluszczak, L. O., Rose, A. W., Kump, L. R (2013), "Geochemical evaluation of flowback brine from Marcellus gas wells in Pennsylvania USA", *App. Geochem.* 28, 55-61, DOI: 10.1016/j.apgeochem.2012.10.002.

628235 Simulation-Based Comparison between Three Different Clean-in-Place Configurations Regarding Their Cleaning Efficiency and Water Use

Abstract:

In many process industries, like chemical, food, process and pharmaceutical industries, large tanks are periodically filled, emptied, and refilled with different products. Prior to tank reuse with new liquid, it is essential to ensure any previous residues are completely removed from the tank's internal walls, as to maintain proper process hygiene. A clean-in-place (or CIP) system is the typical procedure used to clean a vessel after use and before product changeover.

Depending on the product being removed and the equipment size, dismantling of equipment and manual cleaning is usually not attractive economically and a clean-in-place system is desired. A clean-in-place (CIP) process can be automated, standardized, and hence allows for certification. The process involves installing a spray head assembly, figure 1, in which water or cleaning solution jets from the spray head assembly. Different types of spray head assembly exist and are used. The spray head types considered in this study are stationary ball, rotating ball, and orbiting nozzles, figure 2. The stationary ball is the simplest design, but also consumes the most cleaning liquid and takes the most time to cover the same surface. The orbiting nozzles configuration is the most expensive but consumes the least amount of cleaning liquid. The rotating ball is an intermediate design between the stationary ball and the orbiting nozzles.

Typical objectives are obtaining a fully cleaned surface while using the optimal amount of water or solution and within the shortest cycle time. Complete surface coverage with cleaning liquid is essential and "shadowing" effects, i.e., lack of coverage in certain areas, need to be avoided. Due to a wide variety of tank sizes, tank fittings, impeller sizes and their location, among other variables, ensuring the complete coverage and avoiding shadowing can become a lengthy process with each design change. Also, scale up/down rules are nonexistent and engineers rely on previous experiences which changes with time and geographical location making the process difficult to certify and relying on full scale testing which is expensive and time consuming. Computational Fluid Dynamics (CFD) can provide an economical tool to predict film coverage and wetted area. Different tank sizes and accessories (impellers, baffles, dip tubes, internal piping, etc.) can be easily added to the model and their effect on the cleanability can be easily predicted.

For a CIP system, typical input parameters are water pressure (or flow rate), spray head type and its location with respect to the vessel, and finally spray head assembly motion (single-axis rotation for rotating ball, bi-axis or orbiting for orbiting nozzles). In many cases, for larger tanks, there could be multiple spray heads to ensure complete coverage. The engineering challenge is to select and proper place the spray head within the vessel to ensure the complete cleaning, in the shortest amount of time, and using the least amount of cleaning liquid.

Sami et al. [1] studied the effect of inlet water pressure, nozzle design, and distance between the nozzle and target surface on the jet throw (trajectory) and impinging pressure. Metwally et al. [2] studied the effect of nozzle motion on the surface coverage. In this study, different spray heads will be compared, figure 2, in terms of their coverage and time needed to reach that coverage. 100% presented on water jet formation and the impact pressure by a water jet at different nozzle inlet pressure and at various distances between nozzle and surface to be cleaned.

The modeling approach consists of two main ingredients. The first is the water droplet tracking which is done using a lagrangian approach (Discrete Phase Model or DPM for short [3]). This model allows for the accurate tracking of the water droplets and considers its inertia, air drag, and initial droplet conditions from the spray head (location, motion if any). The second ingredient is modeling the water wall film formation [3], spreading, and splashing on the tank surface.

Three different simulations for the same tank were set up, one for each spray head studied. The water flow rate through all the spray heads was identical. Also, the duration for cleaning was identical (one minutes). Figure 3 shows the final film coverage predicted for the three cases studied. It is obvious that the stationary head is suffering from serious shadowing and the top of the tank is not completely cleaned. Both the rotating spray ball and the orbiting nozzles can obtain full coverage. To further quantify the performance of the different spray heads, the % wetted area evolution is plotted in figure 4. The entire cleaning cycle (one minute) % wetted area evolution is shown in figure 4a while the last 40 seconds are shown in figure 4b. Figure 4a shows that the orbiting nozzle starts slower in covering the tank surface, but eventually it overtakes the stationary spray ball. Figure 4b on the other hand further quantifies the shadowing effect when using the stationary spray ball. It also shows that the rotating spray ball is "faster" in cleaning compared to the orbiting nozzle, but the orbiting nozzle is eventually able to catch up.

This modeling methodology is a very powerful tool in expediting the certification and optimization of CIP process. By proper selecting and sizing the system for different equipment configurations, the amount of cleaning liquid used can be significantly reduced and the certification process can be considerably shortened.

[1] M. Sami, H. Metwally, J. Ibrahim, and K. Brown, "Using CFD to Simulate Mixing Tank Clean-In-Place Process", NAFEMS 2019 World Congress, Quebec City, Montreal, Canada

[2] H. Metwally, M. Sami, K. Brown, "Using CFD to Simulate Tank Clean-in-Place Process", AICHE 2020

[3] Ansys Fluent User Manual, "Modeling Discrete Phase," Chapter 24, version 2021 R1

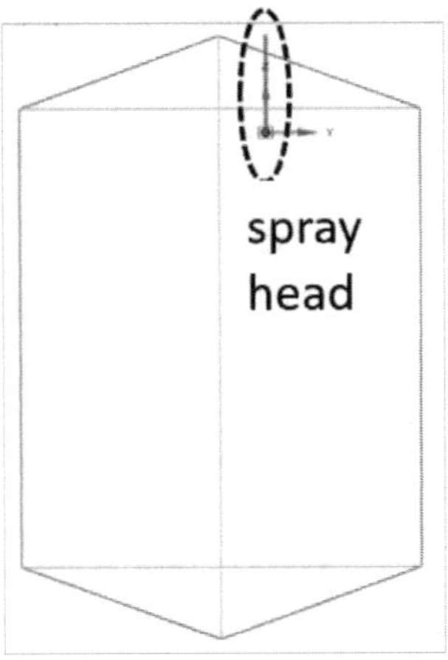

Figure 1, tank with spray head

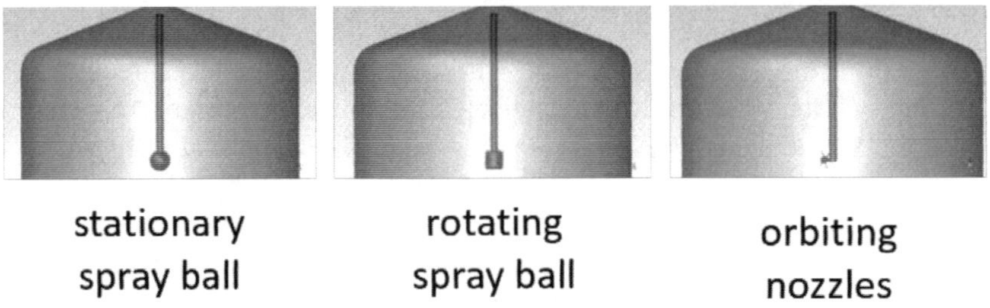

Figure 2, different spray head types

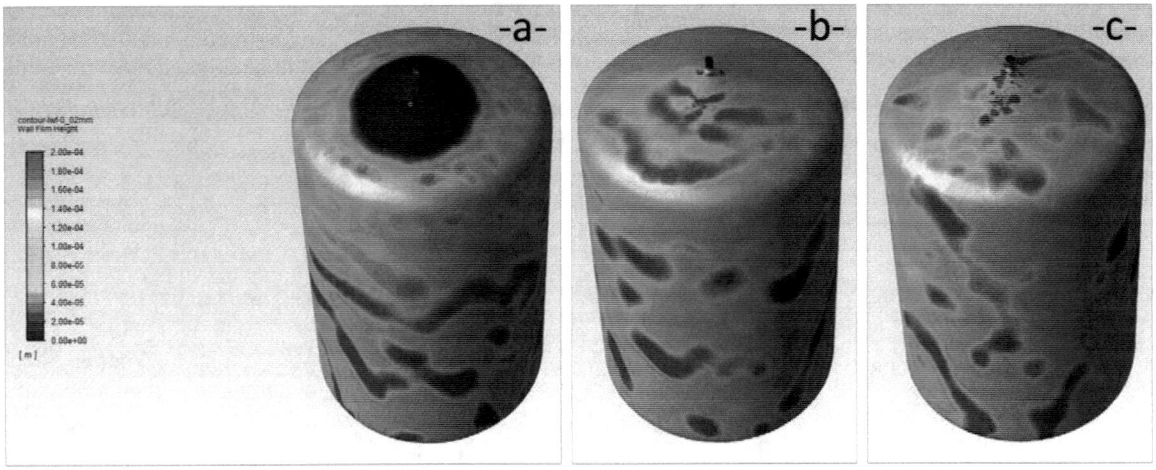

Figure 3, final film coverage comparison
a) stationary ball, b) rotating ball, c) orbiting nozzles

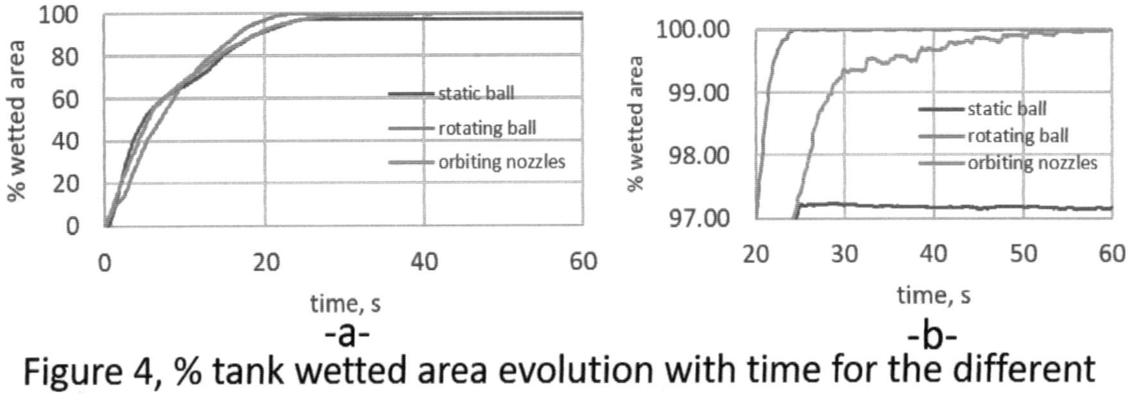

Figure 4, % tank wetted area evolution with time for the different spray head types; entire 60 seconds (a), last 40 seconds (b)

633236 Laser-Irradiated Holey Graphene-Supported Single-Atom Catalyst Towards Hydrogen Evolution and Oxygen Reduction

Abstract:

Single-atom catalysts (SAC) is vital in catalysis due to their ability to boost intrinsic catalytic activity of various chemical reactions, including hydrogen evolution reaction (HER) and oxygen reduction reaction (ORR). However, the challenge remains due to the complex synthesis process, insufficient stability, and massive metal species waste. Herein, we report a sustainable approach to synthesis laser irradiated, mesoporous graphene oxide support (LGO) via Nd:YAG pulse laser irradiation. The porous structure anchor metal-single atoms on nitrogen-doped laser irradiated graphene oxide support (NLG). We utilized nanocarbons 'dangling bonds' to extract metal atoms species with NLG support from its bulk counterparts, such as Co or Fe metal foams. The metal atoms can transfer electrons to NLG support via dangling bonds under optimized conditions to form metal-nitrogen bonds and make Co-NLG or Fe-NLG SAC. Our Co-NLG electrocatalyst requires low potentials of 146 mV to convey a current density of 10 mA cm^{-2} towards HER versus reversible hydrogen electrode (RHE) in acidic solution. Similarly, Fe-NLG electrocatalyst offered onset of 0.79 V towards ORR versus reversible hydrogen electrodes (RHE) in acidic solution. In parallel, the mechanistic understanding is established using density functional theory (DFT) calculations by applying the grand canonical potential kinetics (GCP-K) quantum mechanics model.

623009 Analysis of Transformed Bioprocess Data and Application of Hybrid Modeling Approaches to Enhance Upstream Process Development

Abstract:

Application of data-driven approaches in upstream process development has become a focal point of increased interest in biopharmaceutical industry. This, in part, is due to an increasing volume of on-line and sample data that become readily available during the process development, and upstream bioreactors being the key steps for determining the maximum achievable product quality and quantity. Despite significant advances made in the area (e.g., application of Multivariate Analysis), two key challenges remain: how do the intentional changes made during the process (e.g., feeding) impact the process and cell biology, and how do we use such knowledge to develop more productive processes? To address these questions, this presentation showcases usage of transformed bioprocess data and application of hybrid modeling approach (i.e., combination of data and fundamental-derived knowledge).

623356 Challenges in the Pharmaceutical Development of Lipid Nanoparticle Therapeutics

Abstract:

The discovery of RNA therapeutics has emerged as a promising modality to treat infectious diseases, cancer [1] and undruggable targets [2]. The delivery of RNA is a challenge because these molecules quickly get cleared upon administration and their size, along with negative charge does not allow it to cross into the cellular membrane [1]. As a result, many other platforms including positively charged antibody conjugates, masked endosomolytic agent-dynamic polyconjugates, or lipid nanoparticles offer an oligonucleotide delivery. These platforms also allow for DNA and RNA gene therapy, and vaccines. A recent rise in the development of clinical therapeutic RNA has led to the approvals of siRNA drugs (Patisiran) and mRNA vaccines (Moderna, CureVac, BioNTech) using lipid nanoparticle delivery [1].

The lipid nanoparticle delivery method offers the most mature platform for RNA therapeutics and has gained significant attention over the recent years [1]. However, there remains significant challenges in order to advance to commercialization. One of these challenges, requires efficient delivery of the RNA cargo to the targeted tissue and cellular compartment. Furthermore, the control of the lipid nanoparticle assembly is critical to the understanding the influence of in vivo pharmaceutical properties. A thorough comprehension of the physio-chemical properties to its bio performance is essential to ensuring the best quality commercialized drug product. This includes the molecular size, charge and polarity. The lipid nanoparticle assembly mechanism is also largely unknown [3]. We present an experimental design study of RNA loaded lipid nanoparticle self-assembly via a rapid antisolvent precipitation process, to identify critical parameters in the formation of LNPs. This will also help us understand the mechanism of the LNP assembly driving force to increase robustness for future formulation development and improve reproducibility.

RNA loaded lipid nanoparticles are complex structures of approximately 80-100nm, composed of cationic and neutral lipids. This includes PEGylated lipids in order to improve steric stability. The cationic lipid enables nucleic acid encapsulation and endosomal escape into the cytosol. The neutral lipids include phospholipid and cholesterol, which modulate the physical properties including rigidity, stability and biological membrane interactions. The oligonucleotide cargo, or RNA, is encapsulated at a defined ratio of cationic lipid to oligonucleotide. The process of forming the LNPs utilizes a micromixing device to generate controlled nanometer size, narrow polydispersity, and high RNA encapsulation efficiency [3]. The design of the mixing device is a continuous flow operation allowing for a rapid precipitation process suitable for commercia scale-up [3].

The goal of our developmental strategy is to control the LNP assembly process and reproducibility and correlate physicochemical properties with efficacy and/or toxicity. We use several analytical tools to enable robust and sensitive drug product characterization. For example, small-angle X-ray scattering (SAXS) quantitatively characterizes the structural features of RNA loaded LNPs [3]. Additionally, cryogenic transmission electron microscopy (cryo-TEM) micrographs can be used to identify LNP morphology, either hexagonal, cubic, unilamellar, or bilamellar macromolecular structures. Furthermore, differential scanning calorimetry (DSC) is a thermal analysis tool used to identify phase transitions [4] for LNPs. Both the SAXS and cryo-tem are powerful tools for characterization of LNP structure as a function of formulation manufacture.

In combination of our analytical tools and formulation composition we can generate LNPs of distinct lipid phases. A shift from lamellar to hexagonal shows to correspond to an improvement in potency. A difference in compositional heterogeneity in LNPs is observed to have significant difference in bioperformance. These overall changes in formulation or process are required frequently throughout preclinical, clinical and even sometime post market approval. Henceforth, understanding the key product attributes to in-vivo performance is essential. The analytical tools guide the formulation composition selection, manufacturing process development and ensure manufacturing process consistency.

1. Roces et al., Pharmeceutics, 2020, 12, 1095
2. Gindy et al., Expert Opin. 2012, 9, 171
3. Gindy et al., Langmuir 2014, 30, 16, 4613
4. Eygeris et al., Nanoletters, pending

630409 Dynamic Optimization of an Ultrafiltration System for the Concentration of Monoclonal Antibody Solutions Under Uncertainty

Abstract:

Monoclonal antibodies (mAbs) are a very important type of therapeutic agent with myriad applications in several therapeutic areas, such as immunology, oncology and infectious disease. These immunoglobins owe their success to their higher efficacy, specificity and safety compared to conventional drugs, which is a direct consequence of their mechanism of action. Unfortunately, all these benefits come at a price, namely, mAbs are much more expensive than conventional drugs. In fact, they may cost between 1,000 and 50,000 $/g.

One of the reasons why monoclonal antibodies are extremely expensive is their complex and low-yield production process, which includes several challenging manufacturing steps, namely, an initial fed-batch fermentation followed by centrifugation, depth filtration, several chromatographic purifications, virus filtration and, finally, concentration by ultrafiltration. Therefore, one of the most direct ways to reduce the average cost of monoclonal antibodies is to optimize their production process, in terms of both processing sequence and operating conditions at which the individual manufacturing steps are carried out.

Several authors have proposed optimal design configurations for mAb production processes (Liu et al., 2016) and computed the optimal operating conditions at which the fermentation and chromatographic separations should be carried out (Alhuthali et al., 2019; Papathanasiou et al., 2017). Some of these contributions have also analyzed the impact of model uncertainty on the results of their design and on the associated dynamic optimization calculations (Liu and Gunawan, 2017). Surprisingly, the ultrafiltration step has received very little attention so far, even though the typical yield of this operation may be as low as 90 % (as the monoclonal antibody is being concentrated, a significant amount of it forms a hydrogel that sticks onto the ultrafiltration membrane). Therefore, this contribution investigates the extent to which the systematic use of stochastic dynamic optimization techniques can help increase the efficacy/yield of the ultrafiltration step and reduce its duration.

More specifically, we first develop a hybrid mechanistic-statistical model of the ultrafiltration device, which is comprised of a simplified (yet first-principles inspired) model of the physical system, augmented with a Bayesian neural network (this type of model retains all the benefits of first-principles models, learns from experimental/process data as data-driven models, and provides point estimates of the process variables of interest as well as estimates of their degree of uncertainty). Then, we train this hybrid model on experimental data, collected at several different operating conditions, using advanced optimization-driven uncertainty quantification algorithms. Finally, we make use of the trained hybrid model and of a stochastic dynamic optimization algorithm to optimize the operating conditions of the ultrafiltration system under uncertainty with two principal goals, namely, maximization of the expected process efficacy/yield, measured as final mAb concentration/recovery, and minimization of the ultrafiltration time.

The results of our calculations are extremely promising, in that we can both improve the final mAb concentration/recovery and reduce the ultrafiltration time without violating any operational constraints, even in the presence of significant model uncertainty.

References

Alhuthali, S., Fadda, S., & Kontoravdi, C. (2019). Constrained Optimisation of Cell Culture Feeding Strategy and Temperature Shift Duration to Enhance Monoclonal Antibody Titre and Purity. 2019 AIChE Annual Meeting, Orlando (FL).

Liu, Y., & Gunawan, R. (2017). Bioprocess optimization under uncertainty using ensemble modeling. Journal of biotechnology, 244, 34-44.

Liu, S., Farid, S. S., & Papageorgiou, L. G. (2016). Integrated optimization of upstream and downstream processing in biopharmaceutical manufacturing under uncertainty: a chance constrained programming approach. Industrial & Engineering Chemistry Research, 55(16), 4599-4612.

Papathanasiou, M. M., Quiroga-Campano, A. L., Steinebach, F., Elviro, M., Mantalaris, A., & Pistikopoulos, E. N. (2017). Advanced model-based control strategies for the intensification of upstream and downstream processing in mAb production. Biotechnology progress, 33(4), 966-988.

629569 Confocal Imaging of Protein Gel Layer Formation during Tangential Flow Filtration to Inform Process Conditions Reducing Protein Losses and Increasing Protein Concentration

Abstract:

There are about 570 antibody therapeutics at various phases of clinical trials across the biopharmaceutical industry with 51 mAbs and 11 biosimilars in late-stage trials (Kaplan and Reichart, 2019). Of these 93% are produced in the US and Europe, and half are based on fully human genetic sequences (Grilo and Mantalaris, 2019). Therapeutic levels of monoclonal antibodies (mAbs) lie between 5 to 1000 mg per patient per treatment (Garidel et al., 2017; Maruthamuthu, 2020). Hence, mAbs must be concentrated to high levels (generally in the range of 150 to 200 mg/mL) to achieve therapeutic concentrations and/or reduce dosage volume (Garidel et al., 2017).

Tangential flow filtration (TFF) is the major unit operation in biopharma manufacturing for concentration of proteins. TFF uses shear forces along the membrane surface to limit protein accumulation on the membrane during the ultrafiltration process that occurs at pressures of up to 60 psig. Decreases in flux through the membrane correlate to increases in viscosity and osmotic pressure that occur when bulk protein concentration increases (Baek and Zydney, 2018). In addition, protein accumulation on the membrane's surface results in a gel layer that causes resistance to flux. Mechanisms of gel layer formation are complex and may depend on flow patterns, cross-membrane velocities, and biophysical properties that are a function of protein concentration, molecular weight, and pI.

We report a prototype confocal membrane chamber that enables observation of protein layer development on the surface of the membrane and along the length of the ultrafiltration module. This optical flow cassette was fabricated from a commercially available TFF cassette (Ultracel 30 kD_a pellicon 3) to expose the feed screen and the membrane underneath. A thin, optically transparent plexiglass cover placed over the surface of the feed screen forms a viewable feed channel enabling confocal imaging within a 4mm, optical path length. Flow passes across the membrane through a 1 mm high channel (distance from window to membrane surface). The cover was glued in place along the edges to form a pressure tight seal resulting in an operational replica of a bench-top TFF cassette with a membrane surface area of 44 cm^2.

A solution of Bovine IgG (50 mg/mL; in 10 mM citrate buffer with 50 mM NaCl, pH 5.5) was doped with a fluorescently labeled protein and concentrated in this prototype optical flow cell. This protein was used since its molecular weight of 150 kD is in the same range of molecular weights of many therapeutic mAbs presented in the literature, with a pI of 7.2 that lies in the mid-range of therapeutic mAbs whose pI's range from 6 to 9. The accumulation of protein on the membrane surface was captured with z-stack imaging. Together with separate measurements of flux, retentate concentration, transmembrane pressure, and flow rates, these images are helping to inform development of a computational model for predicting conditions that maximize protein concentration and recovery (Rossi et al., 2021). In this paper, we report visualization and characteristics of a gel layer at TFF membrane surfaces for bulk protein concentrations ranging from 50 to 200 mg/mL for process conditions using confocal microscopy.

References

Baek, Y. and A. Zydney. (2018). Intermolecular interactions in highly concentrated formulations of recombinant therapeutic proteins, Current Opinion in Biotechnology, *53*, 59-64.

Garidel, P., Kuhn, A. B., Schäfer, L. V., A. R., Karow-Zwick, & M. Blech. (2017). High-concentration protein formulations: how high is high? European Journal of Pharmaceutics and Biopharmaceutics, 119, 353-360.

Grilo, A. L., A. Mantalaris. (2019). The increasingly human and profitable monoclonal antibody market, *Trends Biotechnol.*, *37*, 9-15.

Kaplon, H., J. M. Reichart. (2019). Antibodies to watch in 2019, Mabs, *11*, 219-238.

Maruthamutu, M. K., S. R. Rudge, A. Ardekani, M. R. Ladisch, and M. S. Verma. (2020). Process Analytical Technologies and Data Analytics for the Manufacture of Monoclonal Antibodies, Trends in Biotechnology, *38*(10), 1169-1186.

Rossi, F., J. Zuponcic, E. Ximenes, S. Geng, Y. Taa, V. Corvari, M. Ladisch, R. Reklaitis. (2021). "Dynamic Optimization of an Ultrafiltration System for the Concentration of Monoclonal Antibody Solutions under Uncertainty," Abstract submitted to Annual AIChE meeting.

631529 Multiscale Modeling of Spray Coating of Perovskite QDs: Understanding the Role of Molecular Interactions in Particle Aggregation.

Abstract:

The US Department of Energy has set high-priority targets to reduce the cost of photovoltaics from ~ 10 cents/kWh to half of that at 5 cents/kWh by 2030. This has fueled a plethora of research in the discovery, optimization, and commercial production of high-efficiency solar cells. Perovskite solar cells (PSCs), especially $CsPbBr_3$, have received special attention due to their high quantum efficiency and tunable absorption/emission spectra [1]. Often, the perovskite quantum dots (QDs) are synthesized using one of the various batch synthesis routes, and after appropriate post-processing, they are spin-coated and separately cured to make a functional layer for the PSCs [2]. However, the problems of batch-to-batch variation, low-material utilization, and separate curing required in the spin coating process have slowed down and made the commercial production of PSCs difficult. In response, recent studies have demonstrated that spray coating can counter these challenges associated with spin coating by (a) providing a fast-scalable continuous production alternative with better material utilization; (b) offering a large control space for uniform deposition without the need for a separate curing step [3], and (c) integrating it with a continuous crystallizer to result in higher production capacity [4]. That being said, modeling and control of spray coating is a non-trivial task due to a combination of macroscopic phenomena (e.g., transport and thermal), and molecular-level interactions (e.g., dipole-dipole, van der Waals, and steric forces), thereby requiring a multiscale modeling approach.

In spray coating, the colloidal solution from the continuous crystallizer is atomized and sprayed onto a thermally heated silicon substrate through a pressurized nozzle. Consequently, the atomized droplets impinge onto the substrate and form a liquid splat followed by evaporation of the solvent (toluene). In this work, the spray coater is modeled using a system of macroscopic mass and energy balance equations (MEBE), which describe the effect of surface tension, vapor pressure, nozzle velocity, and the size of the impinging droplet on solvent evaporation. Further, during the solvent evaporation, the QD particles undergo particle aggregation due to the presence of various molecular interactions, and can lead to the formation of a non-uniform film, and demonstrate the coffee-ring effect. Thus, a modified form of the discrete element method (DEM), which solves newton's equation of motion for each QD particle, is utilized to accurately model these microscopic interactions [5]. Specifically, the DEM considers four molecular interactions, viz., steric hindrance, dipole-dipole forces, Brownian motion, and van der Waals forces. This is because, (a) the QD particles have a long-chain ligand-shell around them, which results in steric hindrance (repulsion) if two QD particles come very close to each other during aggregation; (b) experimental observations show a non-zero dipole moment for the QD particles [6]; and (c) the QD particles have a size range of 5-10 nm, which makes them very susceptible to undergo Brownian motion. Finally, the continuum model (MEBE) and the DEM are dynamically coupled to develop a high-fidelity multiscale model of spray coating.

The simulation results generated for different nozzle velocities and QD concentrations demonstrate the typical coffee ring formation as seen in the experiments [7]. The results also elucidate the effect of controllable macroscopic variables (i.e., substrate temperature, nozzle velocity, and QD particle concentration) on microscopic film evolution (i.e., aggregate formation, fractal structure and size). Overall, the proposed multiscale modeling framework provides a mechanistic understanding of the role of molecular interactions in QD particle aggregation and film deposition during temperature-controlled spray coating of QD particles for application in PSCs. Furthermore, this work provides a general platform for multiscale modeling of spray coating of different particle systems and will be incorporated with advanced optimization or process control frameworks to ensure adequate set-point (film thickness and smoothness) tracking in future works.

Literature Cited:

1. Protesescu L, Yakunin S, Bodnarchuk MI, Krieg F, Caputo R, Hendon CH, Yang RX, Walsh A, Kovalenko MV. Nanocrystals of cesium lead halide perovskites ($CsPbX_3$, X= Cl, Br, and I): novel optoelectronic materials showing bright emission with a wide color gamut. Nano Letters. 2015 Jun 10; 15(6): 3692-6.
2. Chen LC, Tien CH, Tseng ZL, Ruan JH. Enhanced efficiency of $MAPbI_3$ perovskite solar cells with FAPbX3 perovskite quantum dots. Nanomaterials. 2019 Jan; 9(1): 121.
3. Yuan J, Bi C, Wang S, Guo R, Shen T, Zhang L, Tian J. Spray-Coated Colloidal Perovskite Quantum Dot Films for Highly Efficient Solar Cells. Advanced Functional Materials. 2019 Dec; 29(49): 1906615.
4. Sitapure N, Epps R, Abolhasani M, Kwon JS. Multiscale Modeling and Optimal Operation of Millifluidic Synthesis of Perovskite Quantum Dots: Towards Size-Controlled Continuous Manufacturing. Chemical Engineering Journal. 2020 Dec 7: 127905.
5. Peng Z, Doroodchi E, Evans G. DEM simulation of aggregation of suspended nanoparticles. Powder Technology. 2010 Dec 10; 204(1): 91-102.
6. Wang W, Zhang Y, Wu W, Liu X, Ma X, Qian G, Fan J. Quantitative modeling of self-assembly growth of luminescent colloidal $CH_3NH_3PbBr_3$ The Journal of Physical Chemistry C. 2019 Apr 28;123(20):13110-21.
7. Xu T, Lam ML, Chen TH. Discrete Element Model for Suppression of Coffee-Ring Effect. Scientific Reports. 2017 Feb 20;7(1):1-0.

624233 Findings and Conclusions from a Mobile Worker and Augmented Reality Enabled Continuous Manufacturing Skid Project.

Abstract:

This abstract presents the final conclusions regarding a project initially presented at AIChE 2021 Spring Meeting as an oral presentation (abstract#618949) and poster (abstract#623026).

Background:

Manufacturing USA was created in 2014 to secure U.S. global leadership in advanced manufacturing by connecting people, ideas, and technology. Manufacturing USA institutes convene business competitors, academic institutions, and other stakeholders to test applications of new technology, create new products, reduce cost and risk, and enable the manufacturing workforce with the skills of the future (www.manufacturingusa.com).

MxD encourages factories across the United States to deploy digital manufacturing and design technologies, so those factories can become more efficient and cost-competitive (www.mxdusa.org).

Project detail:

This abstract relates to a continuous process manufacturing skid that has a focus on mobile workers and demonstrates the future of manufacturing by using a digital twin enabled physical skid build. The question the skid answers is "How would you visualize sensor data with the best User Interface/User Experience design to get a seamless experience in continuous manufacturing?"

The findings and conclusions come from a project team consisting of an government owned site personnel, a private chemical processing company, a multi-national technology supplier and multiple service providers. Equipment and assets applied in the project go from sensors to process equipment to pumps/drives to valves to a DCS to a MES and complete engineering and design suite and finally to an IT/OT-integrated and IIoT-enabled setup to support day-to-day operations and mobile workers. Cloud platform with Artificial Intelligence / Machine Learning capabilities is connected to the site and the skid.

Currently there is considerable work done in chemical industry to study the current state and possible future states when it comes to skids, processes, plants, companies, professionals, workers and value chains. Applying novel techniques into operations is in practice an efficient way for human beings to get a real grasp on what is impacted, how it is impacted and what do we want to do with the novel capabilities that come from innovation areas such as the wider digital enterprise or industry 4.0 concept. Concrete studies like this one bring important contributions to the field, especially when introduced as an example in a discussion or presentation at an event such as the AIChE Annual Meeting.

The subject of this abstract was originally planned and divulged in 2019, moved into a contractual obligation in 2020 and is scheduled to conclude in summer of 2021. This timeline allows the authors to go through the whole lifecycle from problem statement to selected methods to real physical results and finally to discuss and expand the implications for future ways of working.

The audience will learn about augmented reality and digital twin use cases around a continuous process testbed. The audience will understand that this skid is available for industry professionals and students to visit at a public facility on Goose Island, Chicago.

Use Cases:

The goal of the process industry test bed skid (physical) and its digital twin (virtual) is to enable field operators, engineers, factory managers, and other key manufacturing roles to gain data-driven insights for increased operational awareness in order to better solve manufacturing problems and optimize processes. A factory manager should be able to communicate with and assign tasks to the mobile worker and view all data necessary to make informed decisions about the assignment of tasks. The solution will be demonstrated using the process test bed and should be agnostic to sensor type, analytics solution, and control type. Visualization of this data should include, but is not limited to, dashboards for factory managers and augmented reality technology for field operation using mobile devices (e.g. tablet and smart glasses).

Conclusions: The three authors of the abstract will each share their findings from their viewpoint, those being the viewpoints of an owner/operator company, the production site and the main technology provider.

632817 Practical Issues in Cybersecurity: From Encryption to Images

Abstract:

Cybersecurity has gained interest for chemical processes in recent years. For example, [1] discusses techniques for considering cybersecurity during traditional chemical engineering procedures such as hazard and risk analysis. One of the main drivers for the cybersecurity discussion is an increased interest in recent years in concepts such as Industry 4.0 [2] and the Industrial Internet of Things [3], and a general increase in data analytics and artificial intelligence approaches. Future work in this domain would benefit from research that enables a more integrated approach to control system cybersecurity in the chemical engineering research field, in which the cyber defenses at a plant are more explicitly considered, as well as advanced control frameworks and concepts. These concepts include resilient frameworks, to analyze how a system functions as a whole and the types of attacks it remains vulnerable to after both, the theoretical requirements for certain resilient control frameworks as well as, practical communication and networking protocols that need to be accounted for. In addition, there are a wide variety of issues becoming more important to consider from a cybersecurity perspective over time. For example, image analysis is important for certain types of decision-making systems, though image data represents another type of information that could be falsified to manipulate such decision-making [4], and computations on the Cloud, where it may be desired to encrypt not only the data but the computations themselves [5, 6], are additional concerns.

Motivated by the above, this talk highlights multiple concepts in the practical aspects of control system cybersecurity as it interfaces with advanced control. For example, using a case study of a benzene hydealkylation process under distributed Lyapunov-based economic model predictive control [7], we explain the role of encryption protocols in enabling safe transfer of information between distributed units, and the extent to which the encryption might enable attacked controllers to be identified on the network. We utilize a process example of a semi-batch polymerization reactor under model predictive control with uncertainty scenarios [8] and offset-free model predictive control [9] to demonstrate the manner in which disturbance-handling is not equivalent to cyberattack resilience, and we also discuss profit targeting as a key metric in distinguishing attacks and faults. We follow up with a discussion of considerations in image analysis, which form part of the increasingly automated world that is more open to attacks, and verification in that domain, and how such concepts relate to cybersecurity and cyberattack resilience. For example, focusing on the zinc flotation process, in which image-based control is considered today [10], we explore the potential use of software such as Blender for creating a preliminary closed-loop simulation methodology for simulating the zinc flotation process, including both system state evolution under control actions, as well as image generation and processing, along with analysis on the impacts of cyberattacks on image data that determine process characteristics. Finally, we will discuss the role of communication techniques, as well as encryption techniques which facilitate communication with the Cloud, in control for nonlinear chemical process systems.

[1] Cormier, A., & Ng, C. (2020). Integrating cybersecurity in hazard and risk analyses. *Journal of Loss Prevention in the Process Industries*, *64*, 104044.

[2] Reis, M. S., & Kenett, R. (2018). Assessing the value of information of data-centric activities in the chemical processing industry 4.0. *AIChE Journal*, *64*(11), 3868-3881.

[3] Sisinni, E., Saifullah, A., Han, S., Jennehag, U., & Gidlund, M. (2018). Industrial internet of things: Challenges, opportunities, and directions. *IEEE Transactions on Industrial Informatics*, *14*(11), 4724-4734.

[4] Bucci, E. M. (2018). Automatic detection of image manipulations in the biomedical literature. *Cell death & disease*, *9*(3), 1-9.

[5] Darup, M. S., Redder, A., & Quevedo, D. E. (2018). Encrypted cloud-based MPC for linear systems with input constraints. *IFAC-PapersOnLine*, *51*(20), 535-542.

[6] Alexandru, A. B., Morari, M., & Pappas, G. J. (2018, December). Cloud-based MPC with encrypted data. In *2018 IEEE Conference on Decision and Control (CDC)* (pp. 5014-5019). IEEE.

[7] Liu, J., Chen, X., Muñoz de la Peña, D., & Christofides, P. D. (2010). Sequential and iterative architectures for distributed model predictive control of nonlinear process systems. *AIChE Journal*, *56*(8), 2137-2149.

[8] Lucia, S., Finkler, T., & Engell, S. (2013). Multi-stage nonlinear model predictive control applied to a semi-batch polymerization reactor under uncertainty. *Journal of process control*, *23*(9), 1306-1319.

[9] Das, B., & Mhaskar, P. (2015). Lyapunov-based offset-free model predictive control of nonlinear process systems. *The Canadian Journal of Chemical Engineering*, *93*(3), 471-478.

[10] Kaartinen, J., Hätönen, J., Hyötyniemi, H., & Miettunen, J. (2006). Machine-vision-based control of zinc flotation—a case study. *Control Engineering Practice*, *14*(12), 1455-1466.

634112 Keynote Talk: Machine Learning and AI Applications in the Chemical Industry

Abstract:

Industrial chemical processes usually come with abundant data and thus provide huge opportunity to leverage machine learning and AI techniques for process improvement. Over the past decade as computational resources have become readily available at much lower costs, together with rapidly growing availability of cloud services, sensor and analytical technology, the number of AI applications adopted for process monitoring, control and optimization have also increased exponentially. Aside from traditional multivariate analysis that is often used for monitoring and fault diagnosis, other analytics fields such as (but not limited to) image analysis, deep learning, sensor fusion, text mining, reinforcement learning, hybrid and physics informed surrogate modeling are also being actively applied to increase reliability, reduce safety events, predict the onset of potential equipment failures and enable real time optimal decision making. This talk will provide examples of AI applications using industrial case studies and also address current challenges as well as outlook.

634113 Keynote Talk: Deploying AI for Automated Monitoring of Physical Infrastructure

Abstract:

This presentation will discuss how AI can be used for automated monitoring of physical infrastructure across various assets and processes. Key topics include:

The purpose and benefits of following an Op-Specific approach to digital transformation based on survey results.

How an operational data platform can reduce resource consumption, improve sustainability goals, and increase revenue.

Steps to implement a specific use-case.

An Operational Data Platform (ODP) can connect the constantly changing realities of the physical world with the power of intelligent software. Plutoshift has built an ODP to unlock the potential of our operational data to enable process monitoring and remote decision making. Plutoshift serves a variety of industries such as chemicals, food & beverage, manufacturing, water treatment, and more. The Plutoshift platform has enabled significant cost reductions ranging from 20-30%. Along with that, it has enabled millions of dollars of revenue increases for customers. The platform is built to help companies achieve sustainability targets as well. For example, the platform helped one global brewer's facility save close to 16 million gallons of water in a single year. These savings are being driven by enhanced visibility of assets and processes across the physical infrastructure.

624540 Kinetic Model and Autothermal Reactor Design for the Oxidative Dehydrogenation of Ethane

Abstract:

Ethylene is a key platform chemical used in the production of a variety of bulk chemicals such as polyethylene, ethylene dichloride, and ethylene oxide. Despite the oxidative dehydrogenation of ethane (ODHE) being extensively investigated as an alternative to steam cracking processes currently used to produce ethylene, a reactor design that can plausibly be implemented on an industrial scale has not yet been proposed in the literature. In this work, we use a combination of theory and experiment to demonstrate that ODHE can be operated autothermally at high pressures with ambient temperature feeds and no external heat input.

A global kinetic model that accurately describes ODHE catalysis between 330- 430°C and 0-6 bar total pressure on a $MoVTeNbO_x$ catalyst was developed using a combination of reaction kinetics experiments and mathematical modeling. Parity plots point to good agreement between model predictions and experimentally measured rates over a wide range of experimental conditions. F-values and t-values as well as the confidence intervals were utilized to evidence the global significance of proposed kinetic models. These global kinetic models were used to conduct bifurcation analysis and identify plausible regions of autothermal operation that were then validated using lab-scale experiments. Our results demonstrate that autothermal reactors can be used to achieve higher ODHE reactor productivity compared to traditional fixed-bed reactors, especially at higher pressures, with their advantages becoming more amplified for highly active catalysts. The results of the study create opportunities for the intensified production of ethene in compact, modular reactors that can operate in concert with renewable sources of electricity that are inherently intermittent in nature.

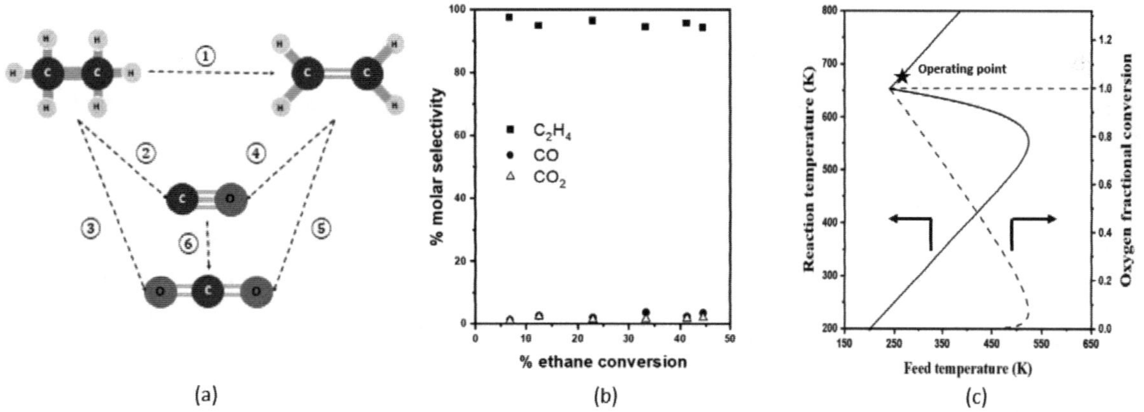

Figure 1. (a) Reaction scheme of the oxidative dehydrogenation of ethane (b) Product molar selectivity as a function of ethane conversion over the $MoVTeNbO_x$ catalyst. Reaction conditions: 643K, 12kPa C_2H_6, 6kPa O_2, balance He. (c) ODHE steady-state bifurcation diagram assuming an LTR model. Reaction temperature and fractional oxygen conversion as a function of feed temperature. The star represents the operating point for the autothermal process. Reaction conditions: $\tau = 1.8s$, $C_2H_6 : O_2 = 6$, $\Delta T_{ad} = 413K$.

624800 Single Reactor Design Concepts for Achieving an Autothermal Operation of Exothermic Oxidative Coupling of Methane and Endothermic Methane Dehydroaromatization Reaction

Abstract:

Need for sustainable processes: Around 3 billion tons of carbon dioxide is emitted every year through the chemical and petrochemical industries. [1] To reduce these emissions and utilize the natural resources better, new sustainable routes are needed to replace the highly optimized and cost-competitive methane conversion to chemicals via the indirect syngas route. Of the proposed directed routes are the oxidative and non-oxidative coupling of methane, which have not yet been commercialized.[2] Among other challenges, these routes require high operating temperatures (600-1000^oC) and strict heat management.[3] An idea to address these challenges is to consider both chemistry routes in the same chemical plant as this can have many potential advantages. One of these potential advantages is reaching autothermal operation which reduces the energy demands and carbon emissions. This work aims to explore options for thermal coupling of exothermic and endothermic reactions in a single reactor but without any mass integration. The first reaction is the oxidative coupling of methane (OCM) which is highly exothermic (ΔH^o_{rxn}= -141 kJ/mol.CH_4) and produces C2+ products like ethane and ethylene. The second reaction is methane dehydroaromtization (MDA) which is endothermic (ΔH^o_{rxn}= +88.4 kJ/mol.CH_4) and yields C6+ products like benzene, toluene, and naphthalene.

Reactors modeling and thermal coupling methodology: As an initial step, each reaction was modeled using a separate reactor with the assumptions of an ideal packed bed reactor. A literature review was carried out to identify available kinetics models for OCM and MDA reactions. Kinetic models provided by Wang and Lin, 1995 for OCM [4] and Zhu et. al, 2018 for MDA [5] were chosen owing to the relatively simple kinetics. A parametric study was then carried out to evaluate the effect of each process variables (pressure (P), temperature (T), feed composition (yi), flow rate (F)), and catalyst weight (mCat) on the reactor performances (conversion (X), selectivity (S), yield (Y)) and most importantly heat duty (Q). Based on this parametric study, an operating window for thermal coupling was identified in between 700 – 850 ^oC temperature, 1 – 5 atm pressure, GHSV 830 – 14,000 h^{-1}, and heat duties of 1–13 MW. This operating window was identified using multiple optimization studies via single-variable-at-a-time optimization to match the heat duties of both reactors as shown in figure 1(a). Owing to many input and output variables, several design options could be proposed. Therefore, we utilized the visual representation of Artificial Neural Network, shown in figure 2, to help navigate between all the different options and provide a structured methodology for performing the optimization for thermal coupling.

Design concepts for thermal coupling of OCM and MDA in a single reactor: This study was carried out by having each reaction in a separate reaction channel divided by a channel wall, across which the heat was exchanged between the channels, as shown in figure 1(b). Heat transfer and pressure drop were estimated using well-established engineering correlations from the literature. The axial temperature profile was assessed using the reaction performance parameters indicated in figure 2. As a starting point, both reaction channels were considered straight, having identical channel length, and filled with spherical catalyst particles, as an ideal packed bed reactor; whereas, the flow rate and the amount of catalyst were varied by adjusting the diameter of the reaction channel. Using these parameters, different design options were considered in this study.

Initial assessment of the study showed that an autothermal operation is achievable. Depending on the chosen reaction conditions and required reactor performances, the ratio of MDA to OCM catalyst varies widely between 1 to 100, which reflects in terms of the reactor geometry, like tube diameter and channel length. Based on the chosen reactor design, the OCM axial temperature profiles present peak temperatures varying from 50 to 300 °C, caused due to excess amount of heat released by OCM, than the amount of heat absorbed by MDA. This is related to the difference in the rates of both the reaction. Therefore, achieving acceptable reactor performances within the operating window draws a great challenge during the thermal coupling. To overcome these limitations, two approaches are utilized which consider non-traditional reactor channel geometry with and without catalyst profiling. Elaboration of these options will be provided while presenting this work.

Towards a global optimum solution and challenges hindering its realization: Several established catalysts are available for both reactions that have their own performances and challenges, and form different products depending on the operating conditions. The current study was performed using simple kinetic models which have a limited validity. In the case of OCM kinetic model, it was only valid for feed composition of 2.5 – 35% CH_4 at 1 atm, 600 – 750 °C and low residence times. It assumes C2 product as C_2H_6 and side product as CO_2. For the case of MDA kinetic model, the validity was limited for a feed of 95% CH_4 at 1 atm, 675 – 750 °C and three components C_2H_4, C_6H_6 and $C_{10}H_8$. Most importantly, this kinetic did not incorporated any coking phenomena, which is a major challenge particularly for the MDA reaction.

Changing these catalysts varies the reaction rates and hence reactor heat duties, which varies the final solution of thermally coupled reactors. Many other design parameters can also drive towards a different optimum, like catalyst shape, geometry and dimensions of the reaction channel. Hence, the presence of multiple design parameters leads to multiple local optima solutions using manual single-variable-at-a-time optimization. Thus, to be able to screen these solutions for achieving a global optimum, an automated multi-objective optimization will be performed. This study shall, therefore, pave the way towards identifying reactor designs for coupling exo- and endo-thermic reactions in a single reactor for achieving an autothermal operation; thereby, making it an energy efficient process with reduced carbon emissions.

References:

[1] H. Ritchie and M. Roser, https://ourworldindata.org/emissions-by-sector. [Accessed: 24-Mar-2021].

[2] S. Majhi, P. Mohanty, H. Wang, and K. K. Pant, *Journal of Energy Chemistry*, vol. 22, no. 4. Elsevier, pp. 543–554

[3] C. Karakaya and R. J. Kee, *Progress in Energy and Combustion Science*, vol. 55. Elsevier Ltd, pp. 60–97

[4] W. Wang and Y. S. Lin, *J. Memb. Sci.*, vol. 103, no. 3, pp. 219–233

[5] Y. Zhu, N. Al-ebbinni, R. Henney, C. Yi, and R. Barat, *Chem. Eng. Sci.*, vol. 177, pp. 132–138

Figure 1 (a)
Independent Reactors

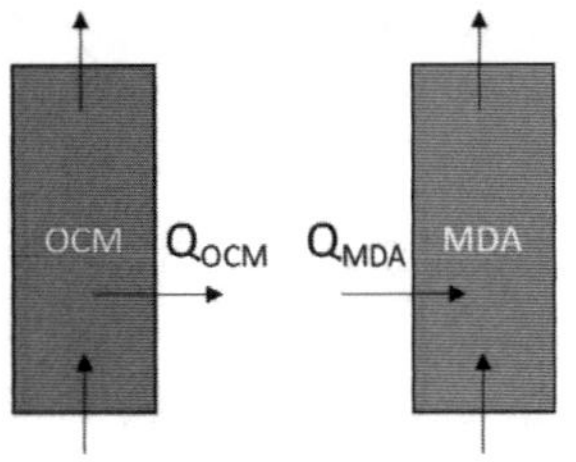

Figure 1 (b)
Dependent Reactors

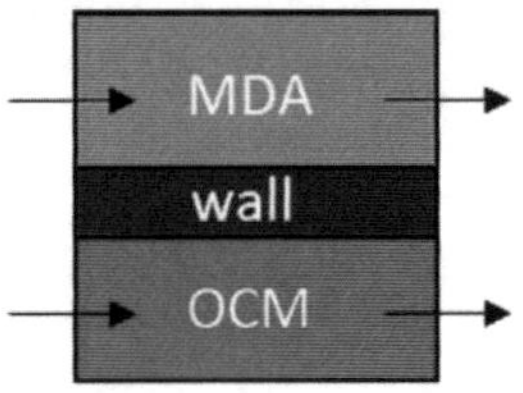

Figure 2
Thermal Coupling Methodology

	Operating Conditions				Feed Composition			Reactor Performance				Heat & Cat. Wt.		
Case Description	F	T	P	GHSV	y_{CH4}	$y_{CH4/O2}$	y_{inert}	$Y_{C2/C6}$	X_{CH4}	X_{O2}	$S_{C2/C6}$	Q	m_{Cat}	Cat R MDA/OCM
	Stdm3/h	°C	atm	h^{-1}	%	-	%	%	%	%	%	MW	tons	-
OCM (Ref. Case)	10,000	800	1	14,919	61.7	3	17.8	33.8	46.1	99.99	73.2	***9.95***	0.185	-
MDA (Ref. Case)	10,000	750	1	1,700	95	-	5	1.1	10.3	-	10.3	***1.15***	3	16.2
MDA *(Opt. Case)*	72,320	800	1	3,688	95	-	5	0.95	12.3	-	7.6	***9.95***	10	54.1

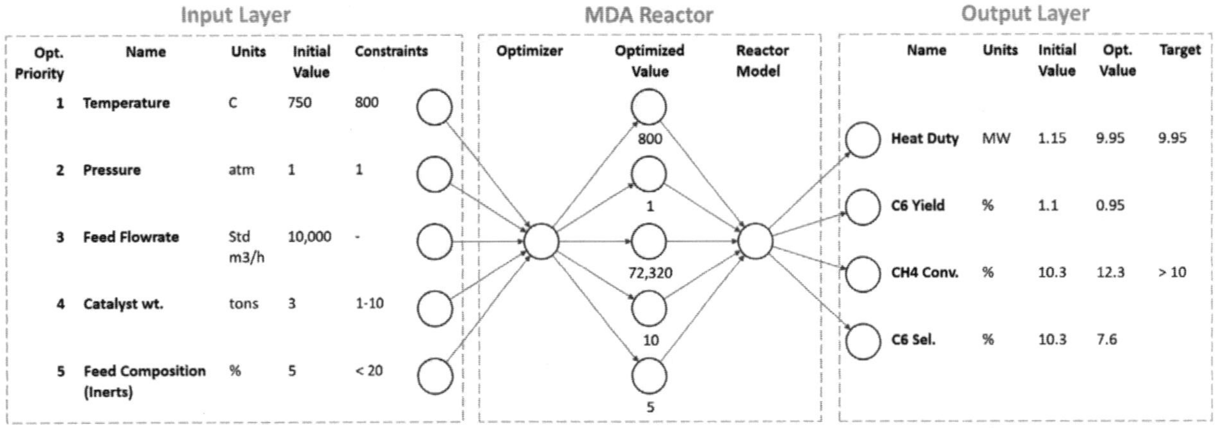

626064 Effect of Cavitation Intensity on Thermal Efficiency and Reactive Performance of a Hydrodynamic Cavitation Reactor

Abstract:

The production of specialty chemicals is to-date conducted almost exclusively in large-volume batch reactors. However, emerging challenges, including supply chain limitations, stricter environmental regulations, and safety concerns, are driving an on-going transition to much smaller, compact, and modular continuous processing units. While this transition itself constitutes a "process intensification" step, the use of novel reactor concepts for continuous processing promises to further enhance the benefits of process intensification for specialty chemicals production terms of energy efficiency, product quality, process safety, and reduced physical footprint of the plant.

In this context, hydrodynamic cavitation reactors (CavR) are promising but to-date rarely utilized. CavR rely on the physical phenomenon of cavitation that involves the rapid formation, growth, and subsequent collapse of vapor micro-bubbles within a liquid which occurs on a very short time-scale (milliseconds) and releases large amounts of localized energy. It hence seems particularly well-suited to the processing of viscous reactive flows, as typical for dispersants production. The onset of cavitation can be predicted using the cavitation number equation ($Cv = (P - P_v) / (0.5 \rho v^2)$), where P is the downstream system pressure, P_v and ρ the vapor pressure and density of the liquid, respectively, and v is the (local) flow velocity of the fluid. Cavitation occurs when Cv is unity and cavitation intensity increases with decreasing Cv.

Here we present an investigation into the effect of cavitation intensity on the thermal efficiency and reactive performance of a rotational CavR in the production of succinimide dispersants. By varying the cavitation number via changes in key experimental parameters, including viscosity and density of the working fluid, downstream system pressure, and rotational velocity, we evaluate the effect of Cv on (average) fluid temperature and reactive conversion. Our results suggest that the observed increase in bulk fluid temperature is mainly due to frictional heating between the rotating cylinder and the fluid, rather than due to cavitation. However, cavitation strongly enhances the observed reaction rates in comparison to a conventional tubular reactor at similar operating conditions. This suggests that either the very high, if short-lived, local temperature maxima during cavitation or the intense local mixing of the poorly miscible reactants result in this strong enhancement beyond reaction rates observed even in a (macroscopically) well-mixed conventional tubular reactor. We complement these studies by a detailed analysis of parameters affecting the cavitation intensity (tangential velocity and fluid viscosity) using stroboscopic videography in order to develop an improved fundamental understanding of cavitation in a rotational hydrodynamic cavitation reactor.

Overall, our results suggest that cavitation reactors constitute a promising, strongly intensified reactor configuration for the processing of viscous mixtures for the specialty chemicals industry.

627009 Continuous Biphasic Microreactor for Production of Hydrogen Peroxide Via Non-Thermal Atmospheric Pressure Plasma

Abstract:

Hydrogen peroxide (H_2O_2) represents an attractive green oxidant due to its high active oxygen content and its clean reduction product (*i.e.,* water). Nevertheless, state of the art production of H_2O_2 comprises catalytic cycles of oxidation and hydrogenation in the presence of an organic substrate (anthraquinone).[1] This process is viable only at an economy of scale, with highly concentrated H_2O_2 requiring special handling for the downstream distribution and storage. Therefore, plasma-assisted production of diluted H_2O_2 can embody an alternative route for localized production, allowing for minimization of waste (only water is produced) and energy demand (electric energy is the only energy source).[2]

In this work, a novel biphasic non-thermal atmospheric pressure plasma microreactor is presented and characterized with respect to its H_2O_2 production. The system features a continuous biphasic stream of helium and water (whose flow rates are regulated by a gas flow controller and a syringe pump, respectively) that passes through a modular plasma region which can be adjusted by changing the reactor geometry. The micro-dimension of the reactor allows for low applied voltage and in turn low power input even for sinusoidal high-voltage signals. The production of H_2O_2 is assessed in relation to varying ratios of gas and liquid flow rate, reactor geometry and power input.

Most importantly, the effect of gas and liquid flow rate in the system is two-fold: on the one hand, high gas flow rate enhances the stability of the plasma and the production of excited species. On the other hand, high gas-to-liquid flow rate ratio has proven to intensify the interfacial area between the two phases with direct effect on the H_2O_2 concentration. In fact, H_2O_2 is expected to be forming on the gas-liquid interface, where recombination of OH radicals formed in gas phase takes place.[3,4] CFD calculations help capturing the different flow patterns linked to the varying flow rates. The effect of all operating parameters on the performance of the reactor is also discussed.

References

1. Strukul, Giorgio, ed. Catalytic oxidations with hydrogen peroxide as oxidant. Vol. 9. Springer Science & Business Media, 2013.
2. Locke, Bruce R., and Kai-Yuan Shih. "Review of the methods to form hydrogen peroxide in electrical discharge plasma with liquid water." Plasma Sources Science and Technology 20, no. 3 (2011): 034006.
3. Zhou, Renwu, Rusen Zhou, Peiyu Wang, Yubin Xian, Anne Mai-Prochnow, Xinpei Lu, P. J. Cullen, Kostya Ken Ostrikov, and Kateryna Bazaka. "Plasma-activated water: generation, origin of reactive species and biological applications." Journal of Physics D: Applied Physics 53, no. 30 (2020): 303001.
4. Wang, Huihui, Robert J. Wandell, Kosuke Tachibana, Jan Voráč, and Bruce R. Locke. "The influence of liquid conductivity on electrical breakdown and hydrogen peroxide production in a nanosecond pulsed plasma discharge generated in a water-film plasma reactor." Journal of Physics D: Applied Physics 52, no. 7 (2018): 075201.

629297 Zero-D Thermodynamic Model As a Simple Tool for Screening Chemical Reaction Candidates and Benchmarking of the Piston Reactor Technology

Abstract:

In the future, electricity derived from renewable energy resources (e.g., solar and wind) will become the most widespread, cheap, and accessible form of energy. This work addresses the potential of utilizing an electrical-driven piston reactor as an alternative technology to conventional reactors to produce value-added chemicals in a more compact, safe, and with less emissions. The reactor concept is based on repurposing conventional engines to be used as chemical reactors. Different than conventional engines, the exhaust gases would become valuable products. One of its essential advantages the ability to achieve high temperatures and pressures upon compression at very short time scales. Therefore, the reactions toward metastable species proceed at relatively high rates. Upon expansion, the fast quenching preserves further conversion of the valuable metastable intermediates, which can be harvested from the effluent stream. The short cycle durations also minimize the energy losses to the surrounding, thus achieving near the adiabatic operation.

The piston reactor technology is still at its early stage of development. So far, the most tested application is the production of synthesis gas by partial oxidation of methane [1-5]. Studies on the production of chemicals other than synthesis gas are limited. [6-8]. Part of this is the multidisciplinary nature and the lack of understanding of this reactor technology and its expected advantages. This work aims to illustrate these advantages by comparing the performance of the piston reactor versus a fixed bed reactor for two relevant industrial reactions. The partial methane oxidation (POX), and methane steam reforming (SRM). These are selected as representative case studies for highly exothermic and endothermic reaction systems to quantitatively analyze the potential of producing hydrogen and mechanical power as in the case of exothermic reactions. Besides benchmarking, this work demonstrates how a zero-D thermodynamic model can be used as a simple tool for screening chemical reaction candidates to guide essential but expensive experimental work.

A zero-dimensional model was developed to describe and simulate the thermodynamic processes inside a single-piston cylinder. The simulation consists of compression and expansion strokes, starting at the bottom dead center (BDC) at the inlet valve closing time equivalent to a crank angle of 180°. A full rotation of the crankshaft is then carried out till the exhaust valve opening time, equivalent to a crank angle of -180°. The governing equations constituting the model's basics are derived based on mass and energy balances dependent on time/crank angle. Therefore, the spatial gradients inside the piston chamber are neglected. The concentration changes due to chemical reactions during the compression-expansion strokes were calculated using time-dependent kinetic models from the literature. Inlet temperature, pressure, RPM, feed composition, and compression ratio are all investigated to elucidate a feasible operating window to carry out the proposed reactions. Similarly, a conventional plug flow reactor (PFR) model was developed to benchmark the piston reactor in terms of conversion and productivity.

A parametric study was carried out using the piston reactor model to identify the conditions needed to achieve hydrogen productivity similar to the PFR reactor model in the case of SMR. A case study of an industrial steam reformer was simulated using a conventional PFR model. The operating conditions were adopted from a publication by Said Elnasaie et al. [9]. It was for a top-fired steam reformer (consisting of 897 tubes) operating at 1130 K and 22 bar. A gas flow rate of 3.953 kg.mol/hr (methane equivalent) containing 20.22% CH_4, 72% H_2O%, 4.92 H_2%, 2.44% CO_2% 0.42% N_2 is typically processed in a single tube pass of 13.72 m length and 0.0978 m inner diameter. The catalyst bulk density for the aforementioned dimensioned was 1360 kg/m^3. Numerical simulations were run using an SRM kinetic model developed by Xu and Froment [10]. Using the aforementioned operating conditions, CH_4% conversion of 87.12% per tube pass was reached, which is very close to that reported by Said Elnasaie et al. from an actual plant output (85.27%). Moreover, the model predicted a hydrogen productivity per tube obtained of 30.87 mol/ (m^3. s). The geometrical parameters of the piston reactor were adopted from a modeling study by Goßler [11]. The total volume of the piston reactor was calculated to be 671.03 cm^3. Figure 1 shows the effect of intake temperature on methane conversion and hydrogen production rate per reactor volume. By increasing the intake temperature, the methane conversion, hence hydrogen production is significantly increased. The latter shows good agreement with thermodynamic studies, suggesting that high temperatures favor the highly endothermic methanation reaction in steam reforming.

To achieve the same hydrogen production rate of 30.15 mol/ (m^3. s) as that of PFR, the piston reactor needs to operate at T_{intake} = 820 K, N=1000 RPM, and only 14.9 kg/m3 of catalyst. This is more than 100 fewer catalyst amounts and 310K lower than the PFR case, which minimizes the preheating energy requirements. This effect is mainly due to the high in-cylinder temperatures that result from compression, which facilitate the methanation reaction to proceed faster. To further enhance the hydrogen productivity, it is possible to increase the engine speed, hence process more gas per piston cycle to achieve a higher production rate. As shown in figure 2, increasing the piston speed from 1000 RPM to 5000 RPM results in lower methane conversions, which was expected since the chemical species' residence time decreases at higher speeds, thus lowering the conversion. However, it is possible to achieve a 30.05 mol/ (m3. s) hydrogen production rate at a lower intake temperate of 690K than the base case scenario. It should be noted that by running the piston reactor at an intake temperature of 1130 K (similar to the PFR) and 5000 RPM, the hydrogen production rate achieved would be approximately five-folds higher compared to the PFR case. Another study was conducted to study the effect of increasing the catalyst weight from 1 to 10 g to further enhance hydrogen production as shown in Figure 2 (c) and (d). It can be observed that using a catalyst weight of 10 g at N = 5000 RPM in the range of the considered intake temperatures, higher methane conversions and hydrogen productivities are achieved compared to the base case scenario (1 g catalyst). In this scenario, a hydrogen production rate of 30.61 mol/ (m^3. s) (similar to the PRF case) is achieved at 489 K (632 K lower compared to the PFR case).

Initial studies on the piston reactor as a future hydrogen production technology (via SRM) have shown great potential in contributing to much less catalyst utilization, energy minimization, compact design, and having a reactor that is driven by renewable electricity. Moreover, the combination of high in-cylinder temperatures and high piston speeds positively affects productivity compared to a conventional PFR reactor. A similar analysis is carried out for POX as potential reaction candidates to produce hydrogen and mechanical work. Full details and finding from these studies will be demonstrated in this work. Although a quick screening and benchmarking tool was developed and demonstrated in this work, two challenges are limiting the utilization of the findings from this work. The limited validity of the kinetic models, which are based on fixed-bed reactor technology for narrow operating window, and the use of heterogeneous catalyst, which is still a challenge to handle in the piston reactor. This work paves the way toward the exploration and development of this promising but less explored reactor technology.

Acknowledgment: This work was made possible by funding from the Qatar National Research Fund (QNRF) project number NPRP12S-0304-190222 and co-funding by Shell Global Solutions International B.V.

References:

1. Karim, G. and N. Moore, *The production of hydrogen by the partial oxidation of methane in a dual fuel engine*. 1990, SAE Technical Paper.
2. Glouchenkov, M., A. Kronberg, and H. Veringa, *Free piston pulsed compression reactor*. Chemical Engineering Transactions, 2002. **2**: p. 983-988.
3. Glouchenkov, M. and A. Kronberg. *Pulsed Compression Technology: A Breakthrough in the Production of Hydrogen*. in *16th World Hydrogen Energy Conference (WHEC)*. 2006. Citeseer.
4. Hiratsuka, Y. *Production of synthesis gas by an internal combustion engine*. in *6th World Petroleum Congress*. 1963. World Petroleum Congress.

5. Anderson, D.M., et al., *Sorption-Enhanced Variable-Volume Batch–Membrane Steam Methane Reforming at Low Temperature: Experimental Demonstration and Kinetic Modeling.* Industrial & Engineering Chemistry Research, 2015. **54**(34): p. 8422-8436.
6. Lowther, F.E. and F.L. Voelz, *Process for producing acetylene using a heterogeneous mixture.* 1986, Google Patents.
7. Charles, E.E., F. Hugh, and M.N.P. William, *Partial oxidation and pyrolysis of saturated hydrocarbons.* 1955, Google Patents.
8. Wulff, R.G., *Method of producing acetylene by compression of natural gas.* 1934, Google Patents.
9. Elnashaie, S.S. and F. Uhlig, *Numerical techniques for chemical and biological engineers using MATLAB®: A simple bifurcation approach.* 2007: Springer Science & Business Media.
10. Xu, J. and G.F. Froment, *Methane steam reforming, methanation and water-gas shift: I. Intrinsic kinetics.* AIChE journal, 1989. **35**(1): p. 88-96.
11. Goßler, H., *Partial Oxidation of Natural Gas in a Piston Engine: Modeling and Simulation.* 2014, Karlsruhe Institute of Technology.

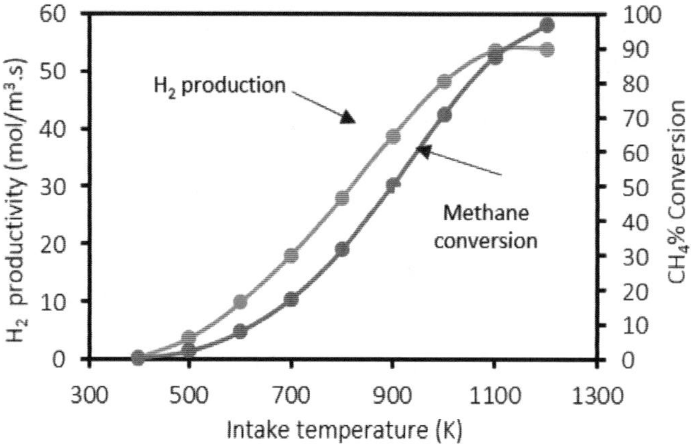

Figure 1: Hydrogen production rate and methane conversion as functions of intake temperature at N = 1000 RPM and P_{intake} = 1 bar

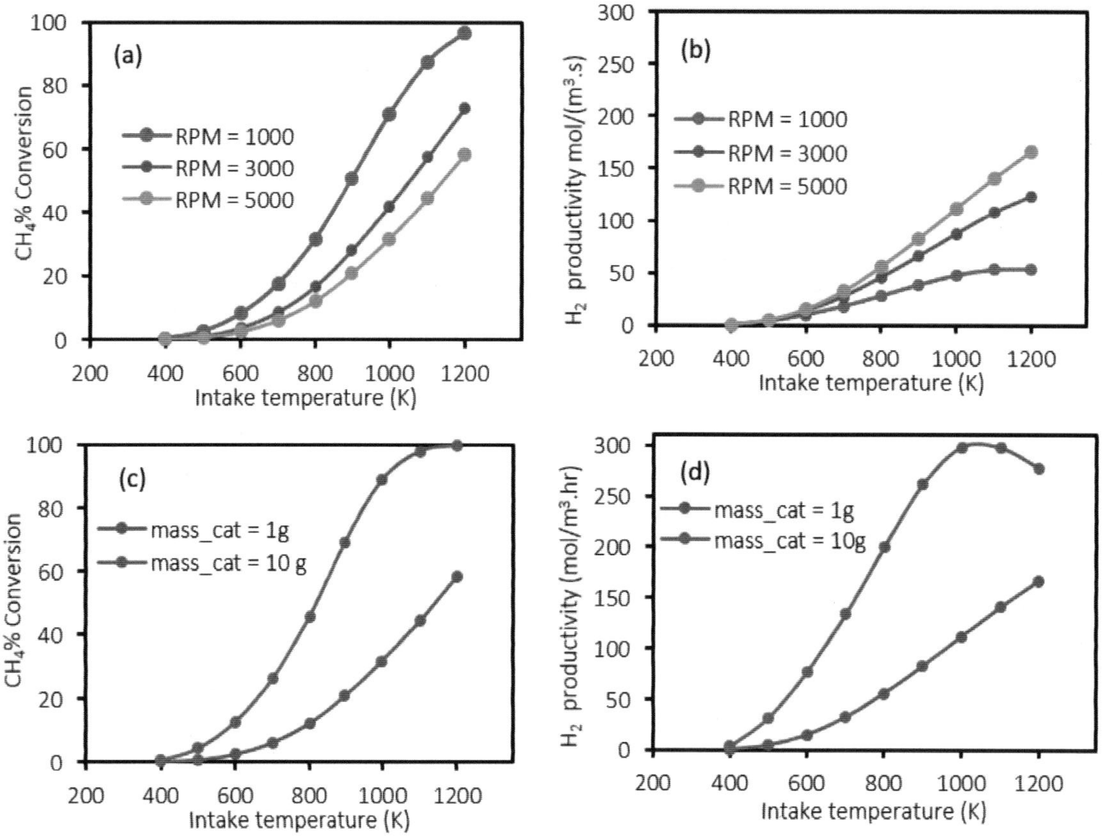

Figure 2: Effect of intake temperature (400-1200K) at RPM (1000-3000) on (a) methane conversion (b) hydrogen productivity (mass_cat = 1g) and at catalyst mass (1-10 g) on (c) methane conversion and (d) hydrogen productivity (RPM = 5000) at P_{intake} = 1 bar

631998 Modeling a Fluidized Bed Reactor for Particle Atomic Layer Deposition with CFD-DEM

Abstract:

Particle atomic layer deposition (ALD) is a vapor-phase coating technique that enables precise surface modification of particulate substrates. ALD coatings are uniquely uniform and controllable, which is desirable for many applications such as catalyst synthesis or cathode coating. Particle ALD is typically performed with a repeated sequence of self-limiting gas-solids reactions in a batch fluidized bed reactor. This configuration is capable of achieving high precursor utilization without agglomerating particles. However, the lack of predictive modeling capabilities for fluidized bed particle ALD reactors presents a challenge for implementing new coating chemistries and limits the commercial application of particle ALD. In this work, we develop a physics-based model for a fluidized bed ALD reactor that aims to predict reactor performance and enable intelligent particle ALD reactor design.

A reactor simulation based on computational fluid dynamics and discrete element method (CFD-DEM) is implemented using open-source multiphase flow modeling software MFiX. A well-characterized glass bead substrate is used to validate model predictions. The packed bed pressure drop and minimum fluidization velocity predicted by the model show good agreement with the experimental fluidization curve. Self-limited ALD reaction kinetics are implemented in the model using a first-order, irreversible Langmuir rate expression. The model can be used to predict the transient gas composition exiting the reactor and inform operating conditions to optimize precursor utilization and coating uniformity. Future work will include measurement of first-order reaction rate parameters for different particle substrates and ALD precursors, and comparison of modeling outputs to an experimental fluidized bed reactor.

634138 Carbon Capture Technology and Natural Gas Treatment

Abstract:

Removal of heavy hydrocarbons (C8+ HHCs) and aromatic (BTX) components from natural gas prior to liquefaction is critical for continuous LNG production. BASF has developed a new, step-change technology that allows for the combined removal of heavy hydrocarbons, aromatics, and water to cryogenic specifications in a single temperature swing adsorption unit to provide treated natural gas for liquefaction. Development of this technology required innovative bed designs to avoid pre-mature molecular sieve degradation, product development to achieve increased BTX capacity, and improved simulation tools for increased model accuracy. Several experiments were designed to prove the molecular sieve could withstand the short cycle times required for removal of HHCs and BTX. These experiments, including the rational behind the innovative bed designs, will be discussed in this paper. Finally, the process of new material development, involving Production, R&D, and Business functions, will be explained.

634137 From Discovery to Deployment of an Aromatic Transalkylation Catalyst

Abstract:

Transalkylation of toluene and C9+ aromatics to produce mixed xylenes is a key technology in modern aromatic complexes. An effective catalyst needs to balance acid function and metal function to achieve optimal performance. Challenging performance targets were set for a next generation catalyst, including increased activity, decreased ring loss and extended cycle length. In this work, we show how researchers took a zeolite discovery through development of a catalyst which was ultimately commercially deployed. The team was able to combine catalyst formulation fundamentals with high throughput testing and data driven tools to rapidly optimize a very broad formulation space. Over 120 unique catalyst formulations, spanning collectively a broad range of variables include zeolite type/combination, metal promoters, zeolite to binder ratios and methods of formulation/synthesis. The resultant catalyst lead was successfully scaled from the 100 gram lab scale to tens of thousands of pounds. In parallel, process experts put the catalyst through its paces to confirm the performance benefits and bring the catalyst to technical readiness and ultimately commercialization and deployment.

622287 Integrated Capture and Conversion of CO_2 into Materials (IC^3M); A Multi-Product Technology for Ccus

Abstract:

The capture, sequestration and utilization of CO_2 are needed to achieve the global goal of decarbonization. The United States Department of Energy (DOE) has recently prioritized utilization of CO_2 by recycling it into fuels and chemicals as a part of a broader negative-emissions strategy, though the energy and capital costs associated with capturing, purifying and compressing CO_2 are often overlooked. In this presentation, we describe a dynamic and flexible technology that couples solvent-based CO_2 capture with catalytic upgrading CO_2 captured in solution to produce low-carbon fuels and chemicals such as methane or methanol with high atom and energy efficiency. The IC^3M approach provides reduced costs while improving efficiency of detached capture and conversion processes. IC^3M allows for multiple products to be made from a single solvent process by simply changing the reagent co-feed and/or catalyst. The ability to produce multiple products allows IC^3M to change to market needs providing superior economic adaptability that single product technologies cannot achieve. We describe here, results from a multidisciplinary study that couples solvent-based capture and catalytic upgrading of CO_2 that is captured in solution to make multiple products. We provide a discussion of the energetics and economics of IC^3M and its viability in the global decarbonization market.

631547 Developing Biomass Valorization Technologies Based on Functionalization Chemistry

Abstract:

The production of various chemicals from biomass usually involves the depolymerization of biomass's three constituent polymers: cellulose and hemicellulose, which are both polysaccharides; and lignin, which is a polymer of phenyl propanoid sub-units. Achieving high yields of useful products from all three fractions is often complicated by low yields both during depolymerization and after the multi-step conversion processes that are often necessary to achieve a useful commercial biobased product. Various degradation reactions of both sugars and lignin intermediates are particular detrimental to achieving products at high yields.

Here, I will discuss how we can use functionalization chemistry, both during lignin extraction and polysaccharide depolymerization, to reversibly "trap" stabilized intermediate molecules, and facilitate their high-yield upgrading. Notably, we can show that, under the right conditions, extracted lignin can be catalytically upgraded at high yields to as few as 1-2 major products. Acetal functionalized lignin and carbohydrates could also serve as new platform molecules for the production bio-based materials including polymers. Altogether, this chemistry offers avenues for simplified scale-up and process development for the production of both high-value and bulk chemicals. I will discuss how these routes offers a realistic path towards de-risking a catalytic biorefining technology and potential commercial development.

632684 Heavy Oil Reforming in a Dual Circulating Fluidized Bed Reactor

Abstract:

TDA Research has developed a patented process to steam reform any hydrocarbon feedstock (ranging from methane to VTB) over nickel-based reforming catalysts while simultaneously regenerating the catalyst in a dual circulating fluidized bed (DCFB) reactor system. In this sense, our process functions somewhat like an autothermal chemical looping steam reformer (a-CLR), preventing deactivation of the catalyst by removing coke and sulfur in the regenerator before it can build up to damaging levels. TDA's technology, called *HyRes*, converts low-value heavy refinery feedstocks into high H_2/CO ratio syngas. Because we use a separate vessel for catalyst regeneration, limited nitrogen is introduced into the syngas, which improves the efficiency of downstream water gas shift (WGS) and hydrogen separation operations such as pressure swing adsorption (PSA).

Refineries that process heavy, sour crudes that contain metals, sulfur, and high molecular weight aromatic hydrocarbons require considerable hydrogen to upgrade these feedstocks. TDA has been granted a patent for this technology that will allow hydrogen to be produced in refineries at a cost that is considerably lower than hydrogen produced from conventional technologies (e.g. SMR) when natural gas prices are high or purchasing hydrogen from a third party.

Under a Phase II Department of Energy SBIR grant, we have successfully designed, built, and operated a bench-scale version (Figure 1) of the DCFB reactor system and have gained important knowledge relevant to the scale-up of our *HyRes* technology. We continue to experiment in our laboratories with various feedstocks, steam-to-carbon ratios, and oxygen content in the regenerator. TDA is also building a pilot-scale skid-mounted reactor system (Figure 2) to demonstrate this technology at the next scale at a refinery or chemical plant field site. Fabrication of the skid and reactor system is underway and we expect to be able to demonstrate our skid-mounted unit in 2022.

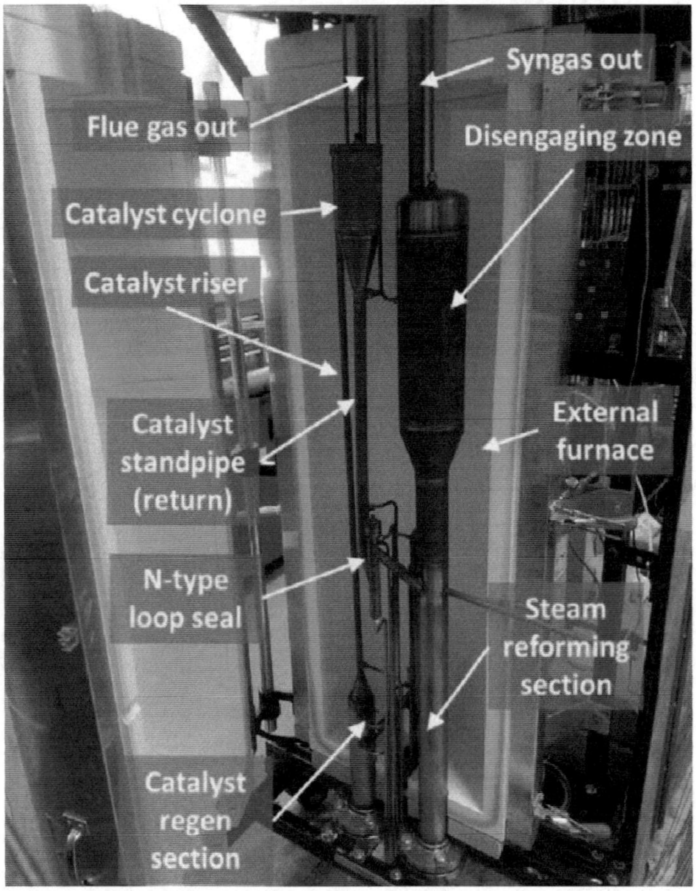

Figure 1: Bench-scale dual circulating fluidized bed (DCFB) reactor system currently operating at TDA.

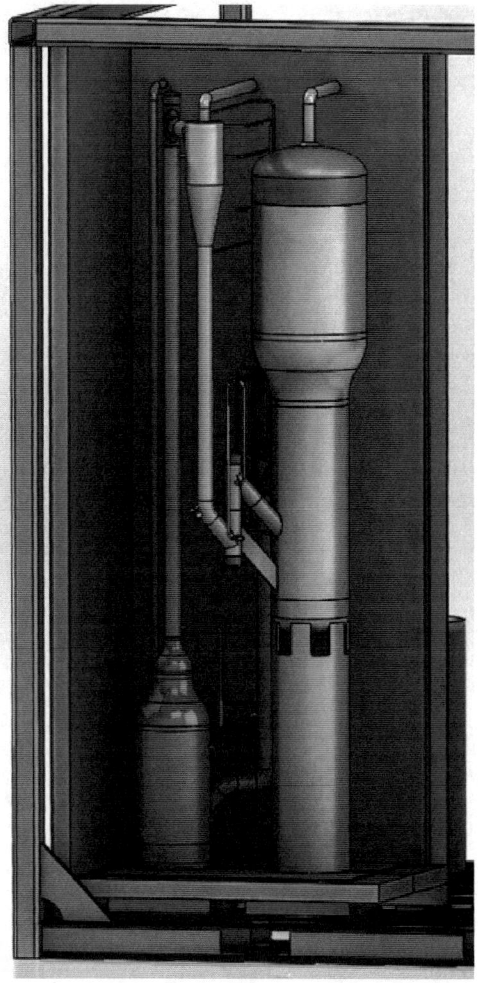

Figure 2: 3D Render of the pilot-scale DCFB reactor currently in fabrication.

629795 Catalyst Development and Process Intensification of a Bio-Renewable Surfactants Platform

Abstract:

Surfactants are the key active ingredient in cleaning products, with long lists of additional builder ingredients added to boost function while maintaining product safety and shelf life. These builder chemicals increase product cost and volume and biodegrade poorly. Sironix Renewables has invented a new class of surfactants, called Oleo-Furan Surfactants (OFS), which eliminate the need for these additional chemicals, reducing volume and resulting in a product that biodegrades readily. The OFS technology links the function of bio-based furan building blocks with natural oils to produce multifunctional and eco-friendly cleaning products. The technology gives improved performance, and by eliminating builder chemicals, reduces volume (and therefore packaging) by up to 30%. These features result in a reduced overall energy consumption while producing a more environmentally friendly product. The joint research project between Sironix Renewables and Los Alamos National Laboratory is designed to leverage the catalytic reaction engineering, catalyst development, and furan chemistry resources of the DOE national laboratories with the surfactants platform of Sironix Renewables to accelerate the DOE-invented and DOE Small Business Innovation Research-funded technology toward market commercialization. This presentation will detail efforts toward process improvements to achieve efficient scale-up of existing surfactants and the development of a new class of furan-based structures to facilitate a greater understanding of their underlying structure-function relationships.

623920 Test Tubes to Tons: De-Risking Pyran's Novel Catalytic Pathway to Bio-Based 1,5-Pentanediol

Abstract:

By employing heterogeneous catalysis, we are able to selectively remove oxygen and effectively manipulate reaction pathways to produce a suite of valuable chemicals from biomass in an economical matter. By combining expertise in heterogeneous catalysis and technoeconomic analysis, Pyran's founding team was able to identify a technically and economically advantaged pathway to produce biomass-derived 5-carbon molecules, including 1,5-pentanediol.

This talk will explore the key factors at the academic and early company stages that can lead to the identification and de-risking of novel catalytic pathways and new-to-market bio-based products. In Pyran's experience, the key to an academic project's potential to generate a commercializable technology is the simultaneous exploration of fundamental catalysis/chemistry and focus on scale-up factors and plant-scale economics throughout the program. It is also wise to take into account product and market considerations at this early stage.

Once a potentially commercializable technology has been identified, or even a startup company formed, it is necessary to consider multiple aspects for de-risking the technology and product; including determination of a cost- and time-effective process and catalyst scaleup strategy, integration of upstream (i.e. feedstock) and downstream (i.e.. polymerization) processes, identification of potential markets and customers, and strategizing early stage financing.

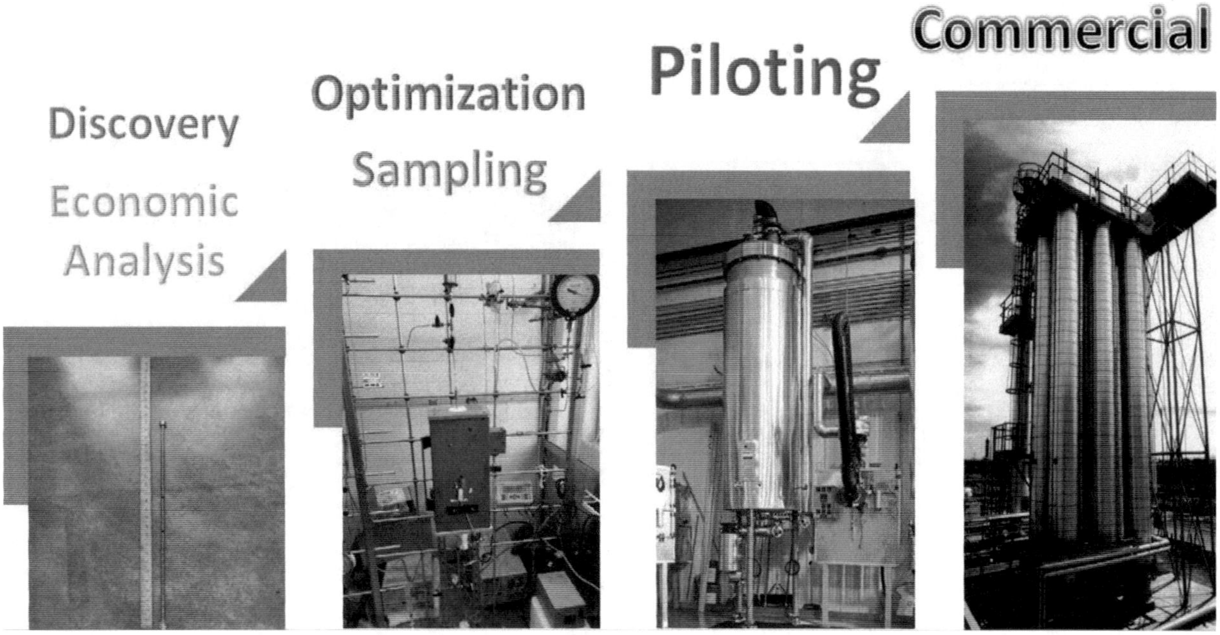

626054 Technology Transfer for Producing Sustainable Aviation Fuels from Wet Waste

Abstract:

With the increasing demand for low-carbon sustainable aviation fuels (SAF), new conversion technologies are needed that can transition from the laboratory to the marketplace while meeting cost and carbon intensity reduction targets. Wet waste is one such promising low-cost, widely available feedstock with the energy potential to displace over 20% of U.S. jet fuel consumption. Recent efforts at NREL have shown that wet-waste derived volatile fatty acids (VFAs) can be catalytically converted into SAF that is suitable for blending into conventional aviation fuels. This talk will examine ongoing work at NREL to partner with industry and advance VFA-SAF catalyst technology towards commercialization. Findings will be presented in regard to i) derisking and scaling VFA ketonization catalyst technology from the bench to gallon-per-day production, ii) navigating the ASTM jet fuel qualification process for new SAF routes, iii) integrating VFA-derived ketones with existing biorefinery infrastructure for hydrotreating fats, oils, and greases, and iv) quantifying VFA-SAF lifecycle carbon intensity reductions for the Low Carbon Fuel Standard. While this talk will focus on wet waste VFA-SAF, the topic of technology transfer holds relevance for advancing other emerging SAF pathways needed for decarbonization.

624449 Enhancing the Dielectric Breakdown Strength of Solid-State Polymer Capacitors By Chain End Manipulations

Abstract:

The need for high power density, flexible, pulsed power, and lightweight energy storage devices requires the use of polymer film-based dielectric capacitors. The maximum energy storage density of a linear dielectric capacitor is directly proportional to the square of the maximum voltage (dielectric breakdown voltage) that can be applied to the dielectric without causing the breakdown. Theoretically, it has been shown that chain ends contribute adversely to the electrical breakdown of polymer dielectrics at high electric fields, resulting in low energy density in polymer capacitors. In this work, we enhanced the dielectric breakdown voltage and hence the energy density of the polymer capacitor by using well-ordered high molecular weight block copolymers (BCP), in which the chain ends are segregated to narrow zones. The well-ordered and easily processable BCP-based capacitors exhibit an energy density of more than 5 J/cm^3, which is higher than that of industrial used biaxially oriented polypropylene. Furthermore, we tested the role of chain ends in the dielectric breakdown of polymer films by comparing the linear polymers with cyclic polymers (no chain ends). The cyclic polymer films demonstrated ~50% enhancement in dielectric breakdown strength and as a result, the capacitive energy density in the cyclic polymer films increased by ~80% as compared to their linear counterparts. These novel insights into manipulating chain end distribution such as in BCPs and with molecular topology to increase the energy density of polymers will help fulfill next-generation energy demands.

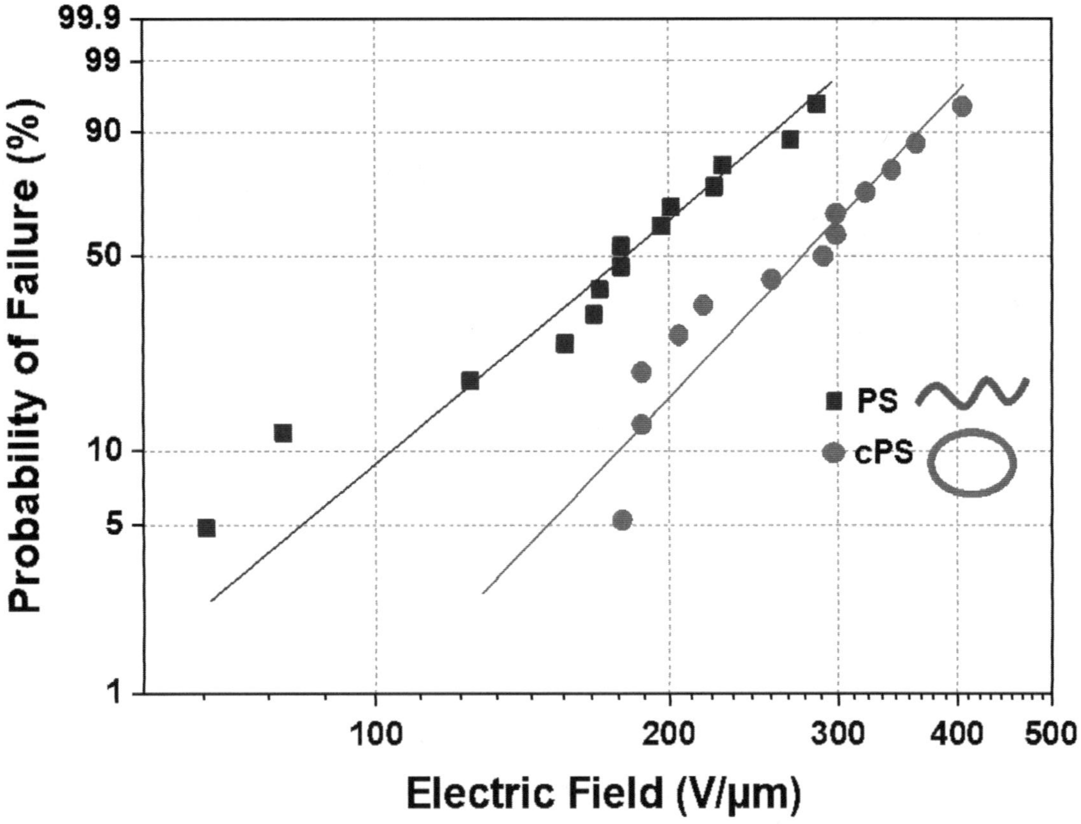

630204 Electropolymerization of Polypyrrole By Photosystem I

Abstract:

The Photosystem I (PSI) protein complex found in all photosynthetic organisms is known for its impressive redox capabilities. PSI has two photoactive redox sites, the oxidizing P_{700} chlorophyll pair and the reducing F_B iron-sulfur cluster with more than 1.1 V of potential difference between them. We are investigating the use of these active sites to perform photo-driven, nanomaterials chemistry directly at the protein donor site. The wide range of reduction potentials places PSI in a position to photooxidatively electropolymerize conductive polymers and photoreduce metal ions into nanoparticles. This presentation will focus on the preparation of PSI-polypyrrole conjugates. A reduction potential of 0.48 V vs. Ag/AgCl at the P_{700} site makes PSI a promising candidate for oxidative polymerization of certain monomers. Because PSI is a renewable and vastly abundant bionanomaterial, electropolymerization via PSI is a sustainable and low-cost alternative to produce polypyrrole. Polymerization occurs with the addition of protein and pyrrole monomer in an acidic aqueous solution under light. Heterogeneous growth of the polymer is also demonstrated from a multilayer film of the PSI protein. The resulting product has been analyzed spectroscopically and electrochemically to confirm the presence of both PSI and electropolymerized polypyrrole. The growth of the polypyrrole alters the photovoltage and photocurrent response of the protein film. The combination of PSI with a conductive polymer can lead to photoactive polymer films that could be utilized in solar energy conversion systems.

624480 Development of Electrically Reversible Ion Exchange (ERIE) Electrodes for Desalination and Aqueous Deionization

Abstract:

Growing demand is straining existing supplies of fresh water, and more efficient and robust methods are needed for removal of mineral or ionic contaminants from other water sources. A new technology employing p-doped conducting polymer electrodes which reversibly electrosorb aqueous ionic contaminants when a small voltage is applied, and regenerate when the potential is removed or reversed has been developed and demonstrated. By electrodepositing doped polypyrrole (PPy) or poly(3,4) ethylenedioxythiophene (PEDOT) onto the porous carbon substrates, we create electrodes with both Faradaic (ion exchange) and non-Faradaic (electrical double layer) charging mechanisms (see Figure 1). Their performance was evaluated electrochemically via Dual Electrode Cyclic Voltametry (DECV) to discern the ion insertion and removal behavior, and via Cyclic Step Chronoamperometry (CSCA) to determine the specific ionic capacities, and the charge/discharge rates of the electrodes. The voltage requirement is only 1.0 and 1.5 volts, and an electrochemical efficiency of up to 90% is achievable. Batch and continuous ion absorption/desorption tests using both symmetric and asymmetric electrode pairs were conducted on saltwater and other aqueous ion solutions to confirm the ionic removal, electrode efficiency and cycling rates, and electrode durability during repeated cycling. These new electrode materials demonstrate ionic capacities (50 to over 100 milliequivalents/gram polymer) that are comparable to or better than ion exchange resins, and greater than the best current Capacitive DeIonization (CDI) electrodes. Our measurements indicate that the ERIE energy requirements are lower than existing CDI or reverse osmosis technologies and offers an attractive alternative for deionization of seawater or brackish waters, and to ion exchange resins for water softening or removal of harmful ionic contaminants.

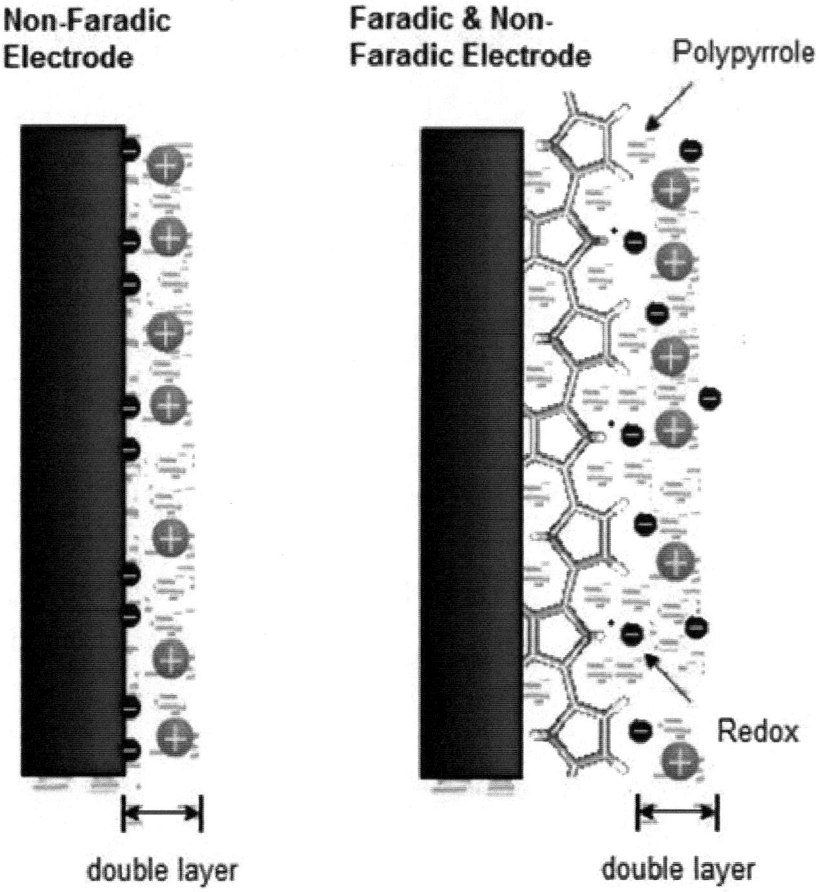
Non-Faradic
Electrode
Faradic & Non-
Faradic Electrode
Polypyrrole
Redox
double layer
double layer

623191 A Multi-Scale Approach to Evaluate the Relationship between Rheological and Textural Properties of Oil-in-Water Cosmetic Emulsions

Abstract:

The production of higher value-added products and consumer-oriented products is a major concern in the chemical industry as a result of market globalization and competitiveness[1]. Consumer-oriented products are used in a wide variety of industrial sectors such as the pharmaceutical, food, personal care, and cosmetic [2]. A major volume of products obtained from those sectors are produced as emulsions [3]. In particular, the cosmetic industry designs emulsified products to satisfy the consumer's sensory expectation since the consumers' selection often relies on product functionality[4]. Hence, it is of significant interest to understand the relationship between the sensations experienced by the consumer and instrumental characterization of the product properties (e.g., rheological and textural characterization).

Previous studies have shown the links between the macroscopic behavior of emulsified systems (i.e., rheology) and their corresponding microscopic structure (i.e., drop size distribution and mean drop size) [5][6][7]. However, these relationships have not been established for emulsified cosmetic products. Thus, it is critical to gain insights into how the product properties at the macroscopic and microscopic scales are correlated and impact product formulation and process operating variables. To understand those relationships between macroscopic properties (i.e., rheology and texture) and microscopic properties (i.e., droplet size and droplet size distribution) for emulsified cosmetic products, the energy consumed during the emulsification process and product formulation needs to be considered. A multiscale approach is a suitable tool to study emulsified cosmetic systems since it can be used to gain insight on the interconnection between the macroscopic and microscopic properties of emulsified products[8].

Often, heuristic rules are frequently used in the design of emulsified cosmetic products. These rules aim to establish the functionality and concentration of some components in the emulsion such as the thickening agents[9]. For instance, previous studies have set a thickener concentration of 1%w/w as a rule of thumb for the design of cosmetic creams[3][10][11]. However, it is not completely understood how to select the thickeners' concentration and type in the design of an emulsified cosmetic product. On the other hand, previous reports have highlighted the influence of synthetic (e.g., Carbopol) and natural (e.g., Xanthan gum) thickening agents on the textural perception of the product[10]. Although synthetic and natural thickeners generate different microstructures in emulsified systems[12], few studies have correlated the rheological and textural properties of emulsified cosmetic systems[4] to the product microstructure. To the authors' knowledge, a study relating the rheology and texture of emulsified cosmetic products to their corresponding microstructure, operating variables, and product formulation (e.g., thickeners' concentration and type) has not been published in the open literature. Those relationships can provide insights on how the product formulation, product properties, and process variables are correlated thus offering a first approach for the efficient and systematic design of emulsified cosmetic products. Those correlations may also provide a deeper understanding of the consumer sensory experience, translating the consumer perception into engineering variables that could be characterized under controlled conditions (e.g., rheological and textural measurements). Accounting for these correlations may represent a reduction in both experimental resources and time for product commercialization.

The present study aims to identify the influence of the product formulation and emulsification energy on the macroscopic (rheology and texture) and microscopic (droplet size distribution) properties of cosmetic emulsions. In particular, this study correlates cosmetic emulsions' rheological and textural properties as a first attempt to link high-controlled instrumental characterization with the sensations that the consumer may experience. Those correlations can provide insights into the role of the thickener on the consumer's perception, which may represent a suitable tool for decision making during the design process. Moreover, this work evaluates the influence of emulsification energy and product formulation on the emulsion's physical stability. To study this phenomena, oil-in-water (O/W) cosmetic emulsions were manufactured varying the thickener type (i.e., Carbopol and Xanthan Gum) and concentration (i.e., 0.1%, 0.73%, 1.37%, and 2% w/w), the dispersed phase concentration (i.e., 10%, 20%, and 30% w/w), and the agitation rate (500, 900, and 1700 rpm). These scenarios were established according to conventional cosmetic formulation proposed in literature[3][10][13].

Results from this study show that the emulsion's macroscopic and microscopic properties are dominated by the thickener type and concentration. The drop size distribution (DSD), rheology, and texture of the prepared cosmetic emulsions remain unaffected once a given thickener concentration is reached. This behavior indicates that there is a critical thickener concentration for the design of emulsified cosmetic products, which depends on the specific thickener used to formulate the cosmetic emulsion (e.g., Carbopol, Xanthan Gum). Our results show that the cosmetic emulsions tested in this work were stable. The cosmetic emulsions' physical stability was improved as the thickener concentration is increased in the formulation. This is mostly due to the polymeric matrix that the thickener generates in the continuous phase of the emulsion. The results also show that changes in the agitation rate and dispersed phase concentration do not affect significantly the product properties when thickener is added into the product. These results indicate that the thickener type and concentration have a strong influence on cosmetic emulsions' sensory performance. Cosmetic emulsions prepared using a natural thickener exhibit notable differences in their DSD, rheological, and textural properties compared to those prepared using a synthetic thickener. This is mostly due to the particular polymeric matrix that natural and synthetic thickeners generate, which causes different rheological and textural behavior of the cosmetic emulsions[10]. Since cosmetic emulsions prepared using thickeners of the same type present similar textural and rheological properties (i.e., natural or synthetic) [10], the correlations proposed in this study could be extended to similar emulsion systems (e.g., moisturizing creams, hand lotions.). Also, the correlations identified in this work can be potentially used to make decisions regarding formulation, product properties, and process variables during the early stages in the design of emulsified cosmetic products. As future work, we aim to expand the analysis for additional formulation scenarios to gain insights on the role of other components such as emollients, exfoliants, surfactants, and humectants.

REFERENCES

[1] E. L. Cussler and G. D. Moggridge, *Chemical product design*. Cambridge University Press, 2011.

[2] K. M. Ng, R. Gani, and D.-J. Kim, *Chemical Product Design Toward a Perspective Through Case Studies*. Elsevier Science, 2006.

[3] C. Wibowo and K. M. Ng, "Product-oriented process synthesis and development: Creams and pastes," *AIChE J.*, vol. 47, no. 12, pp. 2746–2767, 2001, doi: 10.1002/aic.690471214.

[4] L. Gilbert, G. Savary, M. Grisel, and C. Picard, "Predicting sensory texture properties of cosmetic emulsions by physical measurements," *Chemom. Intell. Lab. Syst.*, vol. 124, pp. 21–31, 2013, doi: https://doi.org/10.1016/j.chemolab.2013.03.002.

[5] A. Y. Malkin, I. Masalova, P. Slatter, and K. Wilson, "Effect of droplet size on the rheological properties of highly-concentrated w/o emulsions," *Rheol. Acta*, vol. 43, no. 6, pp. 584–591, 2004, doi: 10.1007/s00397-003-0347-2.

[6] O. A. Alvarez *et al.*, "Influence of semibatch emulsification process conditions on the physical characteristics of highly concentrated water-in-oil emulsions," *Ind. Eng. Chem. Res.*, 2010, doi: 10.1021/ie9020073.

[7] D. Pradilla, W. Vargas, and O. Alvarez, "The application of a multi-scale approach to the manufacture of concentrated and highly concentrated emulsions," *Chem. Eng. Res. Des.*, vol. 95, no. October, pp. 162–172, 2015, doi: 10.1016/j.cherd.2014.10.016.

[8] F. Calvo, J. M. Gómez, L. Ricardez-Sandoval, and O. Alvarez, "Integrated design of emulsified cosmetic products: A review," *Chem. Eng. Res. Des.*, vol. 161, pp. 279–303, 2020, doi: https://doi.org/10.1016/j.cherd.2020.07.014.

[9] J. A. Arrieta-Escobar, F. P. Bernardo, A. Orjuela, M. Camargo, and L. Morel, "Incorporation of heuristic knowledge in the optimal design of formulated products: Application to a cosmetic emulsion," *Comput. Chem. Eng.*, vol. 122, pp. 265–274, 2019, doi: https://doi.org/10.1016/j.compchemeng.2018.08.032.

[10] L. Gilbert, C. Picard, G. Savary, and M. Grisel, "Rheological and textural characterization of cosmetic emulsions containing natural and synthetic polymers: relationships between both data," *Colloids Surfaces A Physicochem. Eng. Asp.*, vol. 421, pp. 150–163, 2013.

[11] P. Dubuisson, C. Picard, M. Grisel, and G. Savary, "How does composition influence the texture of cosmetic emulsions?," *Colloids Surfaces A Physicochem. Eng. Asp.*, vol. 536, pp. 38–46, 2018, doi: https://doi.org/10.1016/j.colsurfa.2017.08.001.

[12] H. M. Ribeiro, J. A. Morais, and G. M. Eccleston, "Structure and rheology of semisolid o/w creams containing cetyl alcohol/non-ionic surfactant mixed emulsifier and different polymers," *Int. J. Cosmet. Sci.*, vol. 26, no. 2, pp. 47–59, Apr. 2004, doi: https://doi.org/10.1111/j.0412-5463.2004.00190.x.

[13] M. Bagajewicz *et al.*, "Product design in price-competitive markets: A case study of a skin moisturizing lotion," *AIChE J.*, vol. 57, no. 1, pp. 160–177, Jan. 2011, doi: 10.1002/aic.12242.

628674 Cyclic Carbonate Plasticizers to Improve Li-Ion Conductivity in Solid Polymer Electrolytes

Abstract:

Solid polymer electrolytes (SPEs) have been the focus of study for decades to address general problems in lithium-ion batteries, such as thermal runaway by removing volatile solvents, uncontrolled mass transport by immobilizing anions to the polymer matrix, undesired side reactions, and lithium dendrite growth by providing a mechanical barrier. Unfortunately, SPEs suffer from poor lithium-ion mobility and conductivity due to the stationary phase of immobilized anions, poor ion pair dissociation, and the slower time scales of polymer diffusivity compared to liquid electrolytes. The focus of this investigation is on improving the conductivity of SPEs, with and without immobilized anions, through the use of organic carbonate-based plasticizers to improve Li-ion dissociation and diffusion. Several solid plasticizer compounds were chosen to avoid highly flammable, liquid plasticizers that harbor the same safety concerns as conventional liquid electrolytes. Studies varying the carbonate structures, the concentration within the SPEs, and the SPE structure were characterized by impedance spectroscopy and coin cell cycling to observe their effect on improving Li-ion conduction.

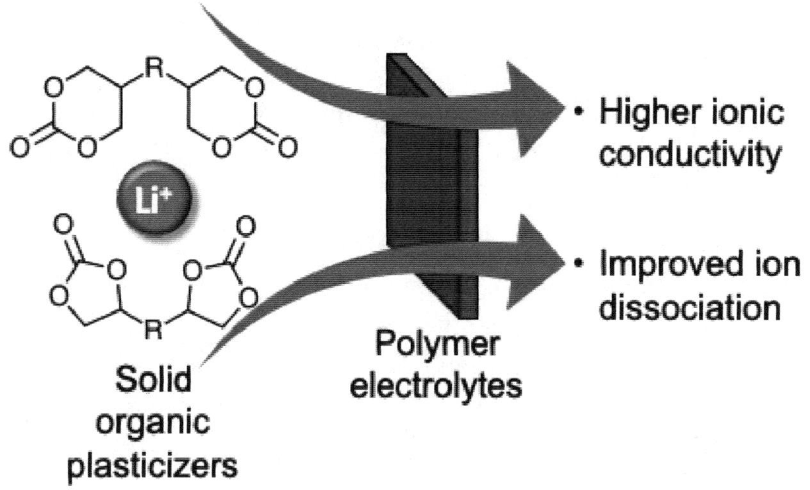

Design and Application of Hydrophobic 2,2,4-Trimethyl-1,3-Pentandiol Deep Eutectic Solvents for Boron Extraction

Narjis Awaja[a], Ghaiath Almustafa[a,b], Ahmad S. Darwish[a], Ioannis Zuburtikudis[b], Hadil AbuKhalifeh[b], Hassan A. Arafat[a,c], Inas AlNashef[a,c].

[a]*Chemical Engineering Department, Khalifa University, P.O. Box 127788, Abu Dhabi, United Arab Emirates.*
[b]*Chemical Engineering Department, Abu Dhabi University, P.O. Box 59911, Abu Dhabi, United Arab Emirates.*
[c]*Center of Membrane and Advances Water Technology, Khalifa University, P.O. Box 127788, Abu Dhabi, United Arab Emirates.*

Introduction

Modern chemistry and chemical engineering trends emphasize the replacement of conventional organic solvents with more stable and less hazardous ones. Thus, new genres of ionic liquids (ILs) and deep eutectic solvents (DESs) are continueously being developed as eco-efficient "designer's solvents". [1–3]. Although ILs are chemically and thermally stable, highly non-volatile, and tunable, most of them suffer from high production costs and potential toxic manifestations [4,5]. Consequently, these challenges impacted their application as green solvents [6].

As analoughs of ILs, DESs combine the practical properties of ILs with attractive features such as simplicity of synthesis, cost-efficency and biodegradability [7,8]. These versatile characteristics of DESs facilitated their use in different fields such as catalysis [9], separation [10–12], biochemistry [13], electrochemistry [14], and nanotechnology [15]. In general, DESs are mixtures of two or more hydrogen-bond donors (HBD) and hydrogen-bond acceptors (HBA) that are associated with each other through hydrogen bond interactions. The resulting eutectic mixture has a lower melting point than the individual components [8].

Different classes of DESs have been reported in the literature since their inception. Initially, only hydrophilic DESs were prepared [16], which were only suitable for non-aqueous applications. More recently, hydrophobic DESs were first reported in 2015 by van Osch *et al.* [17] and Ribeiro *et al.* [18]. Hydrophobic eutectic solvents allowed the separation of different molecules from aqueous media, with various potential applications in the extraction of pharmaceuticals [11], lower alcohols [19], pesticides [20], VFAs [21], phenols [22], heavy metals [23], and more recently, boron [24,25].

Boron removal from aqueous streams is of major interest and is typically conducted for two purposes: either as (a) preventive measure, such as meeting drinking water regulations in desalination plants and protecting boron-sensitive crops and/or (b) for productive purposes, such as the recovery of boron from industrial wastewater or nuclear waste streams.

Thus far, there are few studies discussing the application of DESs for the removal of boron by our group. In the previous work, we screened anumber of potential compounds as HBDs for the formation of DESs with boron extraction capaboilities. We concluded with the 1,3 diols of 2-methyl-2,4-pentanediol (MPD) as the HBD and thymol or menthol were used as HBAs. With those two combinations, boron extraction efficiencies were approximately 90% [24]. As the leachability was considerably hight for MPD. In a more recent work, we further optimized the HBD in the mentioned DESs, by liquifying 2,2,4-trimethyl-1,3-pentanediol (TMPD) into a novel DES with thymol or menthol [25]. This latter combination provided a stable and more efficenct extraction, and the trade-off governing the extraction process was more throguhly examined.

In the aim of enhancing the design of such solvent extraction systems, this work targeted the the investigation of several components as potential HBAs, fixing the HBD as the previously optimized functional extractant, TMPD. The aromatic/cyclic compounds, i.e., thymol and menthol, were replaced with aliphatic compounds such as decyl alcohol or decanoic acid, as well as branched aliphatic compound such as trioctylphosphine oxide. The associated physicochemcoial proepites and their effect on boron extractoiion efficiency were thoughlty tested, and the effect of the HBAs structure on the performance of the DESs was discussed.

Experimental Section

Reagents and Chemicals

Decyl alcohol (Dec) (>98%), decanoic acid (DecA) (>98%), trioctylphosphine oxide (TOPO) (99%), thymol (Thy) (>98.5%) and 2,2,4-Trimethyl-1,3-pentanediol (TMPD) (97%) were purchased from Sigma-Aldrich, USA. Boric acid (analytical reagent grade) was purchased from Fischer Scientific, USA. All chemicals were used as received.

DES Preparation and Characterization

The eutectic mixtures were prepared following a conventional method through agitation and heating at 80 °C for 6 h, using an incubator shaker (KS 4000, IKA®, Germany). The combinations of eutectic mixtures initially prepared for this work are shown in **Table 1**.

Table 1. Combinations of eutectic mixtures

Eutectic Mixture	Molar Ratios
Dec:TMPD	2:1
DecA:TMPD	2:1
TOPO:TMPD	2:1
DecA:Thy:TMPD	1:1:1
TOPO:Thy:TMPD	1:1:1

The physicochemical properties (i.e., density and viscosity) of the prepared solvents were measured using a kinematic viscometer (SVM 3001, Anton Paar, Austria), at 25 °**C**. The cross-contamination between the aqueous phase and the organic DES phase was investigated by mixing equal amounts of both phases in a plastic centrifuge tube for 4 h at 300 rpm and 25 °**C**. Then, the organic phase was analyzed, using a Karl Fischer titrator (Aquamax KF Coulometric, GR Scientific, Germany), to determine the solubility of water in it. A sample was also taken from the aqueous phase to determine the amount of the solvent that transferred to it. This was done by measuring the total organic carbon (TOC) in the water sample, using the cuvette test difference method (LCK 381, Hach Company, Germany), where the total carbon (TC) and total inorganic carbon (TIC) were measured using a portable spectrophotometer (DR 2800, Hack Company, Germany).

Boron Extraction

Equal masses of each of the eutectic mixtures and aqueous solution (1000 ppm boron) were contacted in a plastic centrifuge tube and shaken using a mechanical incubator shaker (KS 4000, IKA®, Germany) at 300 rpm and 25 °C for 4 h. The mixture was then centrifuged (Universal 320, Hettich, Germany) for 5 minutes to ensure complete phase separation. After mixing and centrifuging, a sample was taken from the aqueous phase and diluted in DI water for, and a boron vial test (TNT 877, Hach Company, Germany) along with a portable spectrophotometer (DR 2800, Hack Company, Germany) were used to determine the concentration of boron in the aqueous samples. The boron concentration in the DES was determined by a mass balance before and after extraction. The extraction efficiency was defined as follows:

$$E = (1 - Cw / Cw_o) \times 100\% \qquad (1)$$

where C_{Wo} is the initial concentration of boron in the aqueous phase before extraction and C_W is the concertation of boron in the aqueous phase after extraction.

Results and Discussion

Comparison of Prepared Mixtures

To ensure the effectiveness of the solvent extraction process, the optimization of certain parameters must be done. Such parameters are relevant to the physicochemical properties of the extracting solvent such as density, viscosity, water solubility, and solvent leachability to the aqueous phase, in addition to the effectiveness of the functional extractants represented by the extraction efficiency.

Table 2 below shows the results of the characterization of the prepared eutectic mixtures in addition to the properties of mixtures from the previous two studies. **Figure 1** visualizes the comparison in a radar chart. From **Figure 1** it was observed that the mixture Dec:TMPD offers high boron extraction (97.22%) and its leachability and water solubility are somewhat comparable to the other mixtures, while having the lowest viscosity and density values of 21.479 mPa.s and 0.8601 $g.cm^{-3}$ respectively. This is important in the scope of solvent extraction process as the density of the solvent affects the degree of separation between the aqueous and organic phases, where the larger the difference between their densities the better. Likewise, the viscosity of the solvent influences the mass transport between the two phases, where lower viscosity values allows for better mass transfer across the phases. Furthermore, the amount of water-soluble in the organic phase after extraction (as wt. %) and the amount of the solvent recorded in the aqueous phase (as TOC, $g.L^{-1}$) are also measured and reported, which indicates the degree of the cross-contamination between the organic and aqueous phases. Consequently, the mixture Dec:TMPD was selected for further analysis.

Table 2. Physicochemical properties (at 25 °**C**), thermal properties and boron extraction efficiencies for the selected combinations

Eutectic Mixture	***Density, ρ***	***Viscosity, η***	***Water content[a]***	***Water solubility[b]***	***Leachability, (as TOC)[c]***	***Boron Extraction, E***	***Reference***
	($g.cm^{-3}$)	**(mPa.s)**	**(*wt%*)**	**(*wt %*)**	**($g.L^{-1}$)**	**(%)**	
Thy:MPD (2:1)	0.9591	32. 689	0.1264	4.2568	39.23	90.1	[24]
Men:MPD (2:1)	0.9005	68.3962	0.1866	2.5432	82.55	83.2	[24]
Thy:TMPD (2:1)	0.9635	45.467	<0.1	3.686	10.674	97.77	[25]
Men:TMPD (2:1)	0.9103	124.037	<0.1	2.844	15.963	97.14	[25]
Dec:TMPD (2:1)	**0.8601**	**21.479**	**0.1296**	**4.9106**	**13.63**	**97.22**	**This work**
DecA:TMPD (2:1)	0.9114	21.126	0.6133	3.7617	12.83	94.40	This work
DecA:Thy:TMPD (1:1:1)	0.9361	33.725	0.3053	4.0938	12.54	96.611	This work
TOPO:Thy:TMPD (1:1:1)	0.9058	72.9015	0.2426	1.3103	14.307	94.656	This work

[a] *Water content is the wt% of water in the pure solvent, before contact with the aqueous phase;*
[b] *Water solubility is the wt% of water in the solvent after mixing it with the aqueous phase;*
[c] *Leachability: measurement of the solvent constituents in the aqueous phase via the Total Organic Carbon (TOC)*

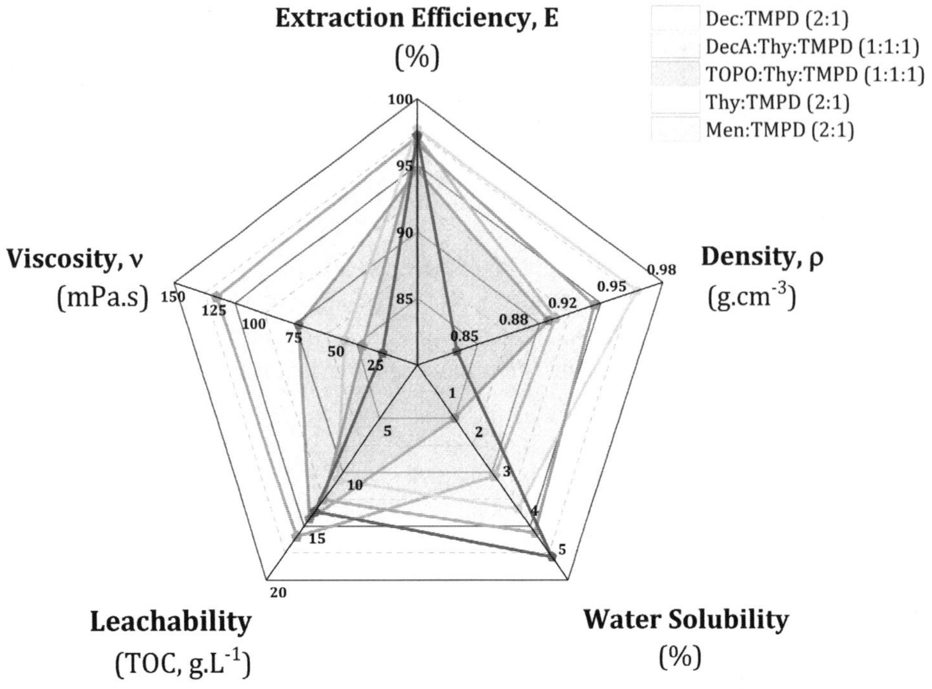

Figure 1. Radar chart comparing the physiochemical properties and boron extraction efficiencies for the prepared eutectic mixtures (Extraction conditions: Boron in aqueous phase = 1000 ppm, pH = 2, T = 25 °**C**, Aqueous/Organic (A/O) mass ratio 1:1)

Stability and Phase Composition of Decanol:TMPD Eutectic Mixture

To investigate the phase behavior of the Dec:TMPD mixtures, different compositions with molar ratios of 8:1, 4:1, 2:1, 1:1, 1:2, 1:4, and 1:8 were prepared. Upon mixing and heating, eutectic mixtures of TMPD and Dec were formed. Initially, all the mixtures yielded clear liquids when cooled down to room temperature (**Figure**). Mixtures with higher ratios of TMPD (1:2, 1:4, and 1:8) solidified after several days.

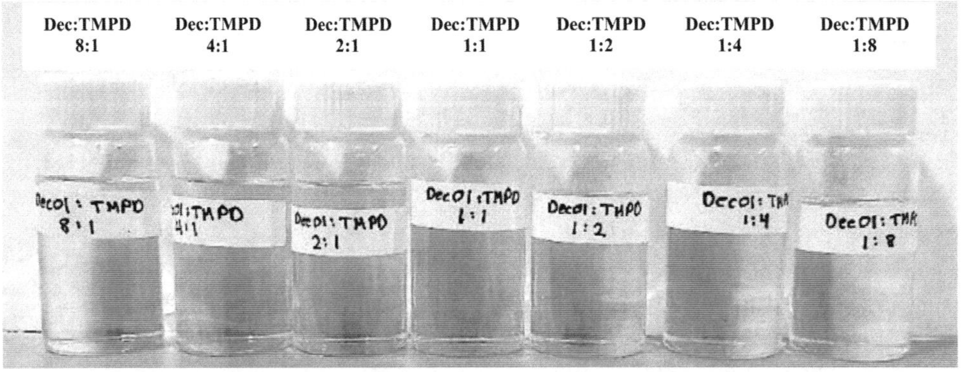

Figure 2. Dec:TMPD eutectic mixtures after cooling down to ambient temperature (~20°**C**); Left to right: Dec:TMPD molar ratio 8:1, 4:1, 2:1, 1:1, 1:2, 1:4 and 1:8

Dec:TMPD Water Phase Separation

The stability of the Dec:TMPD mixtures in the aqueous medium was tested by contacting each of them with an equal amount of DI water. The Dec:TMPD mixtures at a ratio of 1:2, 1:4, and 1:8

where unstable and solidified upon mixing with water, while the rest exhibited a stable aqueous phase separation as shown in **Figure 3**. Thus only the mixtures with molar ratios of 8:1, 4:1, 2:1, and 1:1 were considered for further characterization and boron extraction tests. The amount of water transferred to the solvent phase as well as the amount of solvent leached to the water phase are discussed in the following section.

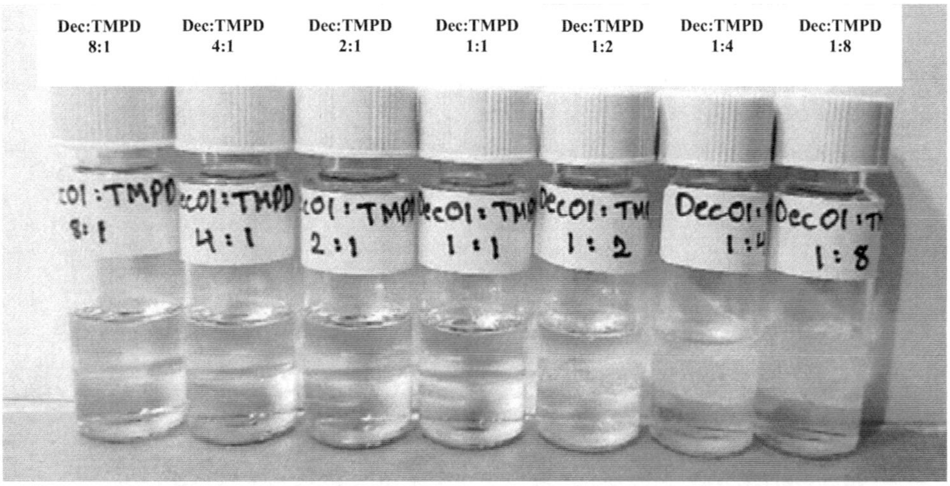

Figure 3. Eutectic mixture-water phase separation for Dec:TMPD mixtures. Left to right: Dec:TMPD molar ratio 8:1, 4:1, 2:1, 1:1, 1:2, 1:4 and 1:8.

Dec:TMPD Characterization and Extraction Efficiency

The Dec:TMPD mixtures were first characterized for their physicochemical properties and the effect of such properties on boron extraction efficiency was also studied, as shown in **Table 3** and **Figure 4**. It can be observed that all four physicochemical properties (density, viscocity, leachbility and water solubility), as well as boron extraction efficiency, increase with higher TMPD molar ratios. The upper and lower density limits are 0.8373 and 0.8787 g.cm^{-3} corresponding to the ratios of 8:1 and 1:1 respectively. Similarly, the corresponding dynamic viscosity limits are 14.222 and 31.637 mPa.s, respectively. Moreover, the general increase in the water solubility with the increase in TMPD is expected due to the nature of TMPD as a water-soluble diol, which is also reflected in the higher leachability values from the organic eutectic mixture to the aqueous phase. On the other hand, increasing the hydrophobic constituent (Dec) results in a decrease in the water solubility and the solvent leachability to the water phase. The solubility of water in the 1:1 mixture and its leachability in the aqueous phase are approximately 5.7 wt%, and 24.25 g.L^{-1}, respectively, while for the 8:1 mixture, they are around 4.1 wt% and 5.72 g.L^{-1}.

Moreover, increasing the amount of TMPD in the eutectic mixture is correlated with higher extraction efficiencies, given its role as the functional extractant. Nonetheless, this increase in the extraction efficiency is accompanied by the increase in density, viscosity, and water solubility of the Dec:TMPD mixture in addition to a noticeable increase in the leachability. While it is desirable to select the solvent that results in the highest extraction efficiency, one must also consider the characteristics discussed above as to not affect the stability of the extraction system in the longer run. As such, the selected optimum ratio of Dec:TMPD is 4:1.

Table 3. Physicochemical properties (at 25 °C), thermal properties and boron extraction efficiencies for the Dec:TMPD mixtures

Molar Ratio	*Density, ρ*	*Viscosity, η*	*Water content*	*Water solubility*	*Leachability, (as TOC)*	*Boron Extraction, E*
	(g.cm^{-3})	(mPa.s)	(*wt*%)	(*wt* %)	(g.L^{-1})	(%)
8:1	0.8373	14.222	0.1214	4.0846	5.72	91.81
4:1	**0.8462**	**16.612**	**0.1528**	**4.7207**	**8.81**	**95.44**
2:1	0.8601	21.479	0.1296	4.9106	13.63	97.22
1:1	0.8787	31.637	0.1496	5.6645	24.25	98.04

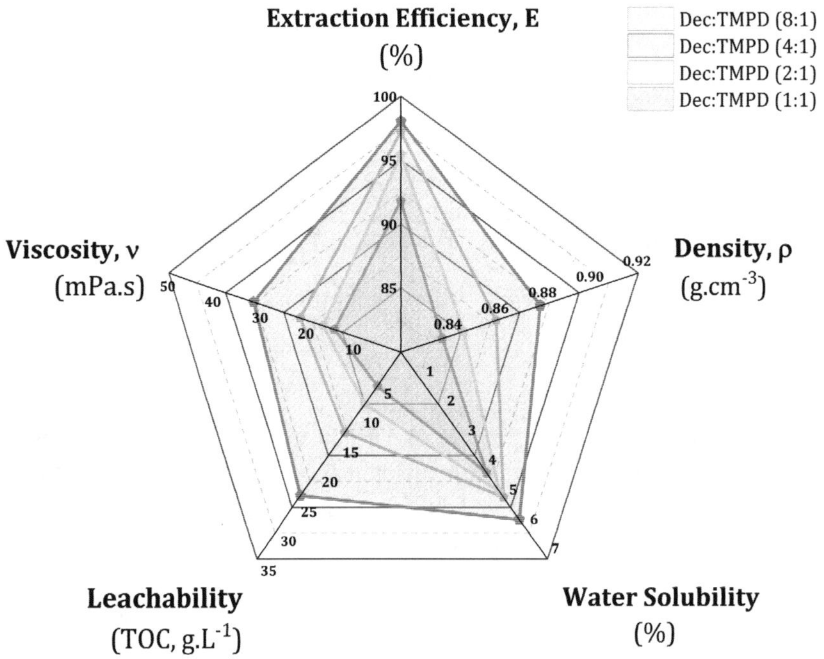

Figure 4. Radar chart comparing the physiochemical properties and boron extraction efficiencies for the Dec:TMPD mixtures at molar ratios of 8:1, 4:1, 2:1, and 1:1. (Extraction conditions: Boron in aqueous phase = 1000 ppm, pH = 2, T = 25 °C, Aqueous/Organic (A/O) mass ratio 1:1)

Conclusion

This work builds on the structured design of TMPD-based DESs presented previously, where the aliphatic compounds (Dec and DecA) and the branched aliphatic compound (TOPO) were investigated as potential HBAs, further enhancing the physcichemcial properties of the eutectic mixtures. The newly developed eutectic mixtures comprising decanol with TMPD featured notably lower density and viscocity values over the mixtures with other HBAs. Furthermore, the properites of the Dec:TMPD system were studied at different molar ratios, and Dec:TMPD (4:1) showed a balanced performance with an excellent boron extractant capability (95.44 %) without compromising the integrity of the liquid-liquid extraction system.

In conclusion, the proper selection of both DES constituents (HBD and HBA) influences the over all performance of the solvent extraction system, where TMPD (as the HBD) targets the selective and efficnet extraction of boron from the aqueous phase, while Dec (as the HBA) contributes toward a stable solvent system for the extraction process.

Acknowledgement

The authors gratefully acknowledge the support of the Center for Membrane and Advanced Water Technology at Khalifa University under grant number RC2-2019-009 and the support of Abu Dhabi University through the ADU Faculty Grant 19300484 to Ioannis Zuburtikudis and Hadil Abu Khalifeh.

References

[1] N. V Plechkova, K.R. Seddon, Applications of ionic liquids in the chemical industry, Chemical Society Reviews. 37 (2008) 123–150.

[2] P. Isosaari, V. Srivastava, M. Sillanpää, Ionic liquid-based water treatment technologies for organic pollutants: Current status and future prospects of ionic liquid mediated technologies, Science of the Total Environment. 690 (2019) 604–619.

[3] I. Krossing, J.M. Slattery, C. Daguenet, P.J. Dyson, A. Oleinikova, H. Weingärtner, Why are ionic liquids liquid? A simple explanation based on lattice and solvation energies, Journal of the American Chemical Society. 128 (2006) 13427–13434.

[4] M. Deetlefs, K.R. Seddon, Assessing the greenness of some typical laboratory ionic liquid preparations, Green Chemistry. 12 (2010) 17–30.

[5] A. Romero, A. Santos, J. Tojo, A. Rodríguez, Toxicity and biodegradability of imidazolium ionic liquids, Journal of Hazardous Materials. 151 (2008) 268–273.

[6] P. Isosaari, V. Srivastava, M. Sillanpää, Ionic liquid-based water treatment technologies for organic pollutants: Current status and future prospects of ionic liquid mediated technologies, Science of the Total Environment. 690 (2019) 604–619.

[7] V. Osch, Design and applications of hydrophobic deep eutectic solvents, Technische Universiteit Eindhoven, 2018.

[8] B.B. Hansen, S. Spittle, B. Chen, D. Poe, Y. Zhang, J.M. Klein, A. Horton, L. Adhikari, T. Zelovich, B.W. Doherty, B. Gurkan, E.J. Maginn, A. Ragauskas, M. Dadmun, T.A. Zawodzinski, G.A. Baker, M.E. Tuckerman, R.F. Savinell, J.R. Sangoro, Deep Eutectic Solvents: A Review of Fundamentals and Applications, Chemical Reviews. 121 (2021) 1232–1285.

[9] L. Sun, Z. Zhu, T. Su, W. Liao, D. Hao, Y. Chen, Y. Zhao, W. Ren, H. Ge, H. Lü, Novel acidic eutectic mixture as peroxidase mimetics for oxidative desulfurization of model diesel, Applied Catalysis B: Environmental. 255 (2019) 117747.

[10] T. Lemaoui, Y. Benguerba, A.S. Darwish, F.A. Hatab, S.E.E. Warrag, M.C. Kroon, I.M. Alnashef, Simultaneous dearomatization, desulfurization, and denitrogenation of diesel fuels using acidic deep eutectic solvents as extractive agents: A parametric study, Separation and Purification Technology. 256 (2021) 117861.

[11] F. Bergua, M. Castro, J. Muñoz-Embid, C. Lafuente, M. Artal, Hydrophobic eutectic solvents: Thermophysical study and application in removal of pharmaceutical products from water, Chemical Engineering Journal. 411 (2021).

[12] J. Kuttiani Ali, C. Maher Chabib, M. Abi Jaoude, E. Alhseinat, S. Teotia, S. Patole, D. Hussain Anjum, I. Qattan, Enhanced removal of aqueous phenol with polyimide ultrafiltration membranes embedded with deep eutectic solvent-coated nanosilica, Chemical Engineering Journal. 408 (2021) 128017.

[13] I. Gállego, M.A. Grover, N. V. Hud, Folding and imaging of DNA nanostructures in anhydrous and hydrated deep-eutectic solvents, Angewandte Chemie - International Edition. 54 (2015) 6765–6769.

[14] M.K. Tran, M.F. Rodrigues, K. Kato, G. Babu, P.M. Ajayan, Deep eutectic solvents for cathode recycling of Li-ion batteries, Nature Energy. 4 (2019) 339–345.

[15] O.S. Hammond, K.J. Edler, D.T. Bowron, L. Torrente-Murciano, Deep eutectic-solvothermal synthesis of nanostructured ceria, Nature Communications. 8 (2017) 1–7.

[16] Q. Zhang, K. De Oliveira Vigier, S. Royer, F. Jerome, Deep eutectic solvents: syntheses, properties and applications, Chem Soc Rev. 41 (2012) 7108–7146.

[17] D.J.G.P. Van Osch, L.F. Zubeir, A. Van Den Bruinhorst, M.A.A. Rocha, M.C. Kroon, Hydrophobic deep eutectic solvents as water-immiscible extractants, Green Chemistry. 17 (2015) 4518–4521.

[18] B.D. Ribeiro, C. Florindo, L.C. Iff, M.A.Z. Coelho, I.M. Marrucho, Menthol-based eutectic mixtures: Hydrophobic low viscosity solvents, ACS Sustainable Chemistry and Engineering. 3 (2015) 2469–2477.

[19] R. Verma, T. Banerjee, Liquid-Liquid Extraction of Lower Alcohols Using Menthol-Based Hydrophobic Deep Eutectic Solvent: Experiments and COSMO-SAC Predictions, Industrial and Engineering Chemistry Research. 57 (2018) 3371–3381.

[20] C. Florindo, L.C. Branco, I.M. Marrucho, Development of hydrophobic deep eutectic solvents for extraction of pesticides from aqueous environments, Fluid Phase Equilibria. 448 (2017) 135–142.

[21] D.J.G.P. Van Osch, L.F. Zubeir, A. Van Den Bruinhorst, M.A.A. Rocha, M.C. Kroon, Hydrophobic deep eutectic solvents as water-immiscible extractants, Green Chemistry. 17 (2015) 4518–4521.

[22] O.G. Sas, M. Castro, Á. Domínguez, B. González, Removing phenolic pollutants using Deep Eutectic Solvents, Separation and Purification Technology. 227 (2019) 115703.

[23] W. Chen, X. Li, L. Chen, G. Zhou, Q. Lu, Y. Huang, Y. Chao, W. Zhu, Tailoring hydrophobic deep eutectic solvent for selective lithium recovery from the mother liquor of Li2CO3, Chemical Engineering Journal. (2020) 127648.

[24] G. Almustafa, R. Sulaiman, M. Kumar, I. Adeyemi, H.A. Arafat, I. AlNashef, Boron extraction from aqueous medium using novel hydrophobic deep eutectic solvents, Chemical Engineering Journal. 395 (2020) 125173.

[25] G. Almustafa, A.S. Darwish, T. Lemaoui, M.J. O'Conner, S. Amin, H.A. Arafat, I. AlNashef, Liquification of 2,2,4-trimethyl-1,3-pentanediol into hydrophobic eutectic mixtures: A multi-criteria design for eco-efficient boron recovery, Chemical Engineering Journal. 426 (2021) 131342.

630637 Nonlinear Predictive Control of an Industrial Selective Catalytic Reduction Unit with Time-Varying Time Delay

Abstract:

Under load-following operation of coal-fired power plants, it is challenging to control nitrogen oxide (NO_x) concentration at the outlet of selective catalytic reduction (SCR) units to comply with environmental regulations. For controlling NO_x, excess ammonia (NH_3) that often far exceeds the stoichiometric requirement is injected in SCRs leading to NH_3 slip that gets released to the atmosphere and increases plant operating costs. In SCR units, NH_3 adsorbs on the catalyst reacting with NOx in the gas phase; it is practically impossible to measure NH_3 loading on the catalyst. Thus, even when NH_3 loading is depleting or increasing, its effect may not be reflected in the real-time outlet NO_x concentration. This leads to a complex, time-varying, time-delay response that is difficult to estimate due to complex adsorption-desorption dynamics of NH_3 on the catalyst coupled with internal and external mass transfer limitations. Thus, the industry-standard feedback-augmented feedforward (FBAFF) controllers and conventional model predicative controllers exhibit poor control performance for SCR control.

In this work, first a comprehensive first-principles dynamic model of the SCR unit is developed. In the existing literature, mass transfer limitations, both internal or external, have been neglected [1]–[3]. In addition, the competing NH_3 oxidation reaction, that can consume available NH_3, particularly when the system experiences thermal disturbances, is often neglected [4]. Additionally, most of these models, including those cited previously, are validated only with lab- or bench-scale data, thus there is a distinct lack of industrial-scale validation for SCR unit models. Data-driven models that have been developed using industrial data lack the predictive capability of first-principles models [5]–[7]. In this work, a first-principles model was developed with consideration for the adsorption-desorption dynamics of NH_3 on the catalyst, the SCR reduction reaction, and the competing NH_3 oxidation reaction, as well as the internal and external mass transfer limitations inherent to the SCR system. Furthermore, the model was validated with industrial data for a wide range of load-following operating conditions to ensure the results were applicable for an industrial-sized SCR unit.

In the power plants, control of the SCR is largely handled via FBAFF control using feedforward models of varying complexity coupled with a feedback trim to control NO_x emissions via NH_3 injection [8]. In the literature, some applications of model predictive control (MPC) have been reported with the goals of limiting both NO_x emissions and NH_3 slip, but considering very simple plant models that lack sufficient details to capture the complex SCR dynamics [9]–[11]. A few papers have considered limiting ammonia slip using a state constraint in their MPC formulations, but these formulations do not necessarily minimize NH_3 slip [12], [13].

In this work, three nonlinear MPC (NMPC) formulation are developed. In the first formulation, tight control of the outlet NO_x concentration with the standard quadratic objective function is exhibited. A second NMPC is proposed using a soft constraint on the outlet NO_x while minimizing the amount of NH_3 slip. Finally, a third NMPC implementation is proposed with explicit accounting for the time delay due to the adsorption-desorption dynamics. Because the time delay is unknown, a state-observer based approach is used whereby an initial estimate of the time delay based on the kinetic parameters of the SCR reactions is updated recursively.

Performance of the NMPC controllers is studied with respect to industrial data for a load-following scenario with comparison drawn against a well-tuned FBAFF controller. Results are discussed from the perspective of controller performance via the standard metrics along with economic considerations in the form of reagent consumption. The study shows that when time-delay is estimated recursively, it leads to considerably superior control performance compared to other approaches especially for rapid changes in the incoming NOx flow and concentration.

[1] I. Nova, L. Lietti, E. Tronconi, and P. Forzatti, "Dynamics of SCR reaction over a TiO 2-supported vanadia-tungsta commercial catalyst," *Catal. Today*, vol. 60, pp. 73–82, 2000.

[2] L. Lietti, I. Nova, S. Camurri, E. Tronconi, and P. Fonatti, "Dynamics of the SCR-DeNO, Reaction by the Transient-Response Method," *AIChE J.*, vol. 43, no. 10, pp. 2559–2570, 1997.

[3] T. Qin, T. Yang, Y. Lv, F. Hong, and Q. Yao, "Dynamic modeling for SCR system of coal fired power plant," in *Chinese Control Conference, CCC*, Jul. 2016, vol. 2016-Augus, pp. 2006–2010, doi: 10.1109/ChiCC.2016.7553660.

[4] E. Muñoz, P. Marín, S. Ordóñez, and F. V. Díez, "The role of reaction kinetics and mass transfer in the selective catalytic reduction of NO with NH3 in monolithic reactors," *J. Chem. Technol. Biotechnol.*, vol. 90, no. 7, pp. 1299–1307, Jul. 2015, doi: 10.1002/jctb.4437.

[5] Y. Lv, J. Liu, T. Yang, and D. Zeng, "A novel least squares support vector machine ensemble model for NOx emission prediction of a coal-fired boiler," *Energy*, vol. 55, pp. 319–329, Jun. 2013, doi: 10.1016/j.energy.2013.02.062.

[6] S. M. Safdarnejad, J. F. Tuttle, and K. M. Powell, "Dynamic modeling and optimization of a coal-fired utility boiler to forecast and minimize NOx and CO emissions simultaneously," *Comput. Chem. Eng.*, vol. 124, pp. 62–79, May 2019, doi: 10.1016/j.compchemeng.2019.02.001.

[7] Y. Lv, X. Lv, F. Fang, T. Yang, and C. E. Romero, "Adaptive selective catalytic reduction model development using typical operating data in coal-fired power plants," *Energy*, vol. 192, Feb. 2020, doi: 10.1016/j.energy.2019.116589.

[8] T. Leopold, "Tuning Ammonia Flow to Optimize SCR Performance," *Power*, Jan. 12, 2010.

[9] J. Zhang, B. Jia, B. Zou, X. Tian, and C. Liu, "A multiple model control method of coal-fired power plant SCR-DeNOx system," in *Communications in Computer and Information Science*, 2018, vol. 925, pp. 306–313, doi: 10.1007/978-981-13-2381-2_29.

[10] K. Zhang, J. Zhao, Y. Zhu, and Z. Xu, "Model predictive control case study: Selective catalytic reduction (SCR) system in coal-fired power plant," in *Chinese Control Conference, CCC*, Aug. 2016, vol. 2016-August, pp. 4300–4305, doi: 10.1109/ChiCC.2016.7554021.

[11] B. Shen, N. Zhao, T. Liu, F. Wu, and C. Zuo, "Modeling and simulation of selective catalytic reduction for flue gas denitration in power plants," in *Advanced Materials Research*, 2012, vol. 383–390, pp. 6580–6586, doi: 10.4028/www.scientific.net/AMR.383-390.6580.

[12] S. Shah, S. Abrol, S. Balram, and J. Barve, "Optimal Ammonia Injection for Emissions Control in Power Plants," *IFAC-PapersOnLine*, vol. 48, no. 30, pp. 379–384, 2015, doi: 10.1016/j.ifacol.2015.12.408.

[13] H. Peng, H. Shioya, and R. Zou, "A predictive control strategy for nonlinear nox decomposition process in thermal power plants," *IEEE Trans. Syst. Man, Cybern. Part A Syst. Humans*, vol. 36, no. 5, pp. 904–921, 2006, doi: 10.1109/TSMCA.2005.855920.

629433 Mechanistic Modelling of Reactive Liquid-Liquid Extraction Towers Using Polar PC-SAFT: Industrial Validation and Optimization of Fat/Oil Hydrolysis

Abstract:

Biorefineries provide an important alternative for petroleum-based refineries to reduce CO_2 emissions and increase the share of renewable feedstocks for the production of chemicals and fuels. Vegetable oils and animal fats are used as renewable raw materials for the production of oleochemicals. Never before in the history of the oleochemical industry have the changes been more dramatic than in the last decades. The rapid growth of palm oil, the rise of oleochemicals production in Southeast Asian Nations and an increased competition with the biofuel industry for feedstocks form a challenge for the traditional oleochemical players in Europe and North America.[1] This increased competition forces European oleochemical companies such as Oleon NV to diversify the feedstocks they process, ranging from low quality animal fat to higher quality vegetable oils. Operating a continuous installation with a variable feed composition poses significant challenges for process operation and control to ensure resource efficiency, high product yields and excellent product quality. Petroleum refineries regularly use mechanistic and statistical modelling to tackle this challenge of feedstock diversity and adapt process conditions for compositional variability of the incoming crude oil.[2] Applying these techniques to biorefineries, such as for the production of oleochemicals, could aid in utilizing variable bio-based feedstock streams more efficiently, enable the use of lower quality-grade feedstocks and improve adaptability toward future changes in demand and supply.[3]

Fat/Oil hydrolysis is the first step in the production of fatty acids. On an industrial scale, oil hydrolysis is mostly performed in large splitting towers via a counter-current reactive liquid-liquid extraction, as shown in **Figure 1**. The oil feed is introduced at the bottom of the tower and forms the continuous organic phase, while water is introduced at the top and forms the dispersed aqueous phase. The reaction takes place in the organic phase, and the formed glycerol is extracted via the aqueous phase. For fat/oil hydrolysis, important considerations are a high degree of hydrolysis, a high concentration of glycerol in the aqueous phase and a minimal colour degradation of the fatty acids. In the past, related process optimization was typically performed via ad-hoc experiments, neglecting possible adjustments for other feedstocks. This lack of accounting for feedstock variability highly limits an oleochemical plant's flexibility to switch between feedstocks. Although the oleochemical industry is mature, modelling research on oil hydrolysis is scarce, due to the complexity of lipid feedstocks and the lack of data on physical properties.[4] In this work, a comprehensive process model of industrial-scale splitting towers that is able to predict the yield for oil hydrolysis is developed and used to investigate optimal processing parameters for different bio-based triglyceride feedstocks.[5]

The splitting tower is represented by a compartmental model, using a cascade of continuous stirred tank reactors and liquid-liquid separators. The developed model extends the state of the art by including a variable glycerol equilibrium ratio, which is a function of the composition and temperature and is calculated using the polar version of the perturbed chain statistical association fluid theory (PPC-SAFT).[6] The group contribution method for PPC-SAFT (GC-PPC-SAFT) is used for all lipid compounds to calculate pure compound properties such as segment number m_i and segment diameter σ_i.[7] The liquid-liquid equilibrium estimated by PPC-SAFT is compared to more traditional property methods UNIFAC_DMD, UNIQUAC and SRK. In addition, the model accounts for the autocatalytic effect of fatty acids in hydrolysis and the isomerization of poly-unsaturated fatty acids.[8, 9] Model validation is performed using process data from three real-life splitting towers covering four feedstock types, i.e., tallow, rapeseed oil, palm oil, and palm fatty acid distillate.

The splitting tower model properly estimates the degree of hydrolysis for tallow, palm, PFAD and rapeseed oil. Composition gradients of the organic phase throughout the tower show that it is crucial to properly account for the changes in the glycerol equilibrium ratio. For the hydrolysis of rapeseed oil, the calculated concentration profiles over the tower show that the reaction is slow at the bottom of the tower due to a low amount of free fatty acids in the organic phase. The reaction rate increases towards the middle, due to a better solubility of water in the organic phase and the autocatalytic effect. At the top, the reaction approaches equilibrium and thus slows down. The importance of feedstock flow rate, water/oil ratio, and temperature profile throughout the tower is analysed and confirmed by sensitivity analysis. Our results show that modifying the temperature profile may shift the reaction equilibrium toward the fatty acid product. In addition, the changed settings result in a reduced steam consumption (-8%), saving water and energy, which is equivalent to a reduction of the climate change impact by 13 kg CO_2eq per tonne hydrolysed rapeseed oil. This knowledge is crucial for improving the energy and resource efficiency of fat/oil hydrolysis, thereby improving its economic and environmental sustainability.

Overall, this work shows how process modelling can deliver a strategic advantage to oleochemical companies for understanding and operating their processes on an industrial scale. For this, the collaboration between academia and industry is essential, as it offers realistic use cases to academia and allows displaying the benefits of new developments and methodologies in process systems engineering to industry.1.

1. Biermann, U.; Bornscheuer, U. T.; Feussner, I.; Meier, M.; Metzger, J. O., Fatty Acids and their Derivatives as Renewable Platform Molecules for the Chemical Industry. *Angewandte Chemie* **2021**, 1-21.

2. Hsu, C. S.; Robinson, P. R., *Practical advances in petroleum processing*. Springer Science & Business Media: New York, 2007; Vol. 1, p 410.

3. Nachtergaele, P.; De Somer, T.; Gelaude, B.; Hogie, J.; Thybaut, J.; De Meester, S.; Drijvers, D.; Dewulf, J., Iterative Lumping Approach for representing Lipid Feedstocks in Fatty Acid Distillation Simulation and Optimization. *AIChE Journal* **2021**.

4. Perederic, O. A.; Cunico, L. P.; Kalakul, S.; Sarup, B.; Woodley, J. M.; Kontogeorgis, G. M.; Gani, R., Systematic identification method for data analysis and phase equilibria modelling for lipids systems. *The Journal of Chemical Thermodynamics* **2018,** 121, 153-169.

5. Nachtergaele, P.; Sin, G.; De Meester, S.; Ruysbergh, E.; Lauwaert, J.; Dewulf, J.; Thybaut, J. W., Simulation of an Industrial-Scale Reactive Liquid-Liquid Extraction Tower using Polar PC-SAFT for Understanding and Improving the Hydrolysis of Triglycerides *ACS Sustainable Chemistry & Engineering* **2021**.

6. Nguyen-Huynh, D.; Passarello, J.-P.; Tobaly, P.; de Hemptinne, J.-C., Application of GC-SAFT EOS to polar systems using a segment approach. *Fluid Phase Equilibria* **2008,** 264, (1-2), 62-75.

7. Rodriguez, G.; Beckman, E. J., Modelling phase behavior of triglycerides, diglycerides and monoglycerides related to biodiesel transesterification in mixtures of alcohols and CO2 using a polar version of PC-SAFT. *Fluid Phase Equilibria* **2020,** 503, 112303.

8. Milliren, A. L.; Wissinger, J. C.; Gottumukala, V.; Schall, C. A., Kinetics of soybean oil hydrolysis in subcritical water. *Fuel* **2013,** 108, 277-281.

9. Gerčar, N.; Šmidovnik, A., Kinetics of geometrical isomerization of unsaturated FA in soybean oil. *Journal of the American Oil Chemists' Society* **2002,** 79, (5), 495-500.

Figure 1. a. Picture of one of the studied industrial-scale splitting towers (Source: Oleon NV). b. Graphical representation of the counter-current reactive liquid-liquid extraction (FA = fatty acids, GLY = Glycerol, TG = Triglycerides).

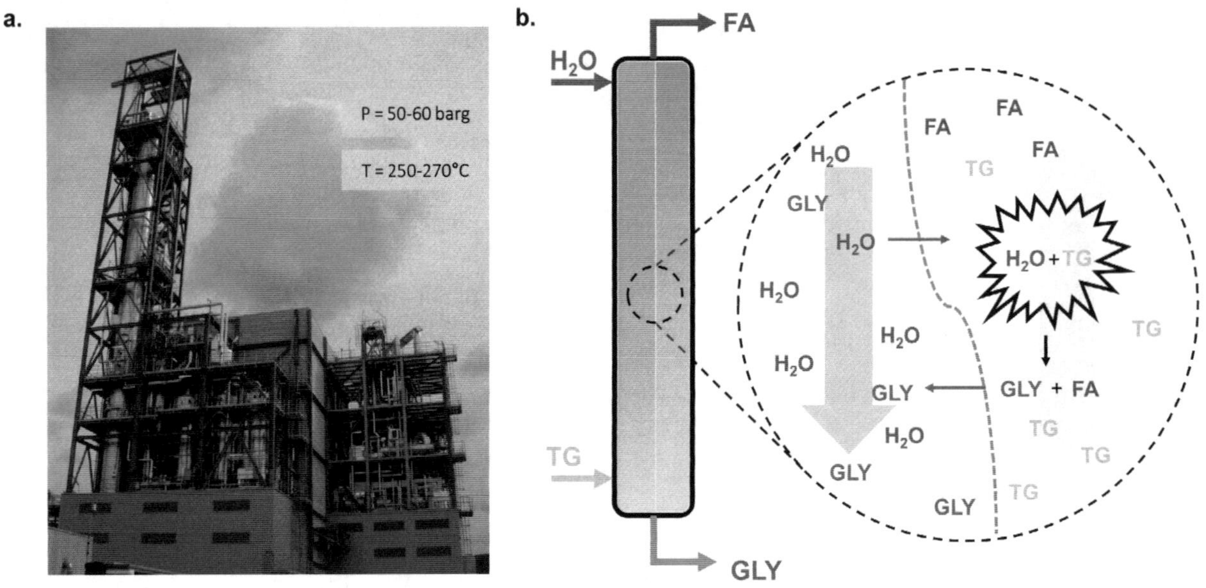
a.
P = 50-60 barg
T = 250-270°C
b.
FA
H2O
TG
GLY
FA
FA
FA
TG
H2O
GLY
H2O
H2O+TG
H2O
H2O
H2O
TG
GLY
GLY + FA
H2O
TG
TG
GLY
GLY
TG

624481 A Multi-Agent and Distributed Cloud Computing Approach for Industrial Production Scheduling Model Development and Deployment

Abstract:

Production scheduling [1] in Dow is a complex process which involves designing production cycles for multiple production trains. A production cycle is a repeating sequence of products that satisfies forecasted demand and balances inventory and product transition costs. Different product cycle sequences and campaigns result in different transition costs. Operationally, these cycles are designed using an excel-based tool and are unrolled into complete schedules which are then adjusted dynamically week-to-week in response to actual customer demand. This process relies heavily on experienced schedulers who build up tribal knowledge over time but also experiences frequent role turnover making the retention of this knowledge a challenge for the business. Furthermore, the current technology does not handle demand allocation of products that are shared across several production trains - this is left to the expertise and judgement of the subject matter expert schedulers. The current work process produces numerous inefficiencies in terms of excess inventory levels and transition costs.

In this talk, we will present a multi-agent [2] based mathematical optimization approach that allocates forecasted demand across multiple production trains and finds the detailed product cycle schedule. The overall optimization model is divided in two independent agents. The role of agent 1 is to determine optimal product demand allocations across the trains and ensure that the allocated products can be cycled through (i.e., the product graph is strongly connected [3]) to ensure feasible product transitions in the production cycle. The agent 1 iteratively loops through solving a MILP based product allocation model and a depth first search (DFS) algorithm [4] to check if the product solution is strongly connected. If the product allocation solution is not strongly connected, its corresponding binary solution is cut off in the next iteration until a strongly connected product solution is found. The role of agent 2 is to use the resulting product demand allocations from agent 1 and determine the product cycle design (such as product sequence and campaign length) and production schedule for each production trains. Simultaneous product cycle design and scheduling is computationally expensive. Agent 2 thus leverages a hybrid MILP scheduling and simulation-based approach to address the high computational complexity. Agent 2 first runs a sampling algorithm to generate candidate product sequences and campaign lengths, which are then solved in the scheduling model to determine the inventory and transition loss cost. The product cycle schedule with the least cost is then implemented in the production site.

Both the demand allocation model (agent 1) and the single train detailed schedule model (agent 2) are deployed to Dow's Azure [5] platform. These are deployed as independent Azure Machine Learning [6] pipeline endpoints. Once these endpoints are called a batch compute cluster is spun up, the models are solved, and the results returned. The demand allocation model will trigger the single train model when the optimal demand allocations are calculated without returning to the user between agent 1 and agent 2 solves. A user may also choose to execute only the single train model when they determine that a multi-train model is unnecessary. Agent 2's sampling solution approach is distributed across the compute cluster using Anyscale's Ray [7] package to solve each sample in parallel thus greatly reducing the solution time to that of a single sample solve time which is on the order of seconds.

The models/agents were tested during a pilot phase deployment where scenarios from different production trains are run and evaluated, and model updates are made to capture all complexities. Having successfully completed this pilot, efforts are underway to scale the solution to over 50 production trains world-wide by the end of the 2022.

This work demonstrates the power of combining distributed and parallel cloud computing technologies with traditional MILP models with a sampling approach which may have been impracticable for sequential solution approaches. Furthermore, the deployment and management of the independent agents is done with the help of MLOps [8] strategies which allows for the seamless and automatic testing and deployment of model updates to end users. The approach shows great promise and potential for expansion to include additional agents, potentially using additional methods like machine learning and artificial intelligence, into this multi-agent system where agents trigger other agents automatically and as needed.

References:

[1] Harjunkoski, Iiro, Christos T. Maravelias, Peter Bongers, Pedro M. Castro, Sebastian Engell, Ignacio E. Grossmann, John Hooker, Carlos Méndez, Guido Sand, and John Wassick. "Scope for industrial applications of production scheduling models and solution methods." *Computers & Chemical Engineering* 62 (2014): 161-193.

[2] Wooldridge, Michael. *An introduction to multiagent systems*. John wiley & sons, 2009.

[3] Nuutila, Esko, and Eljas Soisalon-Soininen. "On finding the strongly connected components in a directed graph." *Information processing letters* 49, no. 1 (1994): 9-14.

[4] Awerbuch, Baruch. "A new distributed depth-first-search algorithm." *Information Processing Letters* 20, no. 3 (1985): 147-150.

[5] Collier, Michael, and Robin Shahan. *Microsoft Azure Essentials-Fundamentals of Azure*. Microsoft Press, 2015.

[6] Klein, Scott. "Azure machine learning." In *IoT Solutions in Microsoft's Azure IoT Suite*, pp. 227-252. Apress, Berkeley, CA, 2017.

[7] https://www.anyscale.com/ray-open-source

[8] *Demystifying XOps: DataOps, MLOps, ModelOps, AIOps and Platform Ops for AI*. Gartner Research. 2021

626543 A Surrogate-Based Topological Compartment Model for Counter-Current Spray Dryers.

Abstract:

Surrogate-based topological compartment models have been widely used to predict the internal distribution of properties in a unit, providing comparable accuracy to CFD models but with less computational cost. In the area of particle technology, the approach has gained attention for aggregating detailed physics into process models in units such as blenders, coaters, and fluidized beds [1]. However, the use of compartment models in spray drying has been limited. To date, only hybrid data-driven and first principles-based models have been proposed for counter-current spray dryers. The two most relevant models have been developed by Process System Enterprise (PSE) [2] and Ali et al. [3]. In both cases, the fluxes were extracted from a CFD-DPM model and used as boundary conditions to a first-principles model incorporating mass and energy balances for the compartment. The main difference between the two approaches lies in the generation of the compartments and the drying model used. PSE uses homogeneous zones with different diffusion-based models available. Meanwhile Ali et al. [3] prosed a fixed compartmentalization with a combined diffusion and boiling model. The results of both showed low error against the CFD-DPM model that was used for the construction. However, a change in the operating conditions modifies the structure and fluxes, making it necessary to rerun the CFD model every time and thus limiting the reapplication of the model. To overcome these drawbacks, in this work a surrogate-based compartment model is proposed as an alternative.

The surrogate-based compartment model has been generated following a step-by-step validation procedure with a decoupling of the momentum, heat, and mass transfer mechanisms as presented in [4]. In the first stage, the distribution and size of each of the zones have been identified based on the volume fraction of the particles. These zones presented in Figure 1 have been made flexible using surrogate models. In the second stage, the fluxes of particles and air between the compartments and the residence time of the particles have been modeled. Unlike previous works, a new numerical procedure has been developed for surrogates based on dimensionless groups. The methodology is based on a three-step algorithm that first selects the necessary dimensionless groups based on minimizing the Bayesian Information Criteria. The second step models the mass fluxes based on these dimensionless groups in the form of Eq. (1) and evaluates the quality of the model; an alternative statistical model is proposed if the model is not good enough. In the same way, the third step of the process generates the surrogate models for the energy fluxes, which have been evaluated under different operating conditions including the heat and mass transfer model defined as in [5]. These surrogate models have been implemented into the final model where the discrete phase is defined by means of a multi-dimensional population balance. The single droplet drying model for the particles includes diffusion and boiling mechanisms as in [5] and it is implemented employing finite differences. The model is finally solved by defining a two-way iterative scheme in Python.

$F = Ar^{a} \cdot Re^{b} \cdot \ldots \cdot St^{c}$ (1)

The model has been validated by decoupling the physics and evaluating it step-by-step. The validation against the momentum of the particles is carried out with the mean residence time of monodispersed injections. The prediction of the mean residence time shows a relative root mean square error of 7%. The validation including the heat and mass transfer is carried out with the temperature at the entrance and exit of every compartment showing an average error of 5 K and a maximum of 11 K in all the compartments evaluated. Even though the differences in predicting the mean residence time and the temperature are low, the errors are still too high to implement the agglomeration and breakage kernels as characterized in the CFD-DPM model. The relative time at which the discrete phase has a wet surface only represents around 1.6% of the total mean residence time. This high level of accuracy required results in too small feasible region for maintaining the surrogate models of the kernels. Thus, the implementation of these kernels into process or compartment models requires at least the use of a data-driven model as an intermediate step and may not yield a significant improvement in the reduction of computational cost.

References

[1] Rogers, A. Ierapetritou, M. (2015) Challenges and opportunities in modeling pharmaceutical manufacturing processes. Computers & Chemical Engineering, 81, 32-39.

[2] Process System Enterprise (2020) Mulizonal tool and Counter-current spray drying model. gProms® Formulated Products.

[3] Ali, M. Mahmud, T. Heggs, P.J. Ghadiri, M. (2020) Zonal modelling of a counter-current spray drying tower. Chemical Engineering Research and Design, 155, 180-199.

[4] Hernández, B. Fraser, B. Martin de Juan, L. Martin, M. (2018) Computational fluid dynamics (CFD) modeling of swirling flows in industrial counter-current spray-drying towers under fouling conditions. Industrial & Engineering Chemistry Research, 57 (35), 11988-12002.

[5] Hernandez, B. Mondragon, R. Pinto, M.A. Hernandez, L. Julia, J.E. Jarque, J.C. Chiva, S. Martin, M. (2021) Single droplet drying of detergents: Experimentation and modelling. Particuology, 58, 35-47.

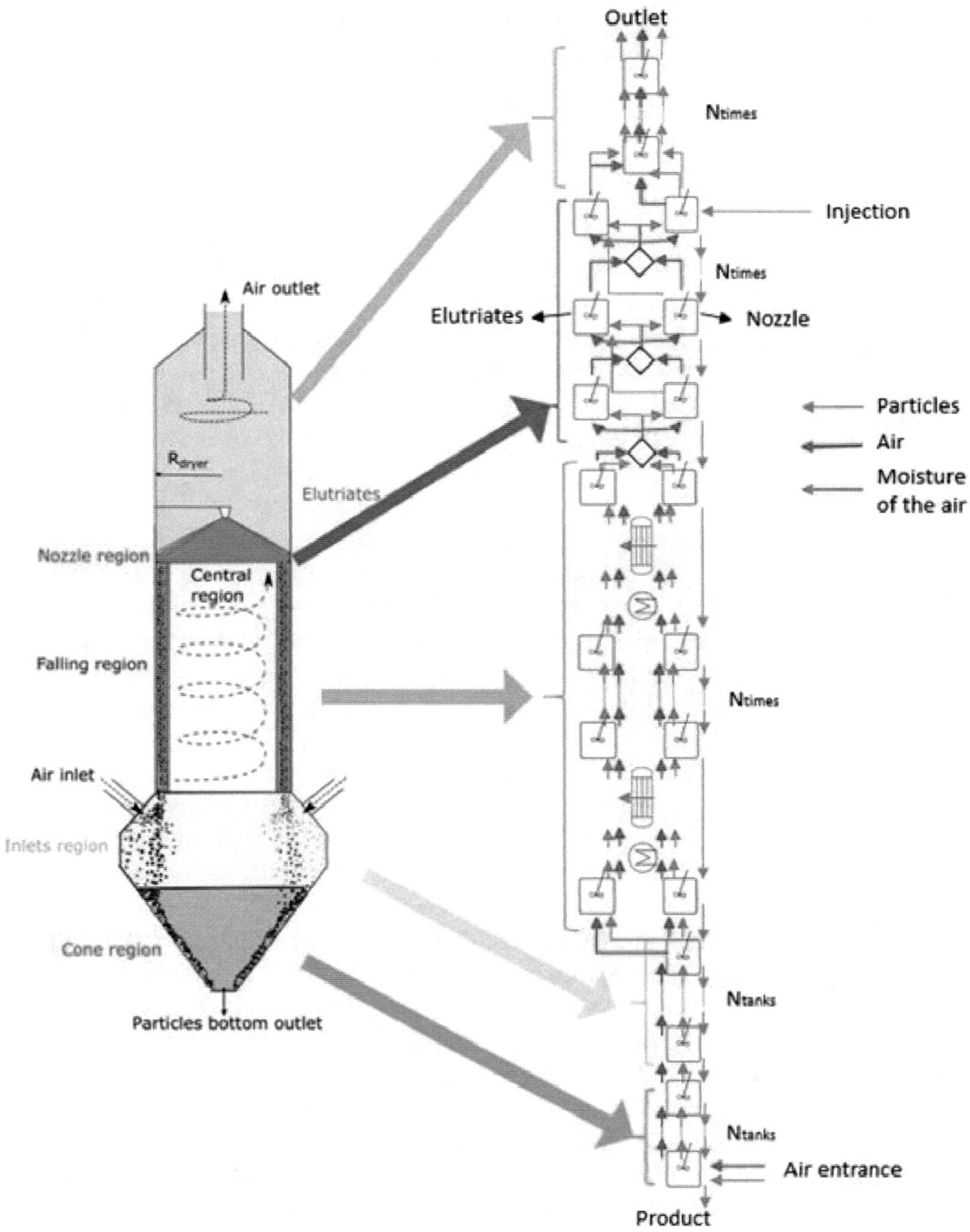

Outlet
Ntimes
Injection
Ntimes
Air outlet
Elutriates
Nozzle
Particles
Air
Moisture
of the air
Rdryer
Elutriates
Nozzle region
Central
region
Falling region
Ntimes
Air inlet
Inlets region
Cone region
Ntanks
Particles bottom outlet
Ntanks
Air entrance
Product

623023 Optimal Design and Operation of an Organic Rankine Cycle (ORC) System Driven By Solar Energy with Sensible Thermal Energy Storage

Abstract:

Since the energy demand is rapidly increasing and climate change deteriorates, the conventional fossil fuel power generation technologies are facing challenges related to greenhouse gas emissions. Environmentally friendly and energy-efficient power generation from renewable energy is a promising way to generate electricity due to the advancements in the renewable energy field beyond hydroelectric power. Among the more recent renewable energy sources, solar energy is the most promising one because of its abundance. However, solar energy has its own weakness such as intermittency and low energy density. To overcome the intermittency of solar energy, the implementation of energy storage allows for more predictable and stable power output systems. Solar energy technologies for electricity production has proven to be a viable option for green energy production. There are a number of different potential power cycles that can convert heat into power. The most versatile and common cycle for this purpose, is the organic Rankine cycle (ORC) for solar energy utilization. Although the conventional steam Rankine cycle dominates in terms of efficiency when utilizing heat sources at temperatures of 400 centigrade or higher, these temperatures are out of reach for conventional solar collector technologies. In addition, the organic Rankine cycle is more compact and less costly compared with conventional steam Rankine cycle power plants.

In this study, we investigate the optimal operation of an ORC system driven by solar energy. The main objective is to develop a model describing an organic Rankine cycle driven by solar energy with constant power output. A sensible energy storage system is configured to overcome the intermittency of solar energy. A circulating fluid that connects the solar collector and the ORC system plays a critical role in this system. The mass flow rate of the circulating fluid determines both the temperature of the circulating fluid and the amount of heat absorbed from the solar collector. Among the concentrating solar collectors, the parabolic trough collector (PTC) has reached the highest level of commercial maturity and accounts for the largest share of the current concentrated solar power market. A simulation-based optimization model is developed in this work. Process simulation of the ORC is performed in Aspen HYSYS, and the mathematical model of the energy storage system and the parabolic trough collector is developed in Matlab. The optimal operation of the system including the mass flow rate of the circulating fluid, operating conditions of the ORC, the inlet and outlet temperature of the circulating fluid are determined jointly based on this model. The control strategy of the system can be determined based on the results of this work. The efficiency of the ORC system can be maximized with the proposed system configuration.

634114 Keynote Talk: Online Measurements for the Petrochemical Industry: Industry 4.0 Trends and Unmet Needs

Abstract:

The primary objectives of the petrochemical manufacturing industry are to operate plants safely, profitably, and reliably. To achieve these objectives, several constraints need to be managed – such as raw material yield, manufacturing costs, and utility requirements. Online measurements via instruments that measure the physical properties and analyzers that measure the chemical properties of the process are critical to the optimized management of these constraints.

Globally, Dow owns and supports > 1 million instruments and > 20,000 online analyzers. Data from these measurements are indispensable to the operation and control of our plants. In this talk we will discuss some of our key requirements for both classes of online measurements, share examples of meaningful advances afforded by Industry 4.0, and articulate key challenges that remain – a call for continued innovation.

628357 Woodchip Moisture Content Estimation Using Short-Range Iot Wi-Fi for the Pulp & Paper Industry

Abstract:

The pulp and paper industry is the third-largest consumer of energy in the US industrial sector thereby leading to tremendous opportunities to improve its energy efficiency and productivity. The pulping process, which converts woodchips into pulp by displacing lignin from cellulose fibers, is one of the most important operations in a pulp and paper mill. For the pulping process, it is important to have real-time knowledge about the moisture content (MC) of the woodchips so that the process can be optimized and/or controlled correspondingly to achieve satisfactory product quality while minimizing the consumption of energy and chemicals. There have been attempts to develop real-time woodchip MC sensors, such as capacitance, resistance, microwave and NIR-based sensors. But these techniques have not been adopted by the industry due to unsatisfactory performance and/or high maintenance requirements that can hardly be met in a manufacturing environment. Also, some of the accurate methods are often lab-based that cannot be implemented online because of time constraint [1]. Due to these limitations, at present, the moisture level is measured only a few times throughout the year in the pulp and paper industry.

To address these limitations, we propose a non-destructive, economic, and robust approach using industrial Internet-of-Things (IIoT) based Wi-Fi and employ channel state information (CSI) to estimate MC in woodchips. We extract CSI by modifying the open-source device drivers for Intel Wi-Fi link 5300 Network Interface Card based on CSITool [2]. CSI contains information about the channel in the form of individual data sub-carriers capturing indoor channel features such as the effect of scattering, fading, and power decay with distance. We hypothesize that when a Wi-Fi signal passes through wood chips, the water or MC in woodchips would cause a corresponding change in the Wi-Fi signal which is then captured in CSI. The goal is to develop a data-driven model to extract the relationship between the change in CSI and the woodchip MC. CSI has been successfully used for moisture classification in wheat [3]. However, compared to wheat moisture classification, woodchip moisture estimation is much more challenging due to the significant heterogeneity in the shape, structure, and arrangement of the woodchips. For example, the woodchip arrangement in the container is expected to have a significant impact on the CSI data. In addition, CSI data have only been used in classifications, which is much easier than regression modeling.

In this work, we propose a regression-based approach to estimate the moisture content in wood chips. To the best of our knowledge, this is the first work to address the continuous estimation of MC using CSI data. Our goal is to develop a robust model that is insensitive to the heterogeneity of the woodchips, particularly the confounding factors such as woodchip shape and the effect of shuffling, to achieve a consistent and accurate MC estimation. While the IIOT devices used are small, low-cost, and rugged to stand for harsh environment, they do have their limitations such as the raw CSI data are very noisy, which can lead to low performing models when directly fed to machine learning algorithms. To address this challenge, our previously developed statistics pattern analysis (SPA) framework [4][5] is used to extract robust and predictive features from raw CSI data. The SPA-based features are then used to develop robust machine learning models for continuous MC estimation in woodchips. In other words, SPA turns messy big data into so-called smart data, which can then be used for accurate sensing. The machine learning techniques studied in this work include artificial neural networks (ANN), support vector machines (SVM), and Gaussian process regression (GPR). We conducted systematic study of these nonlinear regression approaches, and demonstrated that, when combined with SPA, provide a potential solution for real-time woodchip MC estimation for the pulp and paper industry. Specifically, the experimental validation results show highly accurate prediction of more than 20 MC levels in a wide range. In addition to the excellent performance, the proposed approach is also economic and non-destructive, making it a potential technology for industrial implementation in the pulp and paper industry.

[1] R. Govett, A PRACTICAL GUIDE FOR THE DETERMINATION OF MOISTURE CONTENT OF WOODY BIOMASS A Practical Handbook of Basic Information, Definitions, Calculations, Practices and Procedures for Purchasers and Suppliers of Woody Biomass.

[2] D. Halperin, W. Hu, A. Sheth, and D. Wetherall, Tool Release: Gathering 802.11n Traces with Channel State Information, *ACM SIGCOMM Comput. Commun. Rev.*, 2011.

[3] W. Yang, X. Wang, S. Cao, H. Wang, and S. Mao, Multi-class wheat moisture detection with 5GHz Wi-Fi: A deep LSTM approach, in *Proceedings - International Conference on Computer Communications and Networks, ICCCN*, 2018.

[4] Q. P. He and J. Wang, Statistics pattern analysis: A new process monitoring framework and its application to semiconductor batch processes, *AIChE J.*, 2011.

[5] J. Wang and Q. P. He, Multivariate Statistical Process Monitoring Based on Statistics Pattern Analysis, *Ind. Eng. Chem. Res.*, 2010.

633025 Gradient-Weighted Class Activation Mapping (Grad-CAM) Based Explanations for Process Monitoring Results from Deep Neural Networks

Abstract:

In the chemical process industries, high product quality with maximum efficiency and process safety are the critical issues that require effective process monitoring. A fault in a process results in the deviation of process state variables from its normal operating condition. The effect of the fault propagates along the process and disturbs stability of the system. So, it is essential to detect them quickly on their occurrence. To tackle this problem, we motivate to develop the methodology which fulfils the aspect of high accuracy, novel explainability and to make effective use of the multivariate time series process data.

Recent developments in sensor techniques and storage technologies have provided a significant technology push for developing next-generation data driven process monitoring methods [1]. Hence deep learning methods have attracted much attention for fault detection. Deep learning methods like Deep Belief Network (DBN), Convolutional Neural Networks (CNN) and, Hierarchical Deep Neural Networks (HDNN) have been used for fault detection [2] but they lack in providing explanations for their decision. Hence, the eXplainable Artificial Intelligence (XAI) method Integrated Gradients was used by [3] to provide explanations for the results generated from deep learning models. But the Integrated Gradient suffer from one shortcoming that it analyses the samples one by one without accounting for the fact that the data is a time series. In the paper, we develop the XAI method which accounts for the time series nature of the process monitoring problem.

Recently, a new XAI method called Grad-CAM has been developed targeted at providing explanations for time series data in terms of activation maps for CNN [4]. Grad-CAM helps to explore the spatial information that is preserved inside convolution layers. The extracted information from CNN layers helps to understand the important feature of an input data to make classification decision. Grad-CAM uses k feature maps of height h and width u of the last convolution layer, as it contains high level semantics and detailed spatial information. For each feature map A^k, we obtain an importance weight $w_k{}^c$ associated with specific class c. The $w_k{}^c$ is obtained by computing the gradient of final class output y^c with respect to the feature map A^k i.e., $\partial y^c / \partial A^k$. It is further globally averaged over width u and height h. This step is followed by computing a weighted sum of feature maps A^k of class c using $w_k{}^c$. The final heatmap is obtained by applying the *ReLU* operation on the weighted sum to emphasize only positive values. This heatmap highlights the features which are important in making a final decision of the network for class c. In this paper, we use the Grad-CAM method for the explanation of process fault diagnosis.

In this paper, we exploit the real time explainable CNN methodology that classifies the normal and faulty state of the process. The proposed methodology consists of a 2D Convolution layer that is trained to solve a fault detection problem. Then, Grad-CAM is used on a trained model to understand which part of input data contributes more to detect a fault [5]. The contribution of different features visualized using the generated heatmap for each sample helps to identify the key factors in fault occurrence. The performance of the proposed methodology demonstrated on simulated time series classification datasets which contain different input variables and categorical outputs. The CNN was trained and heatmaps were generated using Grad-CAM that provide an accurate explanation about the predicted outcome. In this paper, we will describe the proposed methodology as well as results obtained from the simulated dataset which effectively makes use of the time series dimension to provide intuitive explanations for outputs. The results from fault diagnosis on the Tennessee Eastman challenge problem will also be demonstrated [6].

References

[1] P. Park, P. D. Marco, H. Shin, and J. Bang, "Fault Detection and Diagnosis Using Combined Autoencoder and Long Short-Term Memory Network," *Sensors*, vol. 19, no. 21, p. 4612, Oct. 2019, doi: 10.3390/s19214612.

[2] H. Wu and J. Zhao, "Deep convolutional neural network model based chemical process fault diagnosis," *Comput. Chem. Eng.*, vol. 115, pp. 185–197, Jul. 2018, doi: 10.1016/j.compchemeng.2018.04.009.

[3] V. Pakkiriswamy and R. Srinivasan, "An Explainable Artificial Intelligence Based Approach for Interpretation of Fault Detection Results from Deep Neural Networks," presented at the 2020 Virtual AIChE Annual Meeting, 2020, [Online]. Available: https://www.aiche.org/academy/conferences/aiche-annual-meeting/2020/proceeding/paper/375g-explainable-artificial-intelligence-based-approach-interpretation-fault-detection-results-deep.

[4] R. R. Selvaraju, M. Cogswell, A. Das, R. Vedantam, D. Parikh, and D. Batra, "Grad-CAM: Visual Explanations from Deep Networks via Gradient-based Localization," *ArXiv161002391 Cs*, Dec. 2019, doi: 10.1007/s11263-019-01228-7.

[5] R. Assaf and A. Schumann, "Explainable Deep Neural Networks for Multivariate Time Series Predictions," in *Proceedings of the Twenty-Eighth International Joint Conference on Artificial Intelligence*, Macao, China, Aug. 2019, pp. 6488–6490, doi: 10.24963/ijcai.2019/932.

[6] J. J. Downs and E. F. Vogel, "A plant-wide industrial process control problem," *Comput. Chem. Eng.*, vol. 17, no. 3, pp. 245–255, Mar. 1993, doi: 10.1016/0098-1354(93)80018-I.

630074 A Deep Learning Vision System for Classification of Manufacturing Defects

Abstract:

Manufacturing lines for the assembly of pharmaceutical autoinjectors are highly automated and rely on visual inspection stations to efficiently and safely process a large volume of syringes. As pre-filled syringes enter the inspection line, there are a variety of subtle defects in the flange (e.g., chips, cracks) as well as other processing flaws (e.g., missing components) that can make the assembly unfit for use or otherwise problematic for continued processing. Currently, there is a highly-engineered inspection algorithm that is manually tuned for identifying defects from camera images. The existing approach is able to robustly detect important defects, however, a relatively large percent of rejected products are in fact false positives (e.g., cosmetic defects, lighting irregularities). In this work, we have developed a convolutional neural network approach for classifying defects in syringes with comparable sensitivity and higher selectivity than the previous algorithm. We found that preprocessing and noise reduction were critical for training meaningful models from available images. Additionally, we highlight a series of infrastructure challenges that were overcome during image acquisition and algorithm deployment to transfer our work to and from a good manufacturing process (GMP) inspection line. We also illustrate the importance of algorithm-architecture co-design when developing machine learning models in order for the solution to be practically deployed on the available edge computing device (i.e., the vision processor). This work highlights the use of machine learning in conjunction with traditional computer vision to classify defects beyond what can be identified by expert intuition.

633005 Magnesium Oxychloride Formation Kinetics and Enhanced Water Stability for Sustainable Building Materials Applications

Abstract:

Magnesium oxychloride (MOC) is a ceramic material with significant fire resistant properties and growing potential as a light-weight, structurally strong alternative building material. Despite many advantages, MOC has limitations as a construction material, particularly water stability. While researchers have explored solutions to these shortcomings, fundamental understanding of this material is still urgently needed. Thus we have investigated the fundamentals of the formation cure reaction and mechanisms of water stability enhancement reactions. The formation cure reaction kinetics for magnesium oxychloride 5-phase was monitored from 35 to 55°C using time-resolved quantitative x-ray diffraction and differential scanning calorimetry (DSC). The reaction was characterized as a two-step process: dissolution of magnesium oxide into a gel state followed by crystallization of magnesium oxychloride. Assuming first-order kinetics for both MgO dissolution and MOC crystallization, a kinetic model predicts 42.4 kJ/mol and 26.1 kJ/mol for dissolution and crystallization activation energies respectively. Alternatively, the Avrami nucleation and growth model was fit to DSC measurements predicting diffusion controlled, one-dimensional growth with an activation energy of 72.4 kJ/mol, accounting for both dissolution and crystallization.

For the water stability enhancement, two methods of water stability enhancement have been explored: chlorartinite formation from CO_2 exposure and phosphoric acid addition. The conversion of magnesium oxychloride to chlorartinite by CO_2 forms a protective, semi-insoluble chlorartinite layer on the surface of the magnesium oxychloride crystals, which improves water stability. Phosphoric acid (2.5 to 10 wt. %) was added to the MOC slurry before the cure reaction. Additions of 2.5 wt. % and above had positive impacts on the water stability, preserving ~50 wt. % crystalline MOC after the water stability test, but has a significant impact on the MOC formation kinetics. This is achied via. the formation of an amorphous phase on the MOC crystal surface that contains structural motifs related to insoluble MgHPO4·3H2O (newberyte) and Mg2P2O7·3.5H2O (magnesium pyrophosphate) phases. Results from this work are significantly impactful for the use of MOC in more sustainable alternatives for conventional materials in residential and commercial building applications.

624619 Optimization of TEG Dehydration Process in Natural Gas Processing Under Metamodel Uncertainty

Abstract:

Optimization of TEG Dehydration Process in Natural Gas Processing under Metamodel Uncertainty

Rajib Mukherjee[1,2], Urmila M Diwekar[2*]

[1]*Department of Chemical Engineering, The University of Texas Permian Basin, Odessa, TX 79762, USA*

[2]*Vishwamitra Research Institute, Crystal Lake, Illinois, 60012, USA*

Abstract

Natural gas processing involves removal of acidic gases followed by dehydration that is mainly carried out through absorption in tri-ethylene glycol (TEG). The dehydration process is accompanied by the emission of volatile organic compounds, including BTEX. There are several process parameters associated with major equipment that may impact BTEX emission. In our previous work, a multi-objective optimization is performed for optimal operating conditions identification related to the process parameters to mitigate BTEX emission using data driven metamodeling using machine learning followed by optimization with metaheuristic technique [1]. The metamodel is generated with data obtained from process simulation carried out in ProMax® process simulator. The metamodels are created using limited samples which introduces metamodeling uncertainty [2]. Previous metamodel-based robust designs treated a metamodel as the real model and ignore the influence of metamodeling uncertainty. Thus, the performance of the resulting optimized process parameters may have shortcomings due to the lack of adequately accounting for the uncertainty introduced by the metamodel. In the present work, the bias of metamodel uncertainty has been addressed for parameter optimization. An algorithmic framework is developed to address the challenge of parameter optimization under metamodel uncertainties. We used the better optimization of nonlinear uncertain systems (BONUS) algorithm to solve the problem. BTEX mitigation is used as the objective of the optimization [3]. Our algorithm allows the determination of the optimal process parameters for BTEX emission mitigation from TEG dehydration process under metamodel uncertainty.

Keywords: Natural gas processing, TEG dehydration, BTEX mitigation, metamodeling uncertainty, support vector regression (SVR), BONUS algorithm

[1] Mukherjee, R., & Diwekar, U. M. (2021). Multi-objective Optimization of the TEG Dehydration Process for BTEX Emission Mitigation Using Machine-Learning and Metaheuristic Algorithms. ACS Sustainable Chemistry & Engineering, 9(3), 1213-1228.

[2] Zhang, S., Zhu, P., Chen, W., & Arendt, P. (2013). Concurrent treatment of parametric uncertainty and metamodeling uncertainty in robust design. *Structural and multidisciplinary optimization*, *47*(1), 63-76.

[3] Sahin KH, Diwekar U, M. (2004) Better Optimization of Nonlinear Uncertain Systems (BONUS): a new algorithm for stochastic programming using reweighting through Kernel density estimation. Ann Oper Res 132:47–68

633725 Mapping Environmental and Economic Analysis of Decentralized Cogeneration Energy Management Centers

Abstract:

In the midst of a global conversation on how to reduce GHG emissions, Ontario's electrical grid stands out due to its extensive use of low carbon-intense processes such as nuclear and hydropower that acts as baseload energy sources. However, for peak electricity demands that require a quick ramp up of production, natural gas (NG) turbines are deployed [1], which causes the carbon intensity of Ontario's electrical grid to vary hourly [2]. Another intensive and permanent Canadian energy requirement is heating. Currently, it is provided mostly by NG -fired boilers [3]. The pursuit for better NG utilization is the motivation for this study, which proposes a design of combined heat and power energy management center (CHP-EMC).

CHP-EMC is composed of an NG-powered internal combustion engine connected to an electrical generator (NG-CHP), a hot water tank storage, an NG boiler, and a cooling tower to keep the water temperature within acceptable bounds. EMC is connected to a water loop similar to the ones used in district heating (DH) by a heat exchanger and responds to a community's heating demand using an hourly lumped-demand approach. The simulations are developed in TRNSYS, which is a capable heat transfer and fluid dynamic simulator [4]. The storage tank uses a built-in dynamic model whilst the rest of the components operate in pseudo steady-state. These components are connected by pipe models that account for time delay and flowrate.

The equipment presence and its sizes are decided by minimizing the natural gas consumption while simultaneously maximizing thermal and electrical generation. The optimization strategy used is a black-box approach using modeFrontier software [5]. This optimization and data visualization software can be coupled with any other software that uses input/output files -such as TRNSYS- and it executes several simulations guided by the optimization strategy and its constraints. For this work, classic Multi-Objective Genetic Algorithm (MOGA) is chosen for design variables (including the presence of specific components) and the pareto points are analyzed in more detail. TRNSYS simulation has internal simple controllers that take care of schedule and operational constraints. The result of the optimization aims to create an EMC flexible enough to endure deviations from expected demand profiles.

The emissions of proposed EMC designs are compared to the status quo (electricity supplied by electrical grid and thermal demand supplied by NG boilers) for a year of operation. Total emissions are calculated for each possible value of gCO_{2eq}/kWh to better represent possible scenarios of carbon intensity. The results are summarized in an easy-to-read emissions vs technology map that defines the conditions under which the EMC is less pollutant than status quo (SQ). The optimal design and selection of the EMC (vs. SQ) system is strongly dependent on the carbon intensity of the electric grid during peak-load times, the electric demand profiles, and the cooling demand profiles over the course of its operation. These are best characterized by the combination of greenhouse gas emissions per kWh during peaking operation, the ranges of heating and electricity demand over the course of the year (the maximum minus the minimum), and the kurtosis of the demands.

Following this analysis, an economic evaluation of both systems is carried out, demonstrating the cost of carbon avoided (CCA). CCA is a metric that compares CO_2 and cost of two technological strategies and calculates the cost per ton of GHG emissions potentially reduced [6]. These results are compared to possible carbon tax values, in a similar way to carbon intensity. A second map, cost vs technology, is then developed for easy visualization of the impact of carbon taxes on EMC and SQ operation costs.

The goal of this work is to apply the benefits of decentralized EMCs to reduce GHG emissions from both heating and electricity generation. The results are generalizable and can be applied to any location due to its methodology in representing carbon intensity and carbon tax.

References

[1] IESO, "Power Data," 30 04 2021. [Online]. Available: https://www.ieso.ca/power-data. [Accessed 30 04 2021].

[2] L. Pereira and I. D. Posen, "Lifecycle greenhouse gas emissions from electricity in the province of Ontario at different temporal resolutions," *Journal of Cleaner Production,* vol. 270, no. 122514, 2020.

[3] S. Canada, "Primary heating systems and type of energy," 03 04 2021. [Online]. Available: https://www150.statcan.gc.ca/t1/tbl1/en/tv.action?pid=3810028601. [Accessed 28 04 2021].

[4] TRNSYS, 2019. [Online]. Available: http://www.trnsys.com/index.html. [Accessed 30 04 2021].

[5] ESTECO, "modeFrontier," 2021. [Online]. Available: https://engineering.esteco.com/modefrontier/. [Accessed 30 04 2021].

[6] K. Gillingham and J. H. Stock, "The Cost of Reducing Greenhouse Gas Emissions," *Journal of Economic Perspectives,* 2018.

627430 Development of an Interactive Software Tool for Designing Industrial Solvent Recovery Processes

Abstract:

Abstract

In 2017 the chemical industry was reported to be the world's second-largest manufacturing sector (United Nations Environment Programme 2019). Solvents are used in most operations in chemical and pharmaceutical industries as a reaction medium, selective dissolution and extraction medium, and dilution agents. Thus, there is a sizable amount of solvent waste generated due to process inefficiencies such as inefficient mixing, and inappropriate process equipment being utilized. There are several options for solvent waste disposal, however, the three most common are on-site disposal, off-site disposal, and incineration (J. Raymond, Stewart Slater, and J. Savelski 2010). Even though these methods of solvent waste disposal carry considerable negative environmental impacts they are still widely used since they are readily available and provide an easy method to dispose of large quantities of waste. Solvent recovery is typically not used because of the potential difficulties in achieving a required purity to meet the stringent guidelines imposed on the products in pharmaceuticals and specialty chemical industries as well as the additional infrastructure and investments needed. To this end, this problem needs to be studied carefully by involving all aspects in regard to capital needs, environmental benefits, comparison to traditional disposal methods, while achieving the required chemical purity.

Thus, designing a solvent recovery method is a multi-objective problem focused on minimizing capital investments of the new separation technologies while maximizing the amount of solvent recovery matching the desired purity requirements. This task is further complicated because there is a lack of a guidance system to assist professional engineers in the modeling and design of solvent recovery processes. Such a guidance system or tool that could predict the value of solvent recovery and reuse for industry users, can generate an environmentally friendly trend to shift away from using incineration, on-site disposal, and off-site disposal as primary solvent waste handling methods. As seen in Figure 1A, not only will this tool provide a greener alternative, but also aid in solvent recycling into the chemical plant, which is economically favorable. To this end, if we can develop a user-friendly tool that allows engineers to easily access solvent recovery options, we can provide an economical and environmentally favorable strategy.

To account for all possible waste streams a robust framework is needed. This consists of a maximal process flow diagram that encompassed multiple stages of separations and multiple technologies within those stages. Together this process flow diagram develops the superstructure that provides multiple technology pathway options for any solvent waste stream. We established the different stages in the superstructure as Solid Removal, Recovery, Purification, and Refinement by reviewing the literature on existing materials recovery and separation processes (Yenkie, Wu, and Maravelias 2017; Biegler 1997; Geankoplis 2003). Common technologies are placed in different stages depending on the component they can separate in terms of their physical and chemical properties. A comprehensive chemical database is created using sources, such as the Design Institute for Physical Properties (DIPPR), to store all relevant chemical and physical properties ("DIPPR" 2012). Using General Algebraic Modeling Systems (GAMS) to model the various separation units and stream flows as a multi-objective optimization problem, the amount of solvent recovered can be optimized while minimizing the environmental impacts and total cost of operation (Chea et al. 2020; Yenkie et al. 2016). With the GAMS code as the backbone, a Graphical User Interface (GUI) is created to provide a user-friendly tool to the chemical industry. Figure 1B shows how the GUI connects the superstructure modeled in GAMS model, the chemical database, and user inputs to allow for anyone in the industry to find an optimal technology pathway for their solvent waste.

The GUI will allow for a user to input chemical species and parameters that are specific to the process that is being observed. By connecting MATLAB to the GAMS code, it is possible to model a solvent recovery process solely in the GUI with these input parameters. This tool reports economic, chemical, and environmental factors for the user to consider. The economic factors include an annualized cost breakdown, utility costs, consumable costs, labor costs, and overhead costs. These costs can be seen for the whole selected recovery pathway with details of the individual technologies. The chemical factors report the amount of solvent the predicted pathway can recover and its purity level. The environmental factors show the estimated emissions from the technology pathway and compare these emissions against those for incinerating the waste stream. Thus, this tool is a powerful option for those in the industry who want to look for a cost-effective way of recycling their solvents from the waste streams.

Keywords: Solvent Recovery, Recycling, Graphical User Interface

Acknowledgments

This research was supported by the United States Environmental Protection Agency's Pollution Prevention (P2) Program (NP96259218). The authors thank the Department of Chemical Engineering at Rowan University for their continued assistance with acquiring the tools required for this research.

References:

Biegler, Lorenz. 1997. *Systematic Methods of Chemical Process Design.* 1st ed. Prentice Hall PTR.

Chea, John D., Austin L. Lehr, Jake P. Stengel, Mariano J. Savelski, C. Stewart Slater, and Kirti M. Yenkie. 2020. "Evaluation of Solvent Recovery Options for Economic Feasibility through a Superstructure-Based Optimization Framework." *Industrial & Engineering Chemistry Research* 59 (13): 5931–44. https://doi.org/10.1021/acs.iecr.9b06725.

"DIPPR." 2012. The American Institute of Chemical Engineers. March 29, 2012. aiche.org/dippr.

Geankoplis, Christie J. 2003. *Transport Processes and Unit Operations.* 4th ed. Englewood Cliffs, NJ: P T R Prentice-Hall, Inc.

2010. Raymond, Michael, C. Stewart Slater, and Mariano J. Savelski. 2010. "LCA Approach to the Analysis of Solvent Waste Issues in the Pharmaceutical Industry." *Green Chemistry* 12 (10): 1826–34. https://doi.org/10.1039/C003666H.

United Nations Environment Programme. 2019. "Global Chemicals Outlook II - From Legacies to Innovative Solutions - Synthesis Report (2019)." Global Chemicals Outlook II. United Nations Environment Programme. https://wedocs.unep.org/bitstream/handle/20.500.11822/27651/GCOII_synth.pdf?sequence=1&isAllowed=y.

US EPA, OEI. 2013. "TRI Data and Tools." Data and Tools. US EPA. March 3, 2013. https://www.epa.gov/toxics-release-inventory-tri-program/tri-data-and-tools.

Yenkie, Kirti M., WenZhao Wu, Ryan L. Clark, Brian F. Pfleger, Thatcher W. Root, and Christos T. Maravelias. 2016. "A Roadmap for the Synthesis of Separation Networks for the Recovery of Bio-Based Chemicals: Matching Biological and Process Feasibility." *Biotechnology Advances* 34 (8): 1362–83. https://doi.org/10.1016/j.biotechadv.2016.10.003.

Yenkie, Kirti M., Wenzhao Wu, and Christos T. Maravelias. 2017. "Synthesis and Analysis of Separation Networks for the Recovery of Intracellular Chemicals Generated from Microbial-Based Conversions." *Biotechnology for Biofuels* 10 (1). https://doi.org/10.1186/s13068-017-0804-2.

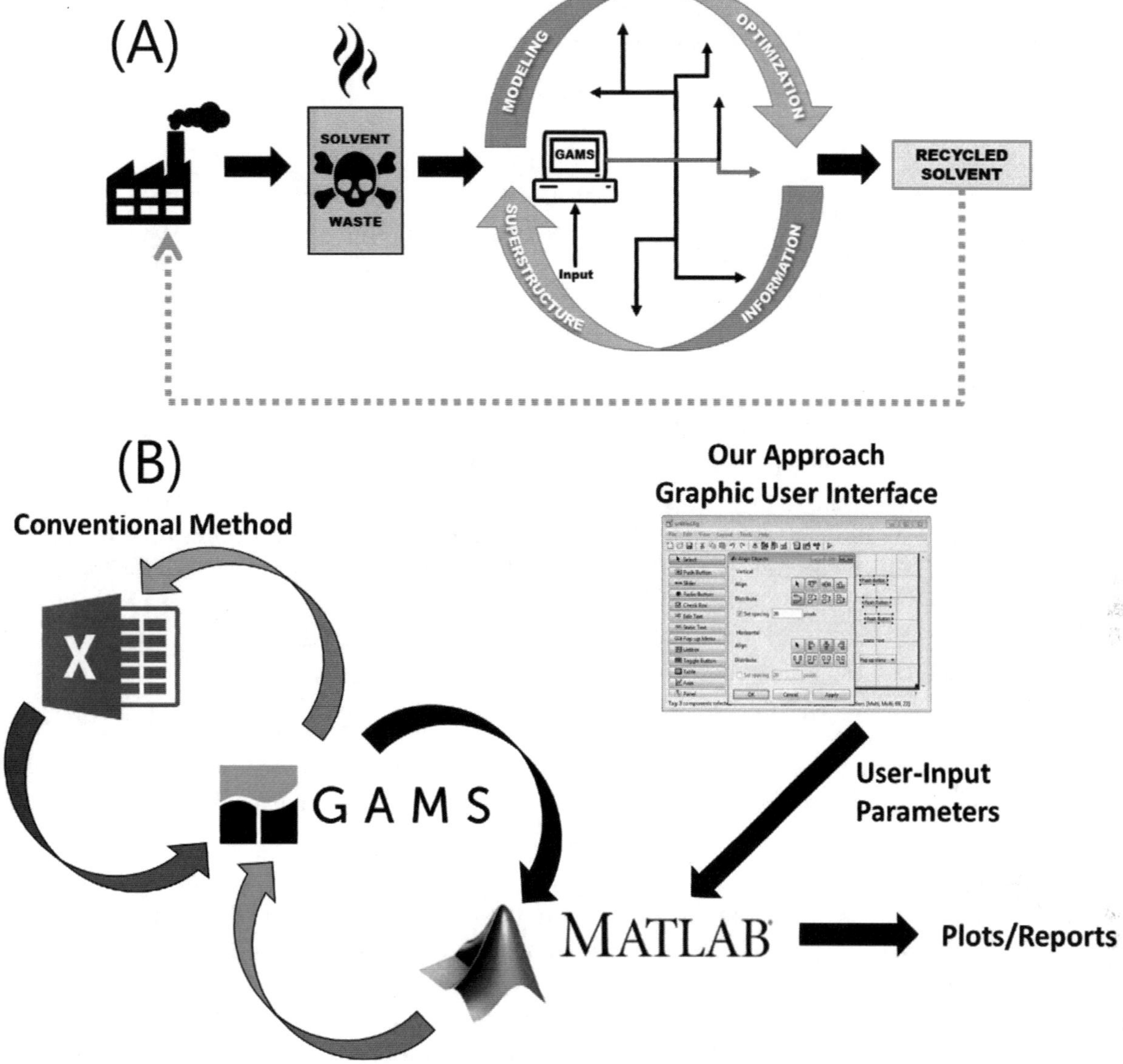

Figure 1(A) Proposed Process for Modeling Solvent Recovery for use in an Industrial setting.
(B) Flow chart of Information Processing of the Proposed Computational Software Tool

625002 High Flux CO2 Selective Membranes for Renewable Natural Gas and CO2 Capture

Abstract:

Reducing carbon emissions is key to preventing further climate change and in correcting the damage that has already been done. Compact Membrane Systems is developing a carbon dioxide selective membrane designed to operate at low pressure with minimal pretreatment. Results from two applications will be discussed: removal of carbon dioxide from flue gas and upgrading of biogas to renewable natural gas.

According to the World Biogas Association, anerobic digestors for biogas production can reduce greenhouse gas emissions by 10% in 2030. Separating carbon dioxide from methane is a key challenge for the renewable gas industry. Biogas is produced largely on farms or small waste centers, not at large centralized plants. While there are existing technologies to separate methane from carbon dioxide, they require multiple stages and high pressures, and are sensitive to process poisons. An efficient and cost-effective solution is needed to bring down barriers that previously prevented small scale operations from transforming their biogas into usable fuel. CMS has developed a membrane aimed at allowing users to capture, separate and upgrade existing biogas streams that would otherwise be flared or released into the atmosphere.

Results from the first field demonstration of Optiperm™ *biogas* will be shared. Working with Energy Projects Manager Matt Steiman at the Dickinson College Farm, a small spiral wound module system has been installed. The system takes raw biogas feed directly from the digester with no pretreatment. The pilot is intended to demonstrate successful operation under field conditions for an extended period of time. Optiperm™ *biogas* technology will be tested for a period of at least 2 months to demonstrate stability in the field and validate performance measured in the CMS laboratory. Optiperm™ *biogas* upgrading streams from 60% to 90% methane in a single stage.

Finally, initial results for utilizing the CMS membranes to capture carbon dioxide from flue gas, including the effect of operating conditions and initial system modeling, will be shared. This application is earlier in development, but results thus far have shown a very high flux membrane that could result in substantially lower capture costs than the current state of the art.

AICHE Fall 2021

Optimizing Energy Efficiency of Ammonia Production via Electrochemical Reaction and Haber-Bosch process.

Gbemisola Ojo and Kyle Camarda
[1]The University of Kansas, Department of Chemical & Petroleum Engineering
4132 Learned Hall, 1530 W 15th Street, Lawrence, KS 66045

Ammonia (NH_3) synthesis is crucial to the production of fertilizers, and accounts for nearly 2% of total global energy consumption. Therefore, finding methods to produce ammonia employing only renewable energy and sustainable feedstocks is paramount. This project uses dynamic simulation to design a 100% renewable ammonia synthesis plant, using only air and water as raw materials. Energy is supplied from a solar source, which is supplemented with industrial batteries. The intermittent nature of solar energy production requires the development of a dynamic production profile for the plant.

In this work, an electrolytic cell powered by a time-varying solar source is employed for hydrogen production. Air is separated via a membrane to provide a nitrogen source, and then the Haber-Bosch reaction is used to produce the product. The degrees of freedom in this problem are the number of batteries purchased, as well as the dynamic production and charging schedules. To determine the optimal production profile, a mixed integer non-linear programing model is formulated and solved. Results demonstrate how varying levels of production and energy storage usage correspond to different levels of energy availability, and show the feasibility of completely sustainable ammonia production, using no raw materials other than water and nitrogen, and creating zero emissions.

632595 Development of Expanded PTFE Structured Adsorbents with Parallel Channels

Abstract:

The overall objective of this work is to develop novel structured adsorbents from expanded PTFE (e-PTFE) comprising activated carbon, carbon molecular sieve and zeolite adsorbents within the expanded e-PTFE framework. Structured columns are being fabricated with various activated carbons and being compared to identical columns fabricated with the corresponding commercial granular or pelleted counterparts for testing under identical conditions. Details about the fabrication technique will be discussed along with representative results from a variety of different experimental tests being conducted. These experimental tests might include single gas adsorption breakthrough curve experiments, volumetric equilibrium adsorption isotherm experiments for column adsorbent mass determination with N_2, 1-bed pressure swing adsorption (PSA) cycling experiments, single gas 10 Hz or 100 Hz volumetric frequency response experiments, and column axial pressure drop experiments. The most recent and compelling result will be presented.

630211 Analysis of Direct Air Capture Process Conditions on Adsorptive Performance of 3D-Printed Aminosilica Monoliths

Abstract:

Abstract

Direct air capture (DAC) is a promising negative carbon technology to extract CO_2 directly from atmosphere. While various sorbent materials have been developed for DAC, they are yet to be tested under realistic process conditions. In our earlier work, we have developed structured sorbents via additive manufacturing and demonstrated their efficacy in selective removal of CO_2 from air. In this present work, we performed a thorough analysis of process performance of 3D-pronted structures comprising of amine-grafted silica monoliths. In particular, 3D printed mesoporous silica monoliths with cell densities of 200, 400, and 600 cpsi graphed with 30 wt% 3-aminopropylsilyl (APS) and tested under temperature swing adsorption (TSA) with various flowrate, desorption temperatures, and feed conditions.[1] The effect of these TSA conditions on several performance metrics including CO_2 purity and recovery, throughput and mass and heat transfer characteristics were investigated systematically to develop a baseline for 3D-printed adsorbent contactors that will be used in DAC processes. Lastly, the energy cost of the TSA system was calculated to provide a cost analysis of the TSA system and the potential scale-up cost of the primary amines DAC adsorbent.

630580 Development of a MOF-Textile Composite for Chemical Defense

Abstract:

Ionic liquid based fiber-welding has been used to attach the metal–organic framework (MOF) UiO-66-NH2 to cotton fibers. The results show that by controlling the extent of the welding process, it is possible to produce fibers that contain a high surface area (approximately 50–100 m2/g), an X-ray diffraction pattern consistent with UiO-66-NH2, and fibers that are chemically reactive to dimethyl 4-nitrophenyl phosphate (DMNP), a common chemical weapon simulant. The ionic liquid/MOF welding solution can be applied by directly placing the fabric in the welding solution or by utilizing an airbrushing technique. Both welding techniques are shown to be scalable with results collected on approximately 1 × 1, 5 × 5, and 15.5 × 15.5 in. swatches. The results are also applicable to weaving methods where the MOF is welded to individual threads and subsequently woven into a textile. The results provide an industrially scalable method of attaching a wide variety of MOFs, and other adsorbent materials, to cotton textiles.

631569 Demonstration of Engineered Structured Sorbents in Various Adsorption Applications

Abstract:

Precision Combustion, Inc. (PCI) has designed, developed, and demonstrated an adsorption technology using beds consisting of structured metal meshes coated with sorbent materials. The engineered structured sorbents, consisting of metal mesh elements which are trademarked and patented as Microlith® by PCI, can be tailored to effectively adsorb a number of species of interest and have been demonstrated to provide beneficial performance for aerospace, building, and fuel cell applications, among others. The resulting sorbents have the potential for direct resistive heating by applying electric current directly to the Microlith® substrate, which offers the potential for rapid, periodic sorbent regeneration as opposed to the longer thermal cycles typical of packed bed adsorbers. As a result, a regenerable adsorber using adsorbent coated on Microlith® structures can reduce the system weight and volume compared to conventional packed bed configurations. Historically, sorbent systems have used some form of packed bed consisting of uniformly shaped pellets or beads, as well as irregularly-shaped granular particles of sorbent media. The downfall of this type of sorbent implementation is the propensity of the material to fluidize, erode, and generate fines. Sorbent media fines have proven problematic in some aerospace applications.

For aerospace (e.g., NASA) and building applications, these structured sorbents were implemented to remove CO_2 and trace contaminants from breathing air. For NASA, these systems must efficiently and reliably control CO_2 and trace contaminant concentrations to levels necessary for long-term crewed space exploration missions beyond low earth orbit. For building applications (e.g., Indoor Air Quality - IAQ or sick-building syndromes), these systems must be capable of providing a long operational window for removing the targeted components. Sorbents such as Zeolites and molecular sieves are excellent candidates for these applications because of their high affinity for capturing molecules of interest along with their chemically inert nature and non-flammable properties. Such properties make them attractive from both performance and safety perspectives. Finally, for fuel cell applications, this Microlith®-based adsorption technology has been implemented for efficient hydrogen sulfide (H_2S) removal from the fuel processor exhaust stream in a fuel reforming-fuel cell system. Our results indicated a significant improvement in performance compared to typical commercial ZnO pellets.

In this paper, the design, development, and optimization of engineered structured sorbent beds based on PCI's Microlith® adsorption technology will be presented. Test results that demonstrate its benefits of increased structural stability gained by eliminating clay bound zeolite pellets that tend to fluidize and erode, and better thermal control during sorption leading to increased process efficiency will be highlighted. We will also discuss the results obtained from various sorption tests, such as removing CO_2 and trace organic contaminants for NASA cabin air cleaning applications and for IAQ applications. Additionally, the development of this technology for sulfur (e.g., H2S) removal in fuel cell applications will be presented.

632661 Tvsa Cycle for Metabolic CO_2 Removal from Spacecraft Cabins Using a Structured Adsorbent: Bench Scale Parametric Study

Abstract:

The objective of this work project is to carry out a parametric study of a temperature vacuum swing adsorption (TVSA) process that utilizes an IntraMicron (IM) structured to determine if more energetically favorable conditions could be used to remove metabolic CO_2 from spacecraft cabin air. The IM adsorbent consists of a copper microfibrous media with entrapped ground and sieved 13X adsorbent particles. This study is being carried out experimentally using the a single-bed apparatus that mimics all the steps of the two-bed TVSA cycle.

This extensive bench scale experimental study is being carried out around the base case separation that meets or exceeds the required CO_2 removal rate for four astronauts. This removal rate is 1 kg of CO_2 per day per astronaut. This corresponds to processing a feed steam containing 2,700 ppm of CO_2 and removing 68.5 % of this CO_2. The effects of the feed flow rate, air save time, regeneration temperature and low vacuum pressure on the process performance are being investigated. The latest results of this project will be presented.

630183 High-Efficient Crystal Particle Manufacture Via Microscale Process Intensification Technology

Abstract:

High-end crystal manufacture has drawn a permanent concern on the high-efficient manufacture of crystal particles, especially in fine chemical, pharmaceutical, electronics, biological and relative engineering fields. In recent years, a series of micro-scale process intensification (MPI) technologies have been widely used in crystal particles preparation via addressing the control of nucleation and growth process. Herein, we review the research progresses of microscale process intensification technology from three aspects, microfluidics devices, microscale force field technology and membrane-based microchannels and interface transfer process. Firstly, the principle of microfluidic and relative microscale device on improving micro-mixing and mass transfer are briefly described. The advantage of microfluidic in continuous nano particle preparation is outlined. Microscale external force field (ultrasonic, high-gravity, electric and magnetic fields) is then introduced as another novel approach for ultrafine nanoparticles and continuous drug crystallization process. Further, in view of the micro-scale intensified mass transfer and micro-scale interfacial force field established on membrane technology, the basic mechanism of membrane crystallization (microscale 2D supersaturation degree control, auto seed detachment, microporous membrane dispersion, etc.) are reviewed. The process coupling and design strategy aiming for enhancing the manufacture capacity is also illustrated. Finally, the developing tendency and key challenges of high-efficient crystal particle preparation technology via microscale processes are overviewed.

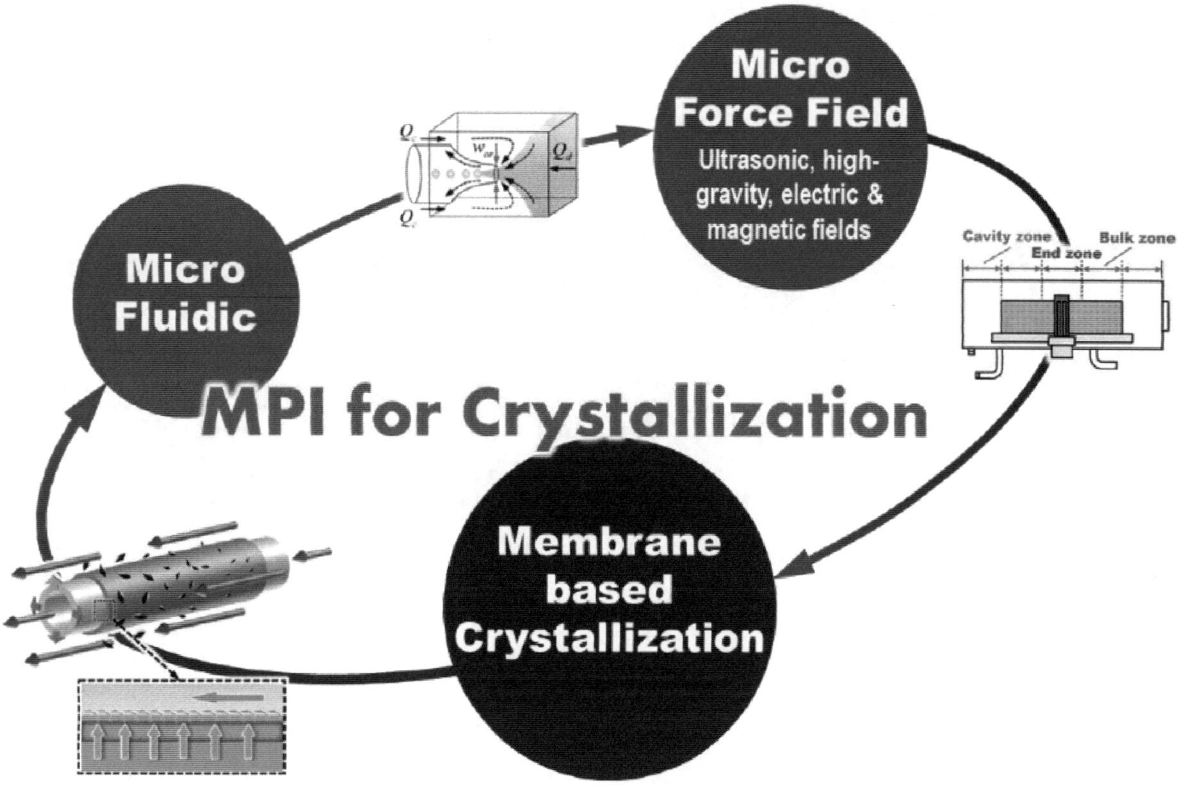

623376 Breakage Facilitated Mixed-Suspension, Mixed-Product-Removal (MSMPR) Crystallization

Abstract:

Continuous crystallization provides advantages of producing solid products with desired quality with smaller footprint. However, the intrinsic nucleation and growth rates of small molecule pharmaceuticals constrain the design spaces of the ranges of robust operation conditions and product properties. In this work, a wet-milling device, used to break large particles into small fragments, was incorporated to MSMSR single stage crystallization. The MSMPR crystallization was controlled by a feedback loop in order to maintain a constant total slurry volume. The approach was applied to paracetamol to demonstrate the effectiveness. Experimental data showed that the introduction of the controllable breakage mechanism offered better particle size distribution control and minimized fouling that resulted in longer operation time than conventional MSMPR.

Ref:

Li, H.; Yang, B.-S. Model Evaluation of Particle Breakage Facilitated Process Intensification for Mixed-Suspension-Mixed-Product-Removal (MSMPR) Crystallization. *Chem. Eng. Sci.* 2019, *207*, 1175– 1186, DOI: 10.1016/j.ces.2019.07.030

629001 Electrochemical Sensor-Integrated, Continuous Flow, Microfluidic Device for Real-Time Measurement of Supersaturation during Salt Screening

Abstract:

Electrochemical sensors can offer quick and robust detection of organic molecules inside microfluidic devices where the amount of sample is small. These sensors can be assembled with a feedback control unit for automatic adjustments of flowrates to obtain higher experimental accuracy. The sensor-integrated microfluidic system can facilitate high-throughput screening of crystalline materials in shorter time by precisely controlling the concentration of organic molecules in each microwell. In this study, a continuous microfluidic mixer is coupled with an electrochemical sensor to enable in-situ measurement of solute concentration in the mixing zone. The sensor consists of three electrodes: working, contour, and reference connected externally to a potentiostat. The electrodes are inserted into the microfluidic mixer horizontally. Here we have used Pt and Ni wires as the counter and working electrodes, respectively. Cyclic Voltammetry (CV) is performed during the continuous operation of the microfluidic device which correlates to the concentration of Naproxen. This sensor integrated microfluidic device is experimentally evaluated for the salt screening of Naproxen. Screening of solid salt forms of Naproxen includes four solvents and four salt formers (SFs) which results in sixteen experimental conditions. All experiments are performed under an optical microscope for in-line measurements of crystals' morphology and growth rates. After the concentration measurement and salt formations, crystallized salts are harvested for PXRD and particle size distribution analysis.

623691 Periodic Forcing Via Temperature Cycles: Model-Based Study of Novel Continuous Configurations for Deracemization

Abstract:

Introduction

Crystallization is an attractive approach to achieve targeted production and isolation of the desired enantiomer in a productive and economically effective way due to its low capital and operating costs. Solid-state deracemization techniques, such as solid-state deracemization via temperature cycles can be applied to convert a racemic mixture of crystals to pure enantiomers by the periodic cycling of the temperature and by the utilization of a liquid-phase catalyst [1]. This process is conventionally operated in batch mode and it has been widely investigated in both experimental and simulation works [1,2,3]. However, its continuous counterpart that can be performed in mixed suspension mixed product removal crystallizers (MSMPRC) has not yet been proposed. This novel process can be more attractive than Preferential crystallization, which can reach only 50% yield.

In this work, we will focus on the model-based design and evaluation of four continuous configurations to perform Periodic Forcing via Temperature cycles (PFTC) and we comparatively assess their performance based on the enantiomeric purity and productivity, i.e., the key performance indicators (KPIs).

Method

The mathematical model of the continuous process configurations was obtained by extending our population balance based model of batch temperature cycles [3]. Deracemization via temperature cycles is a complex interplay of several phenomena and mechanisms, thus the description of size-dependent solubility, crystal growth/dissolution, agglomeration, breakage and racemization are incorporated in the model.

Process configurations

Different configurations could be utilized to perform solid-state deracemization via temperature cycles in an MSMPRC. In all these cases, the MSMPRC initially contains a saturated solution, which is in contact with the seed crystals that are enriched in the target enantiomer and temperature cycles are applied periodically during the process. The configurations differ from each other in the type of the feed: in the first case, we feed racemic crystals and we aim at converting them into the desired enantiomer (configuration (A)). In this case, suspension is withdrawn and the solution is recycled after filtration to keep the volume of the crystallizer constant. For this configuration to be more productive and to reach larger enantiomeric excess, one can attempt to create larger asymmetry between the two enantiomers by partially recycling the obtained solid product (configuration (B)). Apart from feeding crystals, one can feed a suspension to the MSMPRC in an alternative configuration (configuration (C)). One could expect that by feeding not only racemic solid, but solution as well, the concentration of the enantiomers becomes more similar to each other as well, thus the driving force for racemization decreases and the deracemization process is hindered. It is worth noting that the only experimental example of continuous temperature cycling process, which employs spatial thermal oscillations in a tubular flow milli-reactor [4], also utilizes a suspension type of feed.

In continuous crystallization, a mechanism is needed to form new crystals, otherwise complete washing out of crystals from the crystallizer occurs. In the three configurations presented above, the feed contained racemic crystals, thus other crystal-generating mechanisms, e.g., attrition, breakage or nucleation, was not essential. However, in the fourth configuration we propose, only solution is fed (configuration (D)). Therefore, for the sake of simplicity, we have introduced breakage due to wet-milling to the system. It is easy to foresee that with this configuration one can attain larger productivity and enantiomeric purity: as a result of introducing the undesired enantiomer into the system in the solution, one step (the dissolution of the undesired enantiomer) can be spared with respect to the solid-fed configuration.

Results and discussion

With the aid of the mathematical model, we have simulated continuous temperature cycles performed in the proposed continuous configurations [5]. We have analyzed first the solid-fed configuration (configuration (A)) in detail: the enantiomeric excess increases during the startup, and then reaches a periodic steady state, which is bounded by the lower and upper limits corresponding to the enantiomeric excess at low and high temperature, respectively. The average periodic steady state enantiomeric excess was chosen as a key performance indicator, together with the average periodic steady state productivity, to compare the processes. A simulation study on the effect of operating conditions, such as the residence time, feed rate and cooling rate was carried out. We have identified that by increasing the residence time or the feed rate, one can significantly improve the process performance.

Second, we have verified that by partial recycling of the crystalline product, slightly higher average steady state enantiomeric excesss and productivity can be attained (configuration (B)). Third, we have investigated the configuration with the suspension feed (configuration (C)). The average steady state enantiomeric excess was found to be lower than that of the solid-fed process, as expected. We have explored the effect of varying the operating parameters to increase the process performance. Then, we have used the model to simulate the solution-fed operation (configuration (D)). Finally, we have studied an MSMPRC cascade for the production of enantiopure crystals.

Conclusions

Various novel MSMPRC-based approaches have been proposed by means of model-based studies to perform solid-state deracemization via temperature cycles in continuous operation. The advantages and pitfalls of each technique have been assessed, together with the potential performance improvements achieved by carefully selecting the operating conditions. This contribution serves as a guide to transform the conventional batch process of temperature cycles to continuous operation.

References

- [1] Suwannasang, A. E. Flood, C. Rougeot, G. Coquerel, Org. Process Res. Dev. 21 (2017) 623-630.
- [2] Breveglieri, G. M. Maggioni, M. Mazzotti, Cryst. Growth Des. 18 (2018) 1873-1881.
- [3] Bodák, G.M. Maggioni, M. Mazzotti, Cryst. Growth Des. 18 (2018) 7122-7131.
- [4] Cameli, C. Xiouras, G. D. Stefanidis, CrystEngComm, 22 (2020) 3519-3525.
- [5] Bodák and M. Mazzotti, in preparation.

Acknowledgements

This research has received funding as part of the CORE project (October 2016 – September 2020) from the European Union's Horizon 2020 research and innovation programme under the Marie Sklodowska-Curie grant agreement No 722456 CORE ITN

This project has received funding from the European Research Council (ERC) under the European Union's Horizon 2020 research and innovation programme under grant agreement No 2-73959-18.

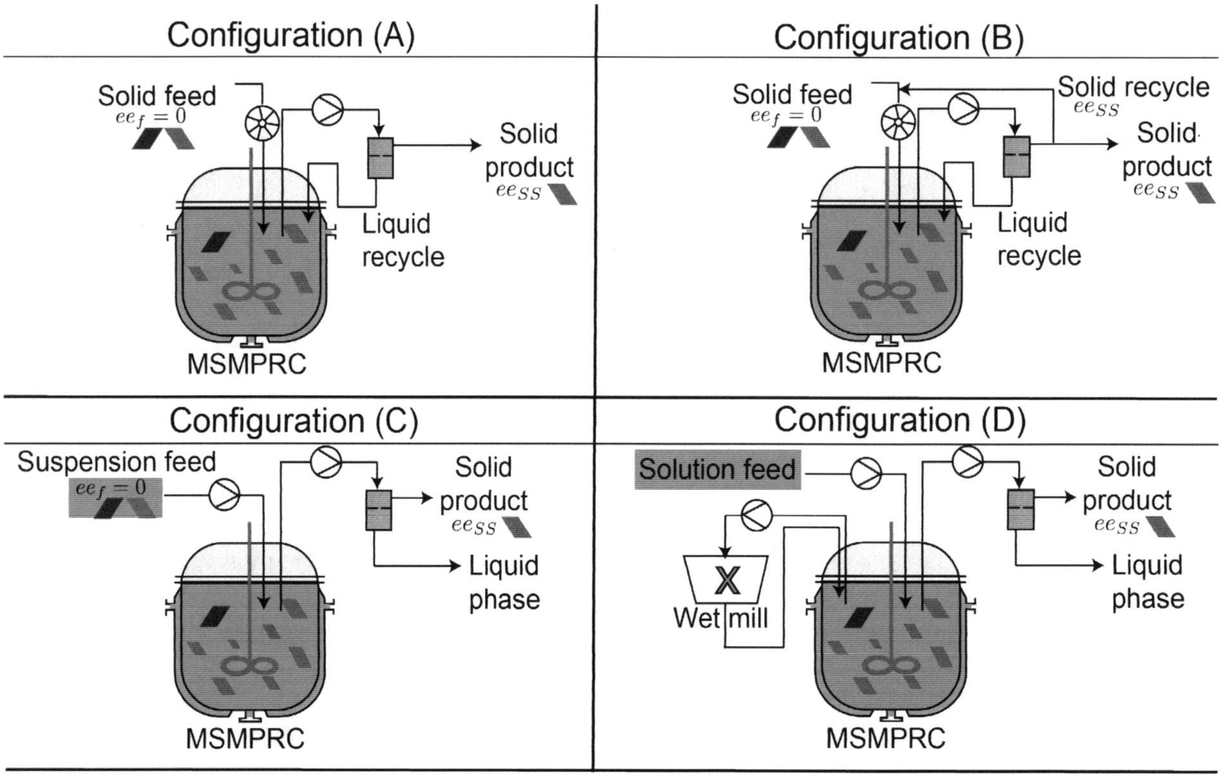

Figure 1. Novel crystallizer configurations are proposed to perform periodic forcing via temperature cycles (PFTC).

621386 Continuous Flow Synthesis of the Metal-Organic Framework HKUST-1 in a Millifluidic Reactor

Abstract:

Metal-organic frameworks (MOFs) are a class of crystalline and porous adsorbents, with wide-ranging applications in gas separations, membrane materials as well as sensors. The basic unit of a MOF consists of metal ions coordinated to organic ligands, which form a coordination polymer with well-defined lattice positions in the crystalline unit cell. The properties of a particular MOF can be finely tuned for a given application through the use of multiple ligands, metal ions and modulators for tuning particle sizes. Conventional lab-scale synthesis procedures for MOFs are autoclave batch reactions with residence times of a day or more. These procedures entail solvothermal synthesis, with expensive and often toxic solvents and starting materials. This explains the reason for a few of them being sold on commercial chemical vendor outlets at exorbitant rates. Thus, given their multifarious applications, economically viable synthesis of MOFs at the industrial scale is the need of the hour.

In this study, we have developed a continuous, droplet-based synthesis procedure of the MOF HKUST-1 in a millifluidic reactor. The droplets consist of aqueous solutions of copper ions, trimesic acid, and various solvent mixtures. Millifluidic reactor comprised of a custom made heating unit including machined metal block and 3D printed parts. Various reaction chemistries for synthesis of HKUST-1 via continuous and batch operation were investigated. We present an ultra-low residence time continuous synthesis of HKUST-1 and compare the physical characteristics and yield with batch synthesis of the same.

631490 Controlling the Effect of Slug-to-Slug Variation on the Crystal Size Distribution of Perovskite QDs: A CFD-Based Approach

Abstract:

Recently, manufacturing of metal halide perovskite quantum dots (QDs), particularly cesium lead bromide ($CsPbBr_3$), has received substantial attention due to its superior tunable optoelectronic properties (related to their size), and because they cater to wide array of applications such as solar cells, next generation displays, and LEDs. [1] Thus, it is of paramount importance to develop a manufacturing route which results in controlled size and size distribution with a desired surface passivation of $CsPbBr_3$ QDs for the targeted application in photonic and optoelectronic devices. Furthermore, device-level applications demand a fast and large scale production of application-ready QDs. However, the large-scale production of these QDs has posed certain engineering and practical challenges. Specifically, an absence of a well-established crystallization model, lack of batch-to-continuous scale-up studies, and unavailability of set-point tracking platforms are the major roadblocks. [2] The fact that slug flow crystallizers (SFCs) offer a large operation range (*i.e.*, precursor concentration, gas-liquid flow rates) has inspired recent studies on proposing SFCs for large scale continuous manufacturing of metal halide perovskite QDs. [3]

Despite the intriguing advantages of SFCs, there are still some key issues with the operation of SFCs. For example, previous studies have not accounted for the effect of slug-to-slug (S2S) variation on crystal size distribution (CSD), and the absence of a modeling and control framework for $CsPbBr_3$ QDs made it challenging to fine-tune the QD size distribution. [4] In response, we have developed a computational fluid dynamics (CFD) model to elucidate the mechanism of slug formation process in a millifluidic SFC, and then combined it with a slug crystallizer model to accurately model QD size distributions and precursor concentration profiles in SFC. Specifically, the slug crystallizer model was constructed by combining a continuum model with a previously developed kinetic Monte Carlo (kMC) model. [5] Based on the CFD-based multiscale modeling framework (*i.e.*, CFD + continuum + kMC model), an optimal operation problem was formulated to ensure adequate set-point (QD size) tracking, resulting in a narrow CSD. Overall, (a) the proposed multiscale model is in good agreement with the experimental results; (b) the CFD simulation highlights the region in which the SFC should be operated (*i.e.*, stable slug flow regime); and (c) the optimizer is capable of effective simultaneous set-point tracking (QD size) and disturbance rejection (S2S variation) while ensuring a narrow CSD.

In summary, the CFD model was used to identify a stable slug regime and quantify the S2S variation in a millifluidic SFC. Additionally, a dynamic multiscale model for a SFC was developed to describe the temporal evolution of average QD size and CSD. Lastly, an optimization-driven controller scheme was implemented to ensure that the QDs match the industry specifications while maintaining a narrow CSD. Overall, the proposed work is a major leap towards continuous nano-manufacturing of metal halide perovskite QDs, which will be crucial in meeting market demands of the optoelectronics industry in the coming decade.

Literature Cited:

1. Protesescu L, Yakunin S, Bodnarchuk MI, Krieg F, Caputo R, Hendon CH, Yang RX, Walsh A, Kovalenko MV. Nanocrystals of cesium lead halide perovskites ($CsPbX_3$, X= Cl, Br, and I):

novel optoelectronic materials showing bright emission with a wide color gamut. Nano Letters. 2015 Jun 10; 15(6):3692-6.

2. Pu Y, Cai F, Wang D, Wang JX, Chen JF. Colloidal synthesis of semiconductor quantum dots toward large-scale production: a review. Industrial & Engineering Chemistry Research. 2018 Feb 14; 57(6):1790-802.
3. Abdel-Latif K, Epps RW, Kerr CB, Papa CM, Castellano FN, Abolhasani M. Facile room-temperature anion exchange reactions of inorganic perovskite quantum dots enabled by a modular microfluidic platform. Advanced Functional Materials. 2019 Jun; 29(23):1900712.
4. Rasche ML, Jiang M, Braatz RD. Mathematical modeling and optimal design of multi-stage slug-flow crystallization. Computers & Chemical Engineering. 2016 Dec 5; 95:240-8.
5. Sitapure N, Epps R, Abolhasani M, Kwon JS. Multiscale modeling and optimal operation of millifluidic synthesis of perovskite quantum dots: towards size-controlled continuous manufacturing. Chemical Engineering Journal. 2020 Dec 7:127905.

630903 Integrated Design and Model Predictive Control of Multiscale Systems Using a Multi-Fidelity Bayesian Optimization Approach

Abstract:

Design and control are interconnected activities that need to be performed simultaneously so that designs with a large degree of operational flexibility can be readily identified [1]. In practice, however, design is typically tackled first (with little-to-no considerations of the dynamics), which can lead to heavily constrained processes with few degrees of freedom available for control. While there is a vast body of literature available on the topic [2]–[4], a tractable and unified framework for optimal integrated design and control (IDC) has remained elusive. This type of framework is needed to enable the systematic design of next-generation (so-called "flexible") manufacturing and energy systems including smart grid technology [5], multiproduct or personalized chemical plants [6], and combined heat and power systems [7]. Although optimal IDC problems can be formulated as multistage stochastic programs (MSPs), they quickly become intractable when the following features are present [8]: (i) relevant dynamics and uncertainties occur on much shorter timescales than the lifetime of the system; (ii) uncertainties are present that are best described by continuous random variables with large variance; and/or (iii) important operational decisions are discrete (e.g., adaptive scheduling and unit commitment). The first two features relate to the multiscale nature of realistic IDC problems, which pose a major challenge for MSPs that suffer from the curse-of-dimensionality (i.e., exponential growth in number of decision variables as the control decisions can differ for each realization of the uncertainty) [9]. The third feature implies the system model is nonlinear with mixed-integer decisions such that a scenario approximation to the MSP is a very large-scale mixed-integer nonlinear program (MINLP) that is far beyond the capabilities of existing solvers, even when an equation-oriented system model is available.

Most previous work on optimal IDC focused on making tractable approximations to the MSP by relaxing features (i)-(iii) (see, e.g., [10]–[13] for several examples including substantially reducing the number of operational stages and uncertainty scenarios). Although these simplifications make the MSP more tractable, they degrade the accuracy of the IDC solution in exactly the ways that need to be captured in order to model the flexibility of the process. To address these challenges, in [14], we proposed a simulation-based optimization (SO) strategy that can tackle a general class of IDC problems consisting of an *outer* optimization over the relevant design variables and an *inner* stochastic simulation that evaluates the expected operating costs and constraints that appear in the outer problem. To reduce dimensionality of the outer problem, we rely on so-called decision rule (DR) approximations [8] to parametrize the recourse/control decisions in a way that scales much more favorably with respect to (i) and (ii) above. In particular, we presented a mixed-integer model predictive control (MIMPC) approach that is able to simultaneously handle scheduling and control decisions mentioned in (iii); this is a significantly more advanced DR than available alternatives that mostly consider PID controllers [15]–[17], with some exploring linear MPC [18], [19]. Due to the multiscale nature of the IDC problems of interest, we utilized Bayesian optimization (BO) that has become a popular tool for solving black-box optimization problems with objective and/or constraint functions that are prohibitively expensive to compute repeatedly and, therefore, must be evaluated as few times as possible.

In [14], we focused on conventional BO, which assumes only a single (computationally intensive) high-fidelity simulator is available. In IDC problems, however, we have access to several cheap approximations to this simulator such as the previously mentioned simplification methods in, e.g., [10]–[13]. As long as the lower fidelity approximations are reasonably correlated to the high-fidelity simulator, they can be used to cheaply eliminate "bad" designs and thus reserve the most expensive simulations for the most promising designs. To take advantage of these approximations here, we extend [14] by replacing traditional BO with a multi-fidelity BO (MFBO) algorithm [20]. To demonstrate its advantages, we apply MFBO to the design of a solar-powered building heating ventilation and air-conditioning (HVAC) system, with grid and battery support, under uncertain weather and demand conditions that vary at the hour scale for a year-long horizon. We show that MFBO repeatedly identifies more profitable designs than single fidelity BO under a fixed computational budget. In addition, we highlight that the choice and sequence of the lower-fidelity models has a strong impact on the rate of convergence of MFBO.

References

[1] V. Sakizlis, J. D. Perkins, and E. N. Pistikopoulos, "Recent advances in optimization-based simultaneous process and control design," *Comput. Chem. Eng.*, vol. 28, no. 10, pp. 2069–2086, Sep. 2004, doi: 10.1016/j.compchemeng.2004.03.018.

[2] L. A. Ricardez-Sandoval, H. M. Budman, and P. L. Douglas, "Integration of design and control for chemical processes: A review of the literature and some recent results," *Annual Reviews in Control*, vol. 33, no. 2. Pergamon, pp. 158–171, Dec. 01, 2009, doi: 10.1016/j.arcontrol.2009.06.001.

[3] Z. Yuan, B. Chen, G. Sin, and R. Gani, "State-of-the-art and progress in the optimization-based simultaneous design and control for chemical processes," *AIChE J.*, vol. 58, no. 6, pp. 1640–1659, Jun. 2012, doi: 10.1002/aic.13786.

[4] P. Vega, R. Lamanna de Rocco, S. Revollar, and M. Francisco, "Integrated design and control of chemical processes - Part I: Revision and classification," *Computers and Chemical Engineering*, vol. 71. Elsevier Ltd, pp. 602–617, Dec. 04, 2014, doi: 10.1016/j.compchemeng.2014.05.010.

[5] A. W. Dowling, R. Kumar, and V. M. Zavala, "A multi-scale optimization framework for electricity market participation," *Appl. Energy*, vol. 190, pp. 147–164, Mar. 2017, doi: 10.1016/j.apenergy.2016.12.081.

[6] M. C. Tang *et al.*, "Systematic approach for conceptual design of an integrated biorefinery with uncertainties," *Clean Technol. Environ. Policy*, vol. 15, no. 5, pp. 783–799, Oct. 2013, doi: 10.1007/s10098-013-0582-x.

[7] R. Madlener and C. Schmid, "Combined Heat and Power Generation in Liberalised Markets and a Carbon-Constrained World," *GAIA - Ecol. Perspect. Sci. Soc.*, vol. 12, no. 2, pp. 114–120, Apr. 2017, doi: 10.14512/gaia.12.2.8.

[8] A. Hakizimana, "Novel Optimization Approaches for Integrated Design and Operation of Smart Manufacturing and Energy Systems," Clemson University, 2019.

[9] M. V. F. Pereira and L. M. V. G. Pinto, "Multi-stage stochastic optimization applied to energy planning," *Math. Program.*, vol. 52, no. 1–3, pp. 359–375, May 1991, doi: 10.1007/BF01582895.

[10] I. E. Grossmann, B. A. Calfa, and P. Garcia-Herreros, "Evolution of concepts and models for quantifying resiliency and flexibility of chemical processes," *Comput. Chem. Eng.*, vol. 70, pp. 22–34,

Nov. 2014, doi: 10.1016/j.compchemeng.2013.12.013.

[11] P. Liu, Y. Fu, and A. Kargarian Marvasti, "Multi-stage stochastic optimal operation of energy-efficient building with combined heat and power system," *Electr. Power Components Syst.*, vol. 42, no. 3–4, pp. 327–338, Mar. 2014, doi: 10.1080/15325008.2013.862324.

[12] D. K. Varvarezos, L. T. Biegler, and I. E. Grossmann, "Multiperiod design optimization with SQP decomposition," *Comput. Chem. Eng.*, vol. 18, no. 7, pp. 579–595, Jul. 1994, doi: 10.1016/0098-1354(94)85002-X.

[13] T. Zhang, R. Baldick, and T. Deetjen, "Optimized generation capacity expansion using a further improved screening curve method," *Electr. Power Syst. Res.*, vol. 124, pp. 47–54, Jul. 2015, doi: 10.1016/j.epsr.2015.02.017.

[14] N. A. Choksi and J. A. Paulson, "Simulation-based Integrated Design and Control with Embedded Mixed-Integer MPC using Constrained Bayesian Optimization," in *Proceedings of the American Control Conference*, 2021.

[15] A. Flores-Tlacuahuac and L. T. Biegler, "Simultaneous mixed-integer dynamic optimization for integrated design and control," *Comput. Chem. Eng.*, vol. 31, no. 5–6, pp. 588–600, May 2007, doi: 10.1016/j.compchemeng.2006.08.010.

[16] V. Bansal, J. D. Perkins, E. N. Pistikopoulos, R. Ross, and J. M. G. van Schijndel, "Simultaneous design and control optimisation under uncertainty," *Comput. Chem. Eng.*, vol. 24, no. 2, pp. 261–266, 2000.

[17] M. Rafiei-Shishavan, S. Mehta, and L. A. Ricardez-Sandoval, "Simultaneous design and control under uncertainty: A back-off approach using power series expansions," *Comput. Chem. Eng.*, vol. 99, pp. 66–81, 2017.

[18] P. Vega, R. Lamanna, S. Revollar, and M. Francisco, "Integrated design and control of chemical processes--Part II: An illustrative example," *Comput. Chem. Eng.*, vol. 71, pp. 618–635, 2014.

[19] N. A. Diangelakis, B. Burnak, J. Katz, and E. N. Pistikopoulos, "Process design and control optimization: A simultaneous approach by multi-parametric programming," *AIChE J.*, vol. 63, no. 11, pp. 4827–4846, 2017.

[20] K. Kandasamy, G. Dasarathy, J. Oliva, J. Schneider, and B. Póczos, "Gaussian Process Bandit Optimisation with Multi-fidelity Evaluations," in *NIPS'16: Proceedings of the 30th International Conference on Neural Information Processing Systems*, 2016, pp. 1000–1008, doi: 10.5555/3157096.3157208.

627233 Adjustable Robust Optimization for the Planning Operations of Integrated Refinery-Petrochemical Site Under Demand Uncertainty

Abstract:

Facing the challenges of tight competition and environmental restriction from global market, petrochemical industry has been forced to seek all potential possible opportunities to improve the overall profit. Therefore, instead of considering planning optimization in refinery and petrochemical plant individually, a comprehensive standpoint of global enterprise optimization has received more attention than ever before [Li et al., 2016; Zhao et al., 2017; Uribe-Rodriguez et al., 2020]. The integrated refinery-petrochemical planning optimization provides a new route to achieve a higher profit margin by producing more valuable products, saving operating cost and raw material purchasing. Furthermore, ignoring the uncertainty in practical manufacturing, such as demand and sale price of final products, can reduce the total profit significantly. To improve the profitability of enterprise-wide refinery-petrochemical planning operations and avoid excess conservativeness derived from the worst-case based classical robust optimization, the framework of adjustable robust optimization under dynamic uncertainty set is proposed for a multi-period problem of integrated refinery-petrochemical site under uncertain demands.

Focusing on the operation planning of a large industrial refinery-petrochemical site which is consist of 24 and 15 processing units in each part, respectively, the nonconvex mixed-integer nonlinear programming (MINLP) is formulated. Complex process operation, such as the scheduling of ethylene cracking process and polymerization process are taken into account simultaneously. Binary variables are defined to denote the type of processed crude oil, purchase decision of raw crudes and selection of unit operation modes. Uncertainty lies in demand of final products and corresponding constraints are reformulated using robust counterpart, affinely adjustable robust counterpart, affinely adjustable robust counterpart with dynamic uncertainty set, respectively. The fluctuation and connection in demand of different periods are drawn from historic statistics. The problems are considered in three scenarios with time length varying in 5, 10, 15 periods respectively and solved by commercial solver BARON. Obtaining from the different models, the respective operational strategies are examined in diverse scenarios with random demand of final products. The computational results show that the method of affinely adjustable robust optimization coupled with dynamic set provides a best average performance on these sampling scenarios, following by the affinely adjustable robust optimization alone. The average profit of static robust optimization is the lowest but fluctuation between scenarios is smallest.

References

1. Li, J., Xiao, X., Boukouvala, F., Floudas, C. A., Zhao, B., Du, G., Su, X., & Liu, H., Data-driven mathematical modeling and global optimization framework for entire petrochemical planning operations, *AIChE Journal,* 62, 3020-3040 (2016)
2. Uribe-Rodriguez, A., Castro, P. M., Gonzalo, G.-G., & Chachuat, B., Global optimization of large-scale MIQCQPs via cluster decomposition: Application to short-term planning of an integrated refinery-petrochemical complex, *Computers & Chemical Engineering,* 140 (2020)
3. Zhao, H., Ierapetritou, M. G., Shah, N. K., & Rong, G., Integrated model of refining and petrochemical plant for enterprise-wide optimization, *Computers & Chemical Engineering,* 97, 194-207 (2017)

629298 A Novel Hybrid Algorithm for Scheduling of Multipurpose Batch Process

Abstract:

As the incremental market demands towards low-volume manufacture and various product categories, a growing number of researchers has been devoted to the investigation on batch processes [1,2]. The inherent flexibility of multipurpose batch process on processing route and shared facilities provides it wide application prospect in industry [2]. As one of the key managerial tools, process scheduling can greatly improve profit and save cost while saving makespan reflected on optimized production schedule. Mathematical programming is the most generally used method to address the batch process scheduling, being classified into sequence-based and time-grid-based models [3]. These formulations show good performance on handling scheduling with various processing features, such as variable processing time, storage policies and intermediate deliveries/demands. Continuous-time formulations [4,5] can effectively address multipurpose process scheduling in small-scale problems but often require large computational time or fail to converge to optimality for industrial-scale applications. The discrete-time formulation with tightening methods [6] is more computationally efficient and can handle intractable large-scale scheduling in reasonable time frames. However, in this approach, the optimal number of time points is not known a priori, which is often identified through the iterative procedure. As a result, the computational time is greatly increased.

Metaheuristics such as genetic algorithm (GA), tabu search and particle swarm optimization, have been proven capable to generate good scheduling solutions in a reasonable computational time even for large-scale batch processes. Many studies [7,8] confirmed that the GA shows good exploration of search space for multipurpose process scheduling attributed to its characteristics on inherent parallelism and strong global search ability. Also, GA can significantly reduce the computational time compared with mathematical programming where excessive CPU time is usually required for complicated scheduling problems. However, most existing GA proposed for multipurpose batch processes in literatures are less generalized and easy to prematurely converge to local optimality. Hybrid algorithm [9, 10] combining the advantages of multiple aforementioned methods can eliminate the limitations of single algorithm and show prominence on solving challenging large-scale problems.

In this work, we proposed a novel hybrid algorithm for scheduling of multipurpose batch process, which combines the strengths of GA on fast convergence and sequence-based mixed-integer linear programming (MILP) formulation on ensuring global optimum. The goal is to determine batching, allocation and sequencing of tasks eventually minimizing the Makespan under demand requirement. The hybrid algorithm comprises two stages: firstly, a new-designed GA is used to quickly generate feasible solutions, which do not require the iterative procedure to identify the optimal number of time points and hence significantly reduce the computational time. The obtained batching and sequence decisions are then used to fix some corresponding binary variables in MILP model. Then, the sequence-based formulation is quickly solved to optimality. Different fixing strategies of binary variables will be compared and identify the best one. In GA, a three-part encoding method is designed based on real value encoding way to direct batching and assigning determination and impact the processing sequence of tasks. Constraints concerning unit/storage capacity, material balance and product demand are handled in the decoding procedure to guarantee feasible solutions. We use elitist strategy and roulette wheel method [11] to select good individuals in each population. And a special crossover method was devised for the three-part real-value chromosome. We develop the sequence-based MILP formulation referring the model of Pan [12] and improve its efficiency by reformulating the big-M constraints concerning sequence and using less big-M terms to monitor inventory level of storage.

To evaluate the performance of the proposed method, four benchmark examples from the literature [5, 6] are solved using the proposed hybrid algorithm. Results show that the proposed algorithm can successfully generate global optimal solutions in all instances. It demonstrates general applicability and strong convergent performance. Compared with existing models, the hybrid algorithm can yield good quality solutions within shorter computational time. For example, in P9 and P11 of the Kallrath example, the method of Vooradi and Shaik [5] cannot converge to optimality in 40000s. By contrast, the proposed algorithm solves these two instances to optimality within 200 s, decreasing the CPU time over two orders of magnitude. In the most complicated instance P0, our approach reports a CPU time reduction by 37.5% compared to that of Velez et al. [6]. More importantly, the hybrid algorithm always generates better solutions than the designed GA.

References

1. Rakovitis N, Zhang N, Li J, Zhang LP. A new approach for scheduling of multipurpose batch processes with unlimited intermediate storage policy. *Chem. Sci. Eng.* 2019; 13(4):784-802.

2. Rakovitis N, Li J, Zhang N. An improved approach to scheduling multipurpose batch processes with conditional sequencing. *Aided Chem. Eng.* 2019; 46:1387-1392.

3. Harjunkoski I, Maravelias CT, Bongers P, et al. Scope for industrial applications of production scheduling models and solution methods. *Comput Chem Eng.* 2014; 62:161-193.

4. Vooradi R, Shaik MA, Gupta NM. Three-index model for Westenberger–Kallrath benchmark scheduling problem. *AIP Conf. Proc.* 2010; 1298:380–385.

5. Vooradi R, Shaik MA. Improved three-index unit-specific event-based model for short-term scheduling of batch plants, *Comput Chem Eng.* 2012; 43:148-172.

6. Velez S, Merchan AF, Maravelias CT. On the solution of large-scale mixed integer programming scheduling models. *Eng. Sci.* 2015; 136(2):139-157.

7. He Y, Hui CW. A binary coding genetic algorithm for multi-purpose process scheduling: A case study. Chem. Eng. Sci. 2010; 65(16):4816-4828.

8. Woolway M, Majozi T. A novel metaheuristic framework for the scheduling of multipurpose batch plants. Chem. Eng. Sci. 2018; 192:678-687.

9. Rakovitis N, Li D, Zhang N, et al. Novel Approaches for Energy-Efficient Flexible Job-Shop Scheduling Problems, *Eng. Trans.* 2020; 20: 823-828.

10. Lee H, Maravelias CT. Combining the advantages of discrete- and continuous-time scheduling models: Part 1. Framework and mathematical formulations, *Comput Chem Eng.* 2018; 116:176-190.

11. Beheshti Z, Shamsudding S. A review of population-based meta-heuristic algorithms, *Int J Adv Soft Comput Appl.* 2013; 5:1–35.

12. Pan M, Qian Y, Li XX. A novel precedence-based and heuristic approach for short-term scheduling of multipurpose batch plants. *Eng. Sci.* 2008; 63:4313-4332.

629370 Impact of Split Delivery in Minimizing LNG Procurement Cost

Abstract:

Historically, LNG has been traded via long term contracts between a producer and a customer. These long term contracts or LTCs, spanned over 15-25 years and provided both, the producer and customer with a stable supply of LNG. With the advent and growth of the LNG spot market in the last decade, many countries have been looking into methods to create a cost effective import portfolio considering both long term contracts and spot market. It is common practice in the LNG industry to load a single cargo for delivery to a single location supplied by a single seller. Since 2017[1], split delivery has also been introduced, which offers a new level of operational flexibility to LNG suppliers and importers. But with a split delivery, importers face a trade-off between increased operational costs and reduced cargo costs. Hence it is important for a buyer to create a procurement plan that takes this into account and optimizes overall procurement costs. In this study, we focus on the optimal procurement of LNG at the import terminal and study the benefits by including split delivery as an option.

In literature, LNG procurement for the customer has been studied primarily as a long term strategic decision with energy security as the priority while minimizing procurement cost. Shaikh et al. (2017)[2] modelled a natural gas import scheme for China, where they developed a mathematical model aimed at minimizing the import cost, transport distance, domestic and political instability associated with each supplier. Later, with a surge of spot market share in LNG trade, it was studied as the portfolio optimization problem that is determining the ratio of LTC and spot market in total imports. Kim and Kim (2018) [3] developed an optimal portfolio for Korea. In this study, they developed a two-step portfolio model mean-variance optimization model which determines the ratio of LTC and spot in imports followed by a linear programming model which chooses the optimal set of contract for LTC procurement. The benefits of split delivery in LNG supply chains have been discussed mainly from the producer's end. Mutlu et al. (2015) [4] considers the problem of developing an annual delivery program with least cost for a supplier. They included penalties for deviating from contractual terms, excess production and maintenance of ships and also a spot market revenue by selling excess LNG. Costs were compared with and without split delivery and they established the cost benefits of split delivery for a producer. Sheiktajian et al. (2020) [5] compared the costs of split and non-split delivery for an LNG manufacturer selling LNG via LTC and spot markets by considering deterministic and uncertain scenarios. Shahrukh et al. (2019) [6] were the first to develop an import strategy for an LNG importer under the terms of an ongoing long-term contract and a spot market available to minimize overall procurement costs for the short term (1-12 months) by proposing a mathematical formulation LNGPM. However they did not allow split delivery in their model. No model in existing literature can establish the benefits of split delivery at the customers end in the short term. In this paper, we seek to fill this gap by considering a scenario with split delivery and compare the costs to the scenario where no split delivery is allowed (LNGPM) to understand the various cost benefits due to splitting of delivery.

We have formulated the above problem as a mixed integer linear programming problem (MILP). In our formulation, we model the LNG available to the importer as a set of LTC and spot cargoes from which the optimal set having least overall cost has to be procured for satisfying the requirements of the importer. We consider multiple terminals at the importer's end, with each terminal having separate demand which must be collectively satisfied by the procured cargoes. We also consider limited berths/jetties and capacities of storage tanks at each terminal. For the cargos, we allow each cargo having its own volume, price, availability window and charter costs. In our model, we consider the option of splitting the delivery between the terminals available only to the LTC cargos and restrict the splitting of delivery between two terminals as seen in the real world. To prevent impractical solutions, the ratio of volume unloaded at each of the terminals is taken to be in reasonable limits . The objective function is to minimize the sum of cargo costs and transport costs under contractual stipulations and constraints based on the mentioned resources and obtain an optimal set of cargos to be procured.

To understand the benefits of split delivery, we present an illustrative example. The example involves an LNG importer having three regionally distributed terminals, having two LNG berths at each import terminal. The planning horizon is taken as 30 days. During the planning horizon, there are sixty spot cargoes available at different time windows. There are also 4 LTC cargos, each corresponding to a particular contract. Each contract has a minimum and maximum amount of LNG that can be procured from that supplier. Each LTC and spot cargo has a cargo volume, LNG price, travel times, and travel costs associated with each cargo. At each import terminal, capacity of storage tanks and a reorder level is specified. Travel time between all combinations of terminals is also defined in case of a split delivery. The illustrative example is solved for both scenarios and their results are compared. We see that there is an additional saving of 1.6 million USD by utilizing the split delivery option. We observe that using split delivery, we procure same number of LTC cargos, but slightly lesser volume of LNG. The other reason for savings is improved flexibility and better scheduling while ordering spot cargos. The additional travel cost incurred due to travel between terminals during a split delivery is offset by the savings in procurement cost. To reduce the effect of additional inventory at the end of the month, we solve the same scenario for 12 months on a rolling horizon basis to give a better comparison of the models. We see that by allowing split delivery, we are able to save 10.67 million USD (0.93%) when compared to the scenario without utilizing split delivery.

In the above example, we have considered benefits of only one customer who avails split delivery. There is a potential to reduce costs further by cooperating with other nearby customers and allowing split deliveries between import terminals of different customers to benefit both customers. In this regard, we have considered three levels of co-operation and compared the benefits among each level. We see that by co-operating and allowing split delivery there is always a reduction in cost seen for the customers. We will also report the observations from various scenarios discussed above.

References:

1. Dhul Qidah (2017, August 21). Qatargas, RasGas complete LNG co-loading for split delivery, LNG global. https://www.lngglobal.com/qatargas-rasgas-complete-lng-co-loading-for-split-delivery

2. Shaikh, F., Ji, Q., Fan, Y., Shaikh, P. H. & Uqaili, M. A. 'Modelling an optimal foreign natural gas import scheme for China'. J. Nat. Gas Sci. Eng. 40, 267–276 (2017).

3. Kim, J. & Kim, J. 'Optimal Portfolio for LNG Importation in Korea Using a Two-Step Portfolio Model and a Fuzzy Analytic Hierarchy Process'. Energies 11, 3049 (2018).

4. Fatih Mutlu, Mohamed K. Msakni, Hakan Yildiz, Erkut Sonmez, Shaligram Pokharel, 'A Comprehensive Annual Delivery Program for Upstream Liquefied Natural Gas Supply Chain'. European Journal of Operational Research (2015), doi: 10.1016/j.ejor.2015.10.031

5. Sheikhtajian S, Nazemi A and Feshari M. Marine Inventory-Routing Problem for Liquefied Natural Gas under Travel Time Uncertainty. International Journal of Supply and Operations Management, 7, no.1, pp 93-111 (2020)

6. Shahrukh, Mohd and Srinivasan, R. (2019). Evaluating the Benefits of LNG Procurement through Spot Market Purchase. Computer Aided Chemical Engineering, 46:1723–1728. doi: 10.1016/B978-0-12-818634-3.50288-5

634117 Keynote Talk: Toward Autonomy for Safe and on-Demand Biomanufacturing on Mars

Abstract:

Biomanufacturing holds promise as a versatile and economic means to transform in-situ resources into products that support long-term crewed space exploration and human mission to Mars. Bioproduction targets for Martian life-support systems primarily include pharmaceuticals, sustenance, and biopolymers that can be used in diverse additive manufacturing applications. A distinguishing feature of a space biomanufacturing factory compared to Earth-manufacturing factories is how versatile and robust a space factory must be to ensure crew survival in unknown environments and under mission uncertainties. In this talk, we will present our vision as to how the autonomy of a Martian biomanufacturing factory will require the safe, robust, and verifiable composition of deep learning, perception, mission planning, and optimal control in a feedback loop.

634118 Keynote Talk: The Roadmap to an Autonomous Chemistry Lab

Abstract:

Current approaches to chemical synthesis rely on time-consuming planning and labor-intensive manual synthesis, which is a rate-limiting step in the discovery of new functional molecules, a significant contributor to the overall cost of preclinical development, and clearly incompatible with the goals of autonomous discovery. In this talk, I will outline the components of a hypothetical autonomous synthesis laboratory designed to facilitate the development and manufacture of new small molecules. We will talk through the hierarchical process of computer-aided synthesis planning, starting from a list of molecules of interest and ending with an optimal schedule of automation-friendly action sequences for their sequential or parallel synthesis. Throughout, I will discuss our and others' progress towards this goal and where the primary technological barriers lay.

632449 Exploring the Physical Aging Behavior of Hpmcas *Via* Thermal Analysis

Abstract:

A major challenge in the design and development of oral delivery drugs is the "solubility challenge," where a substantial amount of newly discovered small molecule drugs exhibits low solubility, resulting in low bioavailability. One strategy to increase the bioavailability of hydrophobic small molecule drugs is to form an amorphous solid dispersion (ASD). In this approach, amorphous active pharmaceutical ingredients (APIs) are typically dispersed into an amorphous polymer matrix. The glassy polymer matrix kinetically hinders the recrystallization of the API, keeping the API in its amorphous state. Amorphous API exhibits increased bioavailability and aqueous solubility in comparison to its crystalline counterpart. However, despite their kinetic stability, ASDs often remain thermodynamically unstable with a preferential phase separation into API-rich and polymer-rich domains under certain environmental conditions. We hypothesize that this unfavorable phenomenon is primarily facilitated by the molecular mobility of the glassy polymer as the system undergoes structural relaxation.

In this study, we examine this hypothesis by analyzing the physical aging behavior of one of the most common polymer excipients in ASDs, hypromellose acetate succinate (HPMCAS), using conventional differential scanning calorimetry and fast scanning calorimetry. Through temperature scans across the glass transition temperature (T_g) and isothermal aging below the T_g, we compute the aging rate and observe the enthalpy recovery behavior on heating. Contrary to theoretical expectations, when it is aged well below T_g, HPMCAS exhibits a unique devitrification mechanism. In this mechanism, the enthalpy lost on physical aging is recovered notably before T_g, indicating a recovery process different from the alpha relaxation. More importantly, this process is not attributed to a secondary relaxation process as it is not discrete but rather continuous such that at longer aging times, the devitrification shifts towards the alpha relaxation process. This unique behavior suggests a relaxation mechanism that may directly influence the stability of drug molecules within the HPMCAS structure and impact ASD design, processing, and storage.

625167 Accurate Temperature Prediction of the Freeze-Dried Cake during Secondary Drying Process in a Laboratory Scale Lyophilizer

Abstract:

Freeze drying is a widely used process that improves the stability of the pharmaceutical products. Conventional freeze-drying consists of two sequential steps: (1) primary drying where ice sublimates from the product to leave behind a porous cake (10-30 hr) and (2) secondary drying where the remaining bound water is removed from the porous cake (2-10 hr). Currently, there is a lack of robust models for secondary drying with comparable accuracy and flexibility as primary drying. To alleviate this issue, we investigated secondary drying using a lab-scale lyophilizer and developed an accurate heat transfer models of this process. Experiments involved lyophilizing various sugars (e.g., sucrose and mannitol) under different operating conditions to determine the effects they have on energy transfer during secondary drying. The overall heat transfer coefficient at the bottom of the vial was determined by using a heat flux sensor to measure the slope of heat flux versus temperature difference between the shelf and the cake. Experimental results show that the heat transfer coefficient is significantly different when compared to the primary drying process, and that the vial's thermal properties play a large role in the heat transfer. Lastly, we develop three theoretical models to describe the heat transfer in the vial based on solving the heat equation with bound water desorption, using published desorption kinetic parameters and equilibrium moisture content data. The models describe the temperature and moisture content in the lyophilized cake with varying degrees of complexity depending on the application of interest. Simulation results are validated by comparison of temperature predictions with experimental data for shelf temperatures $15^{o}C \leq T_{sh} \leq 35^{o}C$. These models have great potential for the precise estimation of the cake temperature and drying time compared to trial-and-error approaches. It is expected that our models will aid in understanding and optimizing secondary drying for the lyophilized product.

621795 Physicochemical Modeling of Drug Stability in Multilayer Polymeric Films Containing an Aqueous Moisture Barrier Layer

Abstract:

Designing a drug dosage form requires the inclusion of excipients that provide enhanced drug solubilization, drug stabilization, controlled release, flavor masking, dilution, among many other properties. The inclusion of these additives can often result in complex interactions requiring significant formulation development to identify the correct combination of additives for a specific active pharmaceutical ingredient. An alternative to this approach is to prepare excipients into discrete layers that provide their relative benefits without needing to identify all blend compatibilization and concerns. Multilayer films provide advantage of incorporating separate additives in separate layers in order to provide different functions such as enhancing aqueous solubility and bioavailability of Biopharmaceutical Classification System (BCS) Class-II and Class-IV drugs in the form of soluble polymeric drug dispersions, and including a beneficial layer of moisture barrier to maintain drug stability in storage. This separation also provides ease of developing mathematical modeling considering only one-dimensional transport for predictive impact during industrial formulation ensuring specific drug release profiles and stability requirements. Here, we model and develop a barrier layer to prevent/minimize water vapor permeation to the active layer so that water sensitive APIs can be protected to a stipulated level during the long-term storage.

Multilayer drug loaded films consisting polyvinyl alcohol (PVA) active layer and tannic acid modified chitosan barrier layer (both layers were prepared by solvent casting) are fabricated using a uniaxial heat press. Active and barrier layers are laminated by putting edible starch-based hot-melt adhesive made from powdered corn-syrup and fructose. Water vapor transmission through the barrier layer was determined by ASTM E96M moisture transmission test at a relative humidity of 70%. Water vapor transmission through the barrier layers is modeled by unsteady-state Fickian diffusion equation coupled with mass transfer at the surface and interface between two layers, and drug impurity formation kinetics. The partial differential equation (PDE) is solved by the method of lines, where ordinary differential equations (ODEs) obtained after spatial discretization are coupled with the ODEs from drug impurity formation kinetics. The water vapor concentration profiles in the barrier layer are used to find mass flux, mass flow rate and cumulative water transmission to the active layer. Data from the experimental water uptake by the active layer and impurity formation were fitted to the model profiles to optimize the transport (diffusivity), mass transfer (surface mass transfer coefficient, partition coefficient) and reaction kinetic constants based on $<1\%$ impurity formation in a year. Finally, relationship between film properties (types, thickness, moisture absorption, delamination upon solvation) and user provided requirements were established in order to develop a generalized procedure for drug formulations.

626788 Development of an HME Based Extended-Release Formulation Using Experimental and in silico Approach

Abstract:

Introduction

Hot melt extrusion (HME) is a continuous manufacturing process primarily carried out using co-rotating intermeshing twin-screw extruders (TSE). In the pharmaceutical context, HME is primarily used to produce amorphous solid dispersions of poorly soluble active pharmaceutical ingredients (APIs). In addition, it can also facilitate the development and production of products with crystalline API embedded into a polymer matrix and nanopharmaceuticals. Being a continuous manufacturing technology, HME-based drug development typically requires prohibitively high amounts of API (kilograms of premix vs grams available in the early development stages) for successful formulation screening, early process development and scale-up. The process setup itself is modular, allowing the process to be tailor-made for any formulation. This might however be problematic in the context of choosing the appropriate and efficient setup for an unknown formulation and it is an additional reason for relatively high quantities of material usually needed for the screening. Effectively solving the multidisciplinary challenge of formulation development, early process screening, effective scale-up and transfer to GMP production represents one the key challenges of the pharmaceutical industry and is especially challenging for novel and imported production technologies like HME.

Methodology

Addressing these challenges, our groups have worked on the development of *in silico* and experimental tools for simpler process development and scale-up. Since to this date most of the process setup and scale-up activities are performed experimentally and empirically, one of the goals was to create *in silico* tools for a rational, science-based process setup and scale-up, while addressing other important aspects, such as API degradation and overall product quality [1]–[3]. The fundamental idea behind process modelling is the breakdown and the detailed analysis of the key process aspects, among which the most prominent might be the analysis of flow patterns developed as a result of the rotation and geometry of the individual screw element pairs [1], [4]–[9]. In this work we´ve followed the development of a novel extended-release product for oral administration. The process development starts with extrusions on the 9mm tabletop extruder (Three Tec ZE9) and continues to 16mm (Leistritz NANO16) and 18mm (Coperion ZSK18) extruders. The experimental development is closely followed with 3D Smoothed Particle Hydrodynamics (SPH) simulations of individual screw elements that comprises the screw configuration, and reduced-order 1D HME process simulations.

Results

The experimental setup on all three extruders is investigated in detail using information collected during the experiments and simulations. The produced extrudates are investigated in term of the material science, release profile, and stability. The extrudate quality is then connected to the prevailing extrusion state with the help of reduced order 1D HME simulations. The process was also scaled from the tabletop 9mm extruder to a triple-flighted 16mm extruder and a 18mm pilot plant size extruder. The differences found in the performance of the individual screw elements, differences in process setup and extruder design and scale are investigated and connected to the resulting extrudate quality.

Literature

[1] J. Matić, A. Witschnigg, M. Zagler, S. Eder, and J. Khinast, "A novel in silico scale-up approach for hot melt extrusion processes," Chem. Eng. Sci., vol. 204, pp. 257–269, Aug. 2019.

[2] J. Matić, A. Paudel, H. Bauer, R. A. L. Garcia, K. Biedrzycka, and J. G. Khinast, "Developing HME-Based Drug Products Using Emerging Science: a Fast-Track Roadmap from Concept to Clinical Batch," AAPS PharmSciTech, vol. 21, no. 5, p. 176, Jul. 2020.

[3] J. Matić et al., "Towards predicting the product quality in hot-melt extrusion: Small scale extrusion," Int. J. Pharm. X, vol. 2, p. 100062, Dec. 2020.

[4] J. Matić *et al.*, "Towards predicting the product quality in hot-melt extrusion: Pilot plant scale extrusion," *Submitted.*

[5] A. Eitzlmayr and J. G. Khinast, "Co-rotating twin-screw extruders: Detailed analysis of conveying elements based on smoothed particle hydrodynamics. Part 1: Hydrodynamics," Chem. Eng. Sci., vol. 134, pp. 861–879, Sep. 2015.

[6] A. Eitzlmayr and J. G. Khinast, "Co-rotating twin-screw extruders: Detailed analysis of conveying elements based on smoothed particle hydrodynamics. Part 1: Hydrodynamics," Chem. Eng. Sci., vol. 134, pp. 861–879, Sep. 2015.

[7] A. Eitzlmayr, J. Matić, and J. G. Khinast, "Analysis of flow and mixing in screw elements of corotating twin-screw extruders via SPH," AIChE J., vol. 63, no. 6, pp. 2451–2463, Jun. 2017.

[8] A. Eitzlmayr et al., "Experimental characterization and modeling of twin-screw extruder elements for pharmaceutical hot melt extrusion," AIChE J., vol. 59, no. 11, pp. 4440–4450, Nov. 2013.

[9] A. Eitzlmayr et al., "Mechanistic modeling of modular co-rotating twin-screw extruders," Int. J. Pharm., vol. 474, no. 1–2, pp. 157–176, Oct. 2014.

[10] R. Baumgartner, J. Matić, S. Schrank, S. Laske, J. Khinast, and E. Roblegg, "NANEX: Process design and optimization," Int. J. Pharm., vol. 506, no. 1–2, pp. 35–45, Jun. 2016.

628373 Multi-Objective Supply Chain Optimization in Personalized Healthcare

Abstract:

Autologous Chimeric Antigen Receptor T cell therapies are a type of immunotherapy, where T cells are taken from a patient's blood and engineered to express the Chimeric Antigen Receptor (CAR). CAR T cells are then capable of targeting and killing cancer cells [1]. The two first commercial products have been approved in 2017 and 2018 by the U.S. Food and Drug Administration (FDA) and he European Medicines Agency (EMA) respectively for the treatment of B-cell lymphoma [2,3]. More recently, the U.S. FDA approved two more treatments for relapsed or refractory mantle cell lymphoma (2020) and for relapsed or refractory large B-cell lymphoma (2021) [4].

Unlike batch produced pharmaceuticals, CAR T cell therapies use patient cells as the starting material, placing the patient schedule in the center of the decision-making process. This unique 1-to-1 production and supply chain model translates into manufacturing lines and distribution nodes designed for and occupied by a single therapy, posing hurdles to the development of robust and responsive networks. Moreover, the predicted increase of patient numbers, the sensitive nature and short shelf-life of these therapeutics pose additional challenges to the coordination of manufacturing, storage and distribution [5].

In this work, we present a Mixed Integer Linear Programming (MILP) model for the optimal design of a supply chain network of CAR T cell therapies under different demand scenarios. The problem is formulated as multi-objective optimization aiming to minimize the average therapy cost (ATC) and the average return time of the therapies (ART). The preliminary results showcased in Figure 1 highlight the trade-off between the two objectives. The results are reported for two annual demand profiles of 200 (triangles) and 500 (squares) patients and two different manufacturing durations (TMFE). The base case scenario represents the current state of the art of 13 days of manufacturing (filled symbols), while a forward-looking scenario where the manufacturing time is reduced to 7 days is also considered (empty symbols).

We can observe that in the base case scenario an increase in the demand leads to an increase of the average therapy cost for the same ATR, which indicates that for prolonged manufacturing time the effect of economies of scale is not significant. On the other hand, in the forward-looking scenario it is observed that for long manufacturing times an increased demand may lead to facilities with capacity in order to satisfy all patients in time.

Future directions include: (i) investigation of scenarios with higher number of patients per year; and (ii) assessment of the effect of patient-specific manufacturing time on the supply chain network. The second point is particularly important, since the patient conditions at the time the cell sample is taken influence significantly cell doubling time and consequently the total manufacturing time required. The results will help identify bottlenecks in the overall manufacturing and distribution process, as well as the impact of a variable manufacturing time on the overall network. Finally, each scenario is simulated for different demand profiles in order to quantify the sensitivity of the patients scheduling for a given total demand and manufacturing time.

Acknowledgments

Funding from the UK Engineering & Physical Sciences Research Council (EPSRC) for the Future Targeted Healthcare Manufacturing Hub hosted at University College London with UK university partners is gratefully acknowledged (Grant Reference: EP/P006485/1). Financial and in-kind support from the consortium of industrial users and sector organizations is also acknowledged.

References

[1] B. L. Levine, J. Miskin, K. Wonnacott, and C. Keir, "Global Manufacturing of CAR T Cell Therapy," Mol. Ther. - Methods Clin. Dev., vol. 4, no. March, pp. 92–101, 2017.

[2] Novartis, "KYMRIAH Treatment Process, Dosing & Administration | HCP," 2018. [Online]. Available: https://www.hcp.novartis.com/products/kymriah/acute-lymphoblastic-leukemia-children/dosing-and-administration/. [Accessed: 24-Mar-2021].

[3] Kite Pharma, "First CAR T Therapy for Certain Types of Relapsed or Refractory B-Cell Lymphoma," 2018. [Online]. Available: https://www.yescartahcp.com/. [Accessed: 24-Mar-2021].

[4] A. Mullard, "FDA approves fourth CAR-T cell therapy". Reviews Drug Discovery 20, 166 (2021)

[5] R. Griffiths and M. Lakelin, "Successfully managing the unique demands of cell therapy supply chains." p. 9, 2017.

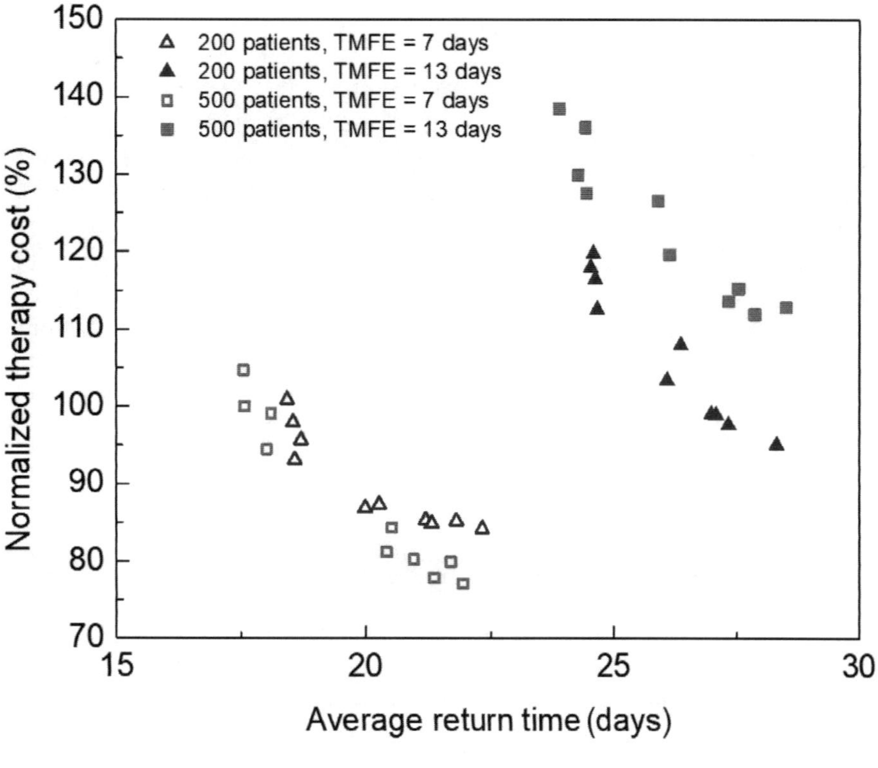

Figure 1: normalized therapy cost versus average return time of a therapy for two demand scenarios and two total manufacturing time excluding quality control.

630367 Robust Bioprocessing of Lignocellulose Using Microbial Tipping Points

Abstract:

Biofuels derived from lignocellulosic feedstocks have been shown to have near zero and in some cases net negative greenhouse gas emissions, driving public and private interest in producing lignocellulosic biofuels on an industrial scale. Anaerobic digestion (AD) is an established and scalable biological process for biogas production from a variety of feedstocks, including lignocellulosic biomass. AD uses a mixed microbial community adapted to the feedstock and operating conditions to produce biogas in four basic steps, hydrolysis, acidogenesis, acetogenesis, and methanogenesis. Biogas is on average 60 vol% methane and 40 vol% carbon dioxide and can be upgraded to renewable natural gas (RNG) by downstream carbon dioxide removal.

Lignocellulosic feedstocks are recalcitrant to hydrolysis, which is often the rate-limiting step for AD of this material. For other feedstocks, e.g. wastewater sludge, food waste, or manure, the rate-limiting step is considered to be methanogenesis. Because of this difference, the process conditions required for microbiomes to achieve high methane production rates during AD of lignocellulosic feedstock may have unique pH, temperature, residence time, and loading rate optima and tolerances. Hydrolysis-limited microbiomes may also have different sensitivities to perturbations of these conditions compared to more prevalent wastewater and manure systems. Understanding the sensitivities of AD to upsets or tipping points is necessary for operating a robust process, with robust defined as being able maintain the key process performance parameters without easily upsetting beyond recovery. A tipping point is defined as the process parameter value at which a biological community becomes unstable in terms of its key product(s) formation. This study proposes a method to define the tipping points for an AD microbiome fed senescent switchgrass. Within those tipping points the variation in process variables avoids irreversible reductions in system performance.

The environmental stresses evaluated in this study are temperature, pH, and overfeeding. These stresses were chosen as those likely to occur during full-scale operation, and that have been known to upset conventional waste digesters. Tipping points will be defined at similar thresholds as those used in inhibitor studies, with tipping points at 50% and 90% decline in methane production after the perturbation is applied. The use of two levels of failure is important as some stresses may not cause catastrophic failure within an operationally relevant range of stress magnitudes. The volatile fatty acid (VFA) concentration will also be monitored as a secondary variable used to identify and characterize a failing digester, as VFAs represent the key intermediates between the potential rate-limiting steps of hydrolysis or methanogenesis. For the pH and temperature tipping point experiments, both positive and negative changes will be evaluated, while for overfeeding only positive changes will be explored. The magnitude of each shock is provided in Table 1. Based on prior AD literature, lower feeding rates are unlikely to result in the dramatic and catastrophic failures that higher feeding rates and both higher and lower pH and temperature conditions can.

Each shock condition will be carried out in duplicate 1-L semi-continuous bioreactors and compared to duplicate controls. The digestate from a pilot scale anaerobic digester fed switchgrass at 55°C, 8.5 pH, and a 10-day retention time will be used to inoculate the 1-L bioreactors. After the shock is administered for a set time the bioreactors will be returned to the process conditions of the pilot reactor and monitored for three retention times for process recovery. Biogas production will be continuously monitored while the VFA concentration, feedstock conversion, and biogas composition are measured at 0.01, 0.1, 0.5, 1, 2, and 3 retention times.

This work aims to define the range of perturbation that a lignocellulosic AD system can experience without significant and irreversible loss of methane production. The method described allows for the sensitivity of a microbial system to be defined in terms of control variables. This then informs how precisely the system must be controlled to avoid biological tipping points leading to loss of product. Defining these tipping points and controlling within them will increase the operating efficiency of microbial systems and enhance the technical and economic feasibility of converting recalcitrant lignocellulosic feedstocks to biofuels.

Table 1. The process variable settings for each tipping point experiment

Process Variable	Settings
Overfeeding (g VS/L)	+10, +20, +30
Temperature (°C)	±10, ±20, ±30
pH	±0.5, ±1.0, ±1.5

624307 Keynote: Multi-Objective Optimization, State Estimation, and Advanced Control of a Semi-Batch Process for the Enzymatic Conversion of Lactose into Value-Added Products

Abstract:

Some of the main challenges experienced when scaling a laboratory enzymatic process to commercial scale include achieving optimal operations, performing online monitoring, and ensuring process controllability. Often times, laboratory-scale enzymatic processes correspond to simply proof of concept experiments that are carried out to characterize the reaction phenomena and develop preliminary process models. For example, in laboratory settings, ideal process conditions, such as negligible spatial variation in concentration, temperature or pH are much more achievable than at industrial scales[1]. As a result, when applying these models to industrial scales, a higher level of uncertainty and error is expected when compared to laboratory scale. Additionally, when developing these models, highly specific offline analytical methods are employed and relied upon due to their increased precision and accuracy over traditional online monitoring methods. However, in many industrial processes, soft sensors or estimation methods must be used online to infer process states from measurements of variables such as the oxygen uptake rate or carbon dioxide production rate. When using these methods, an additional level of uncertainty is introduced due to reliance on calibration ranges and model accuracy[2]. Thus, considering all these factors, it is often challenging to predict if laboratory scale processes can be feasibility scaled up. In this work, a process optimization, state estimation, and advanced control framework is introduced to determine the feasible scaled-up operations of a laboratory enzymatic process, considering a new enzymatic model for the conversion of waste lactose into a series of value-added products known as galactooligosaccharides (GOS).

The enzymatic model used in this work utilizes β-galactosidase to convert lactose into GOS products while distinguishing purely galactose derived di-, tri-, and tetrasaccharides from those containing a glucose group[3]. Due to the structural similarities of these isomers, online monitoring techniques are unable to isolate and uniquely distinguish the specific isomer concentrations. Instead, such techniques are only able to measure the combined concentrations of the isomers and provide one total concentration, thus posing a more challenging monitoring problem. Additionally, this specific model was chosen as it is currently at laboratory scale and could potentially be scaled to a commercial level to address the lactose waste problem plaguing small-scale cheese manufacturers[4].

Due to the many process uncertainties, such as raw feed cost and enzyme cost in this preliminary stage, conducting a robust and accurate economic analysis is challenging and would likely exhibit a wide distribution of economics, ultimately providing unreliable results. Instead, to examine the feasibility of scaling up this process, a multi-objective optimization is carried out considering 2 main factors: maximizing lactose conversion and GOS production rate as these will have the greatest effect on the process economics. Using the results of this multi-objective optimization, a trajectory-based model predictive controller (MPC) is implemented to guarantee that the process adheres to the optimum profile and is able to respond to simulated disturbances that are typical of industrial processes. Finally, a Moving Horizon Estimator (MHE)[5] is implemented to cope with the noise from the online measuring techniques and to decouple the combined isomer concentrations, providing insight into the specific compound concentrations in the reactor for optimal operations and control.

References

[1] Villadsen, J., Nielsen, J., Lidén, G. (2011). *Bioreaction Engineering Principles*, *3*. Springer, Boston, MA.

[2] Stanke, M., Hitzmann, B. (2013). Automatic Control of Bioprocesses. In: Measurement, Monitoring, Modelling and Control of Bioprocesses, *Advances in Biochemical Engineering/Biotechnology*, *1*. Springer, Berlin, Heidelberg.

[3] Schultz, G., Alexander, R., Lima, F. V., Giordano, R. C., Ribeiro, M. P. A. (2021). Kinetic Modeling of the Enzymatic Synthesis of Galacto-Oligosaccharides: Describing Galactobiose Formation. *Food and Bioproducts Processing, 127*, 1-13.

[4] Illanes, A. (2011). Whey Upgrading by Enzyme Biocatalysis. *Electronic Journal of Biotechnology. 14.*

[5] Alexander, R., Campani, G., Dinh, S., Lima, F.V. (2020). Challenges and Opportunities on Nonlinear State Estimation of Chemical and Biochemical Processes. *Processes, 8, 1462.*

631677 Keynote: Virtual-Engineering Software Framework for Integrated Biomass Conversion Modeling

Abstract:

This presentation covers the design and implementation of a software tool to systematically connect computational models of unit operations to simulate an integrated process of low-temperature conversion of biomass to fuel. This virtual engineering (VE) software was designed with the overarching goal of connecting unit models written in various programming languages and requiring different computational resources within a single, flexible framework. The models and features currently considered for the VE library include mechanistic models for pretreatment, enzymatic hydrolysis, and aerobic bioreaction; high-fidelity computational fluid dynamics (CFD) simulations for enzymatic hydrolysis and aerobic bioreaction; and the capability to perform techno-economic analyses (TEA) using Aspen Plus, a commercial software package. The CFD models require access to high-performance computing (HPC) resources, so in addition to handling multiple programming languages and interfaces, the VE software must also be capable of interacting with an HPC scheduler to submit, run, and post-process jobs.

Using the Python programming language, a new VE software package has been developed that contains functionality to manage the input-output communication between various unit models, schedule simulations to run on NREL's HPC and analyze those results, and interface with existing TEA software workflows. A Jupyter-notebook GUI was also created to solicit user input and provide documentation. In cases where multiple models for a particular unit-operation exist, selection between models is accomplished through a simple checkbox, with the appropriate inputs and outputs being parsed and converted seamlessly in the background. Each operation makes use of a different programming language, but the flow of information from pretreatment to enzymatic hydrolysis to bioreaction is managed with an intuitive, centralized file-communication strategy. In this talk, the programming approach and implementation details of the notebook are presented for multiple possibilities of the conversion process, including a demonstration of the ability to manage HPC resources. Additionally, an example of a sensitivity study of treatment parameters governing the overall conversion outcome is shown which highlights the ease of defining new problems using the VE Notebook workflow and leads into a discussion of ongoing work to enable outer-loop optimization studies.

630728 Enabling Low Carbon Hydrogen Production Using Resource Integration Approach

Abstract:

Under the Paris emissions reduction protocol, the global energy sector is expected to shift from carbon-heavy industries to achieve a climate-safe target of 1.5 ^{o}C. To divest from fossil fuel, however, turns out to be more challenging to achieve. Concerns about environmental degradation and dwindling fossil fuel supplies have spurred efforts into adopting renewable energy resources. Hydrogen has been recognized as a key material in the decarbonization of the energy sector and in tackling the global warming challenge. Nevertheless, hydrogen can only exist in combined with other elements in the chemical-bound form. Thus, it has to be produced by means of known energy-intensive processes and thus generates significant amounts of CO_2. One promising alternative in making hydrogen production greener is to incorporate renewable energy in hydrogen production. However, the green hydrogen pathway seems to be more optimistic to see the light in near terms since it suffers from the high cost and the scalability of the hydrogen produced. Therefore, the need for an effective and efficient solution to allow simultaneous production of hydrogen, assess renewable energy use, and incorporate CO_2 management, which is possible through carbon capture utilization and storage. This could serve as the missing link towards a profound energy transformation. This work presents a framework for hydrogen infrastructure design and CO_2 utilization strategies in an industrial park while identifying optimum decisions related to production, storage, and transportation technologies. A techno-economic optimization model based on a superstructure approach has been developed to identify the best possible hydrogen network configurations from a number of technologies, which is achieved by a combination of many possible hydrogen networks (this includes storage and transportation routes) and CO_2 utilization scenarios. A case study that resembles Qatar's industrial sector is set out to investigate the application of the proposed approach.

628182 The Effect of Geological Properties of Underground Storage on Its Suitability for Inter-Seasonal Storage of Hydrogen and Its Role in Achieving Net Zero

Abstract:

Sourcing a sustainable low-carbon energy supply is one of the key requirements towards attaining Net Zero by 2050. Although the use of renewable energies is on the rise, there remain some barriers to their full roll-out, including their intermittency. Underground storage of green hydrogen, generated by electrolysis using renewable energy, across a variety of geological formations and underground features, such as salt domes, rock cavities and aquifers, presents a solution to a year-round national-level supply of renewable energy. Although there are current debates for decentralising the energy grid, the use of inter-seasonal storage provides a good argument for the use of the existing centralised infrastructure.

This study aims to evaluate the role played by underground inter-seasonal storage in decarbonising energy in the UK. It will investigate key factors such as the potential of the UK for inter-seasonal storage, gather data on the properties of the inter-seasonal storage, evaluate what sector is most important for this application, and assess the potential environmental impacts. The effect of properties, such as permeability of surrounding layers, integrity of formation and storage pressure, on the amount of hydrogen that can be stored, how long it can be stored, the rate of loss during storage, and the rate at which hydrogen can be injected into and withdrawn from storage will be investigated. These findings will then be used as inputs to the Value Web Model, a mixed-integer linear programming model for the optimal design and operation of integrated energy value chains. Different optimal scenarios for the use of inter-seasonal storage in the UK will be developed, based on a range of probable costs, efficiencies, and capacities.

The results will be presented in best scenario terms, controlled by factors such as available geologies, demand and other local considerations. These scenarios will be discussed by the role they play for the UK and evaluated on their feasibility. The ultimate goal for the utilisation of inter-seasonal storage is to provide a solution to the current use of non-renewable fuels with the potential in aiding the UK through the transition to renewable energies by 2050.

633237 Stone-Wales Defect-Rich Carbon-Supported Dual-Metal Single Atom Sites for Zn-Air Batteries

Abstract:

This work aims to obtain a fundamental understanding of active sites near stone-wales defects rich nitrogen-doped graphene (DG) with specific coordination of carbon atom rings. It reveals that the stone-wales rich defects (*e.g.,* pentagon (5), pentagon—octagon—pentagon (i.e. 585), or pentagon-heptagon-heptagon-pentagon (5775) rings, appears correspondingly with carbon rings that brought active sites during catalytic reactions. Moreover, we anchored dual isolated metallic atoms (Ni/Fe) on DG support via linkers (O/N) called NiFe-DG. X-ray absorption spectroscopy indicates Ni/Fe metal single atoms are embedded via $Fe\text{-}N_4$ and $Ni\text{-}N_4$ coordination on DG surfaces. It exhibits high catalytic activity for oxygen reduction reaction (ORR) with an onset potential of 0.97 V, a half-wave potential of 0.86 V, and diffusion current density of 5.7 mA/cm^2, which is at par with commercial Pt/C. The catalyst shows superior stability, retained 82% of the initial current density even after 12 hours under an applied potential of 0.86 V. Similarly, the oxygen evolution reaction (OER) overpotential of 358 mV was achieved at 10 mA/cm^2 with a lower Tafel slope value (76 mV/dec) than commercial Pt/C. It maintains 85% stability for 12 h at a constant potential of 1.588 V, shows better stability than commercial Pt/C.

629975 Determination of Safety Parameters in Lithium-Ion Batteries

Abstract:

Lithium-ion batteries (LIBs) have raised increasing interest due to high energy density and efficient energy storage. LIBs are used not only in portable electronics, such as cell phones and computers, but also for energy storage systems (UPS and Grid applications.), and electric vehicles. LIB's offer; excellent performance, high energy density, high specific capacity & voltage, no memory loss, very little self-discharge, and a wide operating temperature range. This has made LIBs increasingly popular. However, safety is a major concern, and a large number of LIBs have been recalled due to explosion and fire accidents. Consequently, increasing attention is being paid to LIB safety issues and a large number of safety strategies are being developed.

The safety of a battery is determined by the battery chemistry, its operating environment, and its abuse tolerance. The internal failure of a LIB is caused by electrochemical system instability. Thus, understanding the material properties, electrochemical reactions, and side reactions occurring in LIBs is fundamental in assessing battery safety. Temperature and voltage are the two controlling factors of the battery reactions. Safety accidents are accompanied by continuous heat and gas generation, which causes battery rupture and ignition of the combustible materials. Cells are equipped with safety features such as separator shut down to stop the flow of current and opening of the vent to help prevent the cell from bursting and/or catching fire in addition to others.

In this study, a testing method was developed using accelerating rate calorimeter (ARC®) to determine thermal behavior of a cell under thermal abuse conditions. Onset temperature of self-heating of the cell, separator shutdown temperature, leak/vent temperature, and ignition temperature of the cell were all determined. The two types of cells that were tested were cylindrical 18650 cells and polymer pouch cells. All tests were run with 100% state of charge. Attached below is an ARC test result from a polymer pouch cell (4.2V, 1800 mAh). It was found that the separator shutdown temperature was 140°C, the primary self-heating onset temperature was 134°C, and ignition temperature was 160°C. Cell leak temperature was found to be in the range of 141°C to 147°C.

Silica Gel/ $MgSO_4$ Hybrids for Thermal Energy Storage

Suboohi Shervani, Curtis Strong, and F. Handan Tezel*

Department of Chemical and Biological Engineering, University of Ottawa, 161 Louis-Pasteur, Ottawa, Ontario K1N 6N5, Canada.

* Handan.Tezel@uottawa.ca (author for correspondence)

Abstract

In the past few years, fossil fuels usage have been increased significantly, which produce large amount of carbon-dioxide, a greenhouse gas, and causes climate change. Therefore, a clean technology is needed to fulfill the global energy demand. Renewable energy sources, such as solar thermal, can be used as potential alternatives to the fossil fuels. However, thermal energy storage is required to fully utilize the solar energy.

Thermo-chemical energy storage technology is a promising technology used in space heating and cooling applications. Salt hydrates show high energy storage density and appeared as encouraging candidate for thermal energy storage. Nevertheless, lack of stability/ repeatability are the main issue. The current study focuses on $MgSO_4$-based hybrids for thermal energy storage. The structural properties of the hybrid is measured by SEM, and the adsorption/regeneration characteristics are measured by a lab scale thermal energy storage apparatus. The hybrid shows high energy storage density ~ 150 kWh/m^3at 50% relative humidity and 120°C regeneration temperature.

Keywords: Salt Hydrates, Adsorption, Thermal Energy Storage, Energy Storage Density, Hybrids

632187 Thermochemical Heat Recuperation for Compressed Air Energy Storage

Abstract:

Compressed Air Energy Storage (CAES) suffers from low energy and exergy conversion efficiencies (ca. 50% or less) inherent in compression, heat loss during storage, and the commonly employed natural gas-fired reheat prior to expansion. Previously, isothermal, and adiabatic (or 'advanced' adiabatic) compressed air energy storage have been proposed to enhance the efficiency of the process. Among the adiabatic schemes, thermal storage can be used to store heat from compression process. Prior to expansion, the air can recuperate the stored thermal energy, which could reduce or even eliminate the need for natural gas-fired reheat. Although sensible heat storage (SHS) and latent heat storage (LHS) have been previously investigated for this purpose, there are very few published studies looking at the use of thermochemical energy storage (TCES) for the CAES application. Here, a direct heat transfer scheme with combined thermochemical energy storage and sensible energy storage is proposed. The present study develops a mathematical model to address the conceptual design of the proposed idea. A detailed model and analysis are performed for TCES-rock-filled PBTES (packed bed filled with both barium oxides and rocks) and rock-filled PBTES. Results showed that similar round-trip efficiency is achieved between rock-filled and TCES-rock-filled PBTES due to the relative low heat capacity and heat of reaction for the barium oxides materials used in the current study. However, with TCES material placed on top of the packed bed, more stable turbine air inlet temperature (less than 200 K temperature variation) is obtained compared with all rock-filled packed bed (more than 300 K temperature difference) if idle time after charge is less than 20 h. In addition, with future advanced materials, the proposed idea could potentially improve the roundtrip efficiency of the process, have longer duration of storage and stable turbine air inlet temperature under suitable operating conditions. The concept could also be applied to Liquid Air Energy Storage.

623606 Exploring the Role and Value of Grid-Scale Energy Storage in Deep Decarbonisation

Abstract:

As the power system in the UK undergoes decarbonisation, the transformation of the capacity mix remains a topic of discussion. Not only novel generation technologies, such as power plants with carbon capture and storage (CCS), but also storage technologies are to be considered. The increasing share of intermittent renewables in the energy mix creates the need for higher storage capacity and/or more flexible plants. Furthermore, the additional requirements of other sectors of the energy system, specifically the added demand from electric vehicles (EVs) and from heat pumps (HP), may impact the power demand profile and not only raise overall demand but also increase peak demand. This raises the question which storage technologies and how much storage capacity is required to satisfy the power demand at all times.

In the UK context, pumped hydro storage (PH) is already a part of the system capacity. Interconnection with neighbouring countries is also used to balance demand peaks and renewables fluctuations. These technologies, however, are mainly used for short-term storage, as in, storage within a day or between days. It is of interest if long-term storage, such as power-to-gas storage technologies, could also have a place in the capacity mix. In this work, we therefore examine under which circumstances power-to-gas storage has value in the future UK energy system under a net zero carbon target by 2050. Specifically, we model a route of power to synthetic methane to salt cavern storage to CCGT (P2M) with recycling of CO_2, which has been determined to be a promising power-to-gas storage route based on techno-economic analysis by Yao *et al.*[1]. We use the electricity systems optimisation model with capacity expansion (ESO-X) developed by Heuberger *et al.*[2]. The model constraints include build rates, ancillary services requirements, plant flexibility constraints, etc. We use one year of full-hourly renewables availability (solar, onshore wind and offshore wind) and demand profiles. This allows the analysis of seasonal storage. The demand is considered to be comprised of three parts, power demand, added demand from electric vehicles, and added demand from heat pumps as shown in Figure 1. Owing to the high amount of uncertainty regarding the level of electrification for both transport and heat, we consider four scenarios for both with varying levels of electrification ranging from minimal to total. The same applies to the generating technologies allowed in the system. We analyse scenarios with and without gas-fired power plants with CCS (CCGT-CCS) as a form of stable, low-carbon generation, and bioenergy with CCS (BECCS), as negative emissions technology, since the deployment of these technologies is uncertain. Furthermore, we analyse the value of key technology parameters of P2M such as input-to-output power ratio and round-trip efficiency and perform sensitivity analysis regarding the CAPEX of the technologies.

Excerpts of the dispatch schedules for scenarios with and without CCS and P2M are shown in figure 2. Our results suggest that in a scenario where all technologies are available for deployment with historical build rate limits, the increasing seasonality of the demand curve and the fluctuation of the increased share of renewables would optimally be absorbed by a combination of all archetypes of technologies. Renewables contribute a large share of power on days with high availability. CCGT-CCS and CCGT provide load-following and flexibility on peak days, respectively. The continuous operation of BECCS offsets emissions from gas-fired power plants. Using our central numbers for technology features and costs, P2M would have value in a cost-optimal technology mix by absorbing excess renewable energy and releasing it on peak days and hours (around 6 GW with 3.6 TWh storage capacity in 2050).

When CCS is removed from the capacity mix, P2M becomes a key technology. In our scenario without CCS or P2M, meeting demand with renewables and short-term storage (PH and battery storage) requires large amounts of curtailment, incurs lost load, and hence doubles the total system cost. When P2M as seasonal storage is added (26 GW and 13 TWh installed in 2050), it is used in the system to absorb power from renewables instead of curtailing it and providing flexible generation on low-availability days. This reduces the necessary amount of renewables capacity substantially. Notably, in scenarios with P2M available, it is deployed in greater amounts than battery storage despite the higher CAPEX and lower round-trip efficiency, indicating the increased need for seasonal as opposed to short-term storage.

Our sensitivity analysis shows a link between the deployment of onshore wind and power-to-gas storage. Lower seasonal storage CAPEX enables the deployment of both in conjunction, raising the renewables capacity as shown in Figure 3. Furthermore, our model suggests that renewables in combination with seasonal storage could be cost-competitive with CCGT-CCS in case of continuously high renewables build rates. We further identified the ratio of charging power to discharging power of the power-to-gas technology as a parameter with impact on its deployment potential and found that cost reductions of the technology may provide more value to the system than technology improvements.

In summary, grid-level power-to-gas storage may have a place in the future energy system in the UK. Whether or not it has value depends heavily on the build rates of renewables and on the relative cost of power-to-gas storage and low-carbon dispatchable power such as CCGT-CCS. Our results underscore the value of a diverse energy system with a mix of multiple technologies for generation and storage.

References

1. Yao JG, Bui M, Mac Dowell N. Grid-scale energy storage with net-zero emissions: comparing the options. *Sustainable Energy & Fuels* 3.11 (2019): 3147-3162.
2. Heuberger CF, Rubin ES, Staffell I, Shah N, Mac Dowell N. Power capacity expansion planning considering endogenous technology cost learning. *Appl Energy*. 2017;204:831-845.

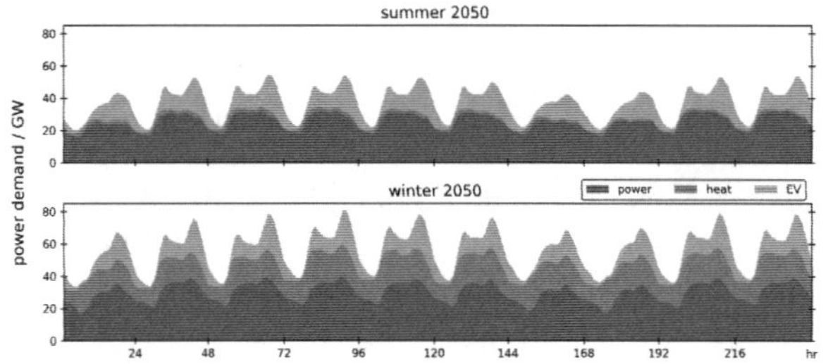

Figure 1: Example of the demand profile used in the model. Hourly power demand, added demand from heat pumps and electric vehicles, for ten days in summer and winter.

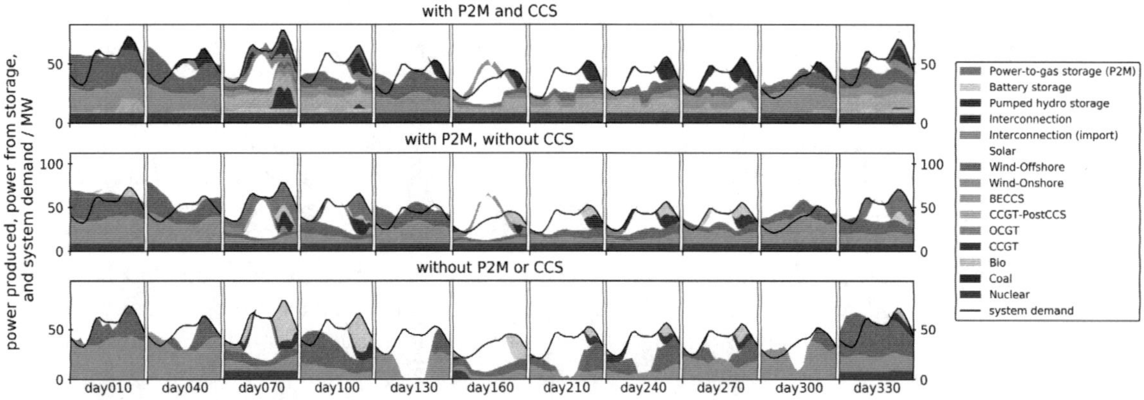

Figure 2: Excerpts of hourly dispatch schedules throughout the year (power produced and power from storage) for central electrification scenario. Stack higher than demand indicates power is fed to storage.

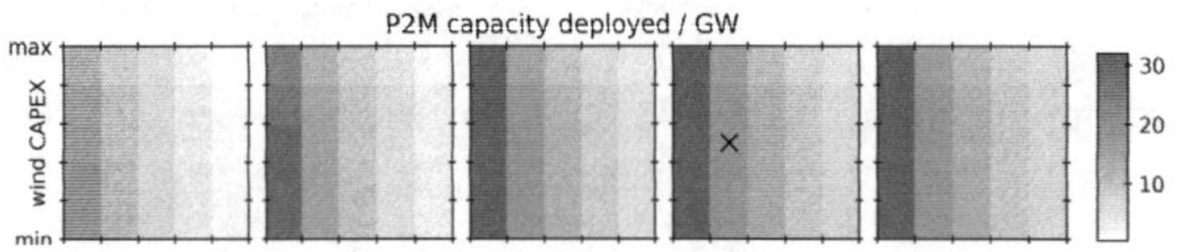

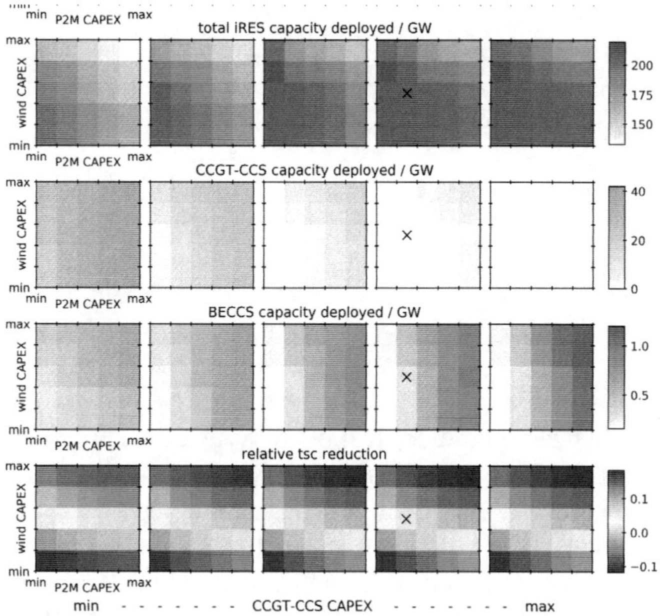

Figure 3: Deployment of power-to-gas storage (P2M), intermittent renewables (iRES), CCGT-CCS, and BECCS, and relative reduction of total system cost (tsc) depending on technology CAPEX. "min" values are estimated lower bounds, "max" values are 1.5 of the current CAPEX. Build rate limits set to 2x historical values. Cases marked with (x) indicate model did not converge within allocated time; values are interpolated based on neighboring cases.

629997 Data-Driven Incipient Fault Management for Proton Exchange Membrane Fuel Cell

Abstract:

In the efforts of establishing cleaner energy systems, hydrogen is receiving increasing attention for transportation and energy storage applications. Among various hydrogen systems of interest, fuel cell vehicles (FCVs) are well studied as the next-gen energy vehicles. Specifically, proton exchange membrane fuel cell (PEMFC) is considered the leading technology for handling hydrogen as fuel. It converts chemical energy from hydrogen to electrical energy via an electrochemical reaction and leaves water as the only byproduct. Therefore, PEMFC is attracting the interest of both the fuel cell community and the automobile industry. However, there are challenges associated with large-scale commercialization of PEMFC for electric vehicles such as operational safety, reliability, performance, and cost [1]. Though several prognostics and health management (PHM) [2, 3] and fault diagnosis [4, 5] studies have been performed to overcome these challenges, the problem of incipient fault diagnosis in PEMFC has not been fully examined. Incipient faults are usually masked by disturbances or other system uncertainties due to their small magnitudes and thus, are unnoticeable and difficult to detect [6, 7]. If no corrective action is taken, incipient faults can evolve into functional faults and lead to catastrophic failures. Therefore, early detection and isolation of incipient faults is highly valuable for the on-board safety of automobiles.

Motivated by the aforementioned considerations, we propose an incipient fault management framework that uses an adaptive data-driven modeling method, OASIS [8] (to detect faults), along with contribution analysis (to isolate faults) through a moving window strategy [9]. First, we obtain a data-driven process model using OASIS for the current operating condition. The model is used to predict the real-time dynamics. Based on this prediction, we continuously evaluate and monitor process risk associated with the fuel cell operation. Here, we incorporate moving window approach for risk assessment to improve the detection accuracy of incipient faults. Also, the adaptive model obtained through OASIS can cope up with any changes in the system, which further helps with decoupling incipient faults from process uncertainties. Next, after the detection step, we initiate contribution analysis to identify the location of the detected incipient fault. Finally, we isolate the process variable with the maximum contribution to the process risk from the remaining components of PEMFC. In this work, we investigate a scenario of hydrogen leakage with a small flowrate. As hydrogen is highly inflammable, it is important to diagnose any small traces of hydrogen leakage to facilitate safe operation of the PEMFC system. To conclude, the proposed framework has the ability to 1) predict the fuel cell operation under varying operating conditions, 2) detect and isolate an incipient fault before it propagates as a functional fault, and 3) assess the process risk to ensure safe operation of the PEMFC system.

Literature cited

[1] Zhang, X., Zhang, T., Chen, H. and Cao, Y., 2021. A review of online electrochemical diagnostic methods of on-board proton exchange membrane fuel cells. *Applied Energy*, *286*, 116481.

[2] Jouin, M., Gouriveau, R., Hissel, D., Péra, M.C. and Zerhouni, N., 2013. Prognostics and Health Management of PEMFC–State of the art and remaining challenges. *International Journal of Hydrogen Energy*, *38*(35), 15307-15317.

[3] Meraghni, S., Terrissa, L.S., Yue, M., Ma, J., Jemei, S. and Zerhouni, N., 2021. A data-driven digital-twin prognostics method for proton exchange membrane fuel cell remaining useful life prediction. *International Journal of Hydrogen Energy*, *46*(2), 2555-2564.

[4] Tian, Y., Zou, Q. and Lin, Z., 2020. Hydrogen Leakage Diagnosis for Proton Exchange Membrane Fuel Cell Systems: Methods and Suggestions on its Application in Fuel Cell Vehicles. *IEEE Access*.

[5] Li, Z., Outbib, R., Giurgea, S. and Hissel, D., 2018. Fault diagnosis for PEMFC systems in consideration of dynamic behaviors and spatial inhomogeneity. *IEEE Transactions on Energy Conversion*, *34*(1), 3-11.

[6] Safaeipour, H., Forouzanfar, M. and Casavola, A., 2021. A survey and classification of incipient fault diagnosis approaches. *Journal of Process Control*, *97*, 1-16.

[7] Chen, H., Jiang, B., Zhang, T. and Lu, N., 2020. Data-driven and deep learning-based detection and diagnosis of incipient faults with application to electrical traction systems. *Neurocomputing*, *396*, 429-437.

[8] Bhadriraju, B., Bangi, M.S.F., Narasingam, A. and Kwon, J.S.I., 2020. Operable adaptive sparse identification of systems: Application to chemical processes. *AIChE Journal*, *66*(11), p.e16980.

[9] Ji, H., He, X., Shang, J. and Zhou, D., 2016. Incipient sensor fault diagnosis using moving window reconstruction-based contribution. *Industrial & Engineering Chemistry Research*, *55*(10), 2746-2759.

626145 Efficient Process for the Production of High Conductivity, Carbon-Rich Materials from Coal

Abstract:

Abstract: Domestic coal can be used to manufacture high value carbon products for multiple applications. The market value of these high performance materials often exceeds the fuel and heat value of coal, which illustrates there are sustainable market forces for manufacturing carbon materials from coal. Current processes to produce high performance carbon materials from coal pose significant challenges due to the high temperatures and corrosive reagents required to substantially modify the native coal structures. Under a Department of Energy sponsored program, Physical Sciences Inc. (PSI) developed and demonstrated an innovative approach for producing high value, carbon-based products from coal feedstocks for electrochemical applications. The key innovation is a two-step process that generates both a high conductivity carbon material (HCCM) and valuable byproducts (mineral fractions for trace elements recovery and low emission fuels). PSI demonstrated: (1) a highly robust and scalable processes (multi-grams to kilograms) to produce the target carbon product with appropriate surface area (>50 m^2/g) and low mineral content (Fe<100 ppm); (2) comparable electrochemical performance of PSI's coal-derived HCCM in battery electrodes compared to a benchmark commercial product; and (3) economic feasibility for scale-up and commercialization. This presentation will provide a comprehensive overview of the overall process that provides economical production of high conductivity carbon materials for electrochemical applications. In addition, the technical approach, key results including detailed analysis, conclusions, and the next steps will be discussed in this presentation.

Acknowledgement: This material is based upon work supported by the U.S. Department of Energy, Office of Science, Office of Chicago Ops, under Award Number(s) DE-SC0018837.

628927 Modular Processing of Flare Gas for Hydrogen and Carbon Nanofibers

Abstract:

In the United States, the "shale boom" and wide implementation of hydraulic fracturing has dramatically reduced the price of natural gas. As a result, oil producers increasingly choose to flare or vent associated natural gas rather than sell it, as investing in expensive transportation infrastructure, such as pipelines, is less economically prudent. Moreover, tight oil wells exhaust quickly as production from a single well depletes rapidly, thus imposing difficulties in predicting the optimal pipeline location[1]. A profitable alternative to harnessing this waste of natural resources is needed.

A one-step Catalytic-Chemical Vapor Deposition (CCVD) process to produce carbon nanofibers (CNFs), carbon nanoparticles (CNPs), and hydrogen is a financially promising repurposing of the flared natural gas. An economically viable CCVD process largely depends on the production of a cheap catalyst. Particle Atomic Layer Deposition (ALD) enables the formation of highly dispersed metal nanocatalysts on a given support. The highly dispersed low-cost catalyst is essential as this process contains a *sacrificial catalyst*; the catalyst, support, CNFs and CNPs combine as one product to improve the strength and durability of concrete. The sacrificial catalyst circumvents the well-known disadvantages of rapid catalyst deactivation in methane thermal cracking[2] and avoids the costly step of catalyst/product separation. To combat the challenges of rapid well depletion, this process will be scaled up to a modular unit that can be transferred between hydraulic fracturing sites. If each flare site in the United States transitioned to this modular CCVD process, over 30 million metric tons of CO_2 emissions per annum could be eliminated[3].

In this work, transition metals are deposited onto a silica fume support via ALD. CNFs and CNPs are then produced via CCVD of methane with the ALD nanocatalyst. The carbon product is evaluated for its improvements to concrete durability via various compression and deformation tests. The carbon product evaluation and empirical reaction data inform both the modular unit design and technoeconomic analysis.

[1] Energy, B. *The Williston Basin: Greasing the Gears for Growth in North Dakota*; BENTEK Energy: 2012, 2012

[2] Abanades, A.; Ruiz, E.; Ferruelo, E. M.; Hernandez, F.; Cabanillas, A.; Martinez-Val, J. M.; Rubio, J. A.; Lopez, C.; Gavela, R.; Barrera, G.; Rubbia, C.; Salmieri, D.; Rodilla, E.; Gutierrez, D., Experimental analysis of direct thermal methane cracking. *International Journal of Hydrogen Energy* 2011, *36* (20), 12877-12886.

[3] U.S. Energy Information Administration. Natural Gas Gross Withdrawals and Production.

http://www.eia.gov/dnav/ng/ng_prod_sum_a_epg0_vgv_mmcf_a.htm (accessed April, 2021)

632244 Modeling Particle Atomic Layer Deposition in a Fluidized Bed with CFD-DEM

Abstract:

Particle atomic layer deposition (ALD) is a vapor-phase coating technique that enables precise surface modification of particulate substrates. ALD coatings are uniquely uniform and controllable, which is desirable for many applications such as catalyst synthesis or cathode coating. Particle ALD is typically performed with a repeated sequence of self-limiting gas-solids reactions in a batch fluidized bed reactor. This configuration is capable of achieving high precursor utilization without agglomerating particles. However, the lack of predictive modeling capabilities for fluidized bed particle ALD reactors presents a challenge for implementing new coating chemistries and limits the commercial application of particle ALD. In this work, we develop a physics-based model for a fluidized bed ALD reactor that aims to predict reactor performance and enable intelligent particle ALD reactor design.

A reactor simulation based on computational fluid dynamics and discrete element method (CFD-DEM) is implemented using open-source multiphase flow modeling software MFiX. A well-characterized glass bead substrate is used to validate model predictions. The packed bed pressure drop and minimum fluidization velocity predicted by the model show good agreement with the experimental fluidization curve. Self-limited ALD reaction kinetics are implemented in the model using a first-order, irreversible Langmuir rate expression. The model can be used to predict the transient gas composition exiting the reactor and inform operating conditions to optimize precursor utilization and coating uniformity. Future work will include measurement of first-order reaction rate parameters for different particle substrates and ALD precursors, and comparison of modeling outputs to an experimental fluidized bed reactor.

636257 Study and modeling the wettability contact angle and area measurement for EOR in 2D imaging technology by using python algorithm

Abstract:

Wettability contact angle & area are studied and modeled by 2D image technologies obtaining the python algorithm. The investigated approach by using 2D image capturing and image analyses that are applied for EOR mathematically. as well as wetting contact angle and area trends in the rock sample. And then, generate all possible pore contact angles. In this study, the oil recovery prediction model is also developed by constructing a mathematical model based on statistical grouping and averaging for all measured pore-walls available in the rock sample. Pore parameters such as pore-area, pore-count, and pore wall, and pore shape distribution are all captured and measured experimentally, and data can be accurately generated.

631576 Some Features of Ion Exchange Membranes in Complex Environments for Electrochemical Conversions

Abstract:

Ion exchange membranes are frequently used in redox flow batteries and electrochemical flow cells in general. However, there are few general guidelines on how to select membranes. Uptake and transport of ions and the factors underlying them are critical determinants of the behavior of conducting membranes. Substantial work was carried out on issues such as partitioning of ions into membranes in the early studies of Nafion. The heavy focus on fuel cell applications narrowed the scope of many studies to an emphasis on proton conduction. However, with the increased emphasis on redox flow batteries and other, similar electrochemical devices, there is a need to build a stronger base of studies of ions other than protons or hydroxide in various membrane situations. Furthermore, the operating conditions for practial applications typically entails exposure of membranes to concentrated electrolytes.

We will discuss recent studies of the uptake, conduction and transport of ions in various membranes that are exposed to both aqueous and non-aqueous environments. In this presentation, we will describe the separation of factors determining permeation through ion exchange membranes and the molecular basis of these parameters based on our recent work. Several highlights of the work include:

- Comparisons of the measured uptake and transport, in many different membrane types and with different treatment protocols, for materials exposed to typical VRB electrolytes.
- Interpretation of those data in the context of Donnan potentials, with implications for use of the Donnan potential itself as a descriptor.
- Studies of the conductivity and other transport (e.g. NMR diffusion) in membranes ion exchanged with different cations (aqueous) to probe the Nernst-Einstein equation and its interpretation in light of various concentration references.
- Studies of conductivity and other transport (e.g. NMR diffusion) in membranes ion exchanged with different cations in non-aqueous electrolytes, showing some surprising results for conductivity as a function of cation type. This is further investigated in detail using various spectroscopic methods, culminating in a description of a simple membrane modification that will lead to decreased permeation while maintaining conductivity.

By comparing these behaviors across different membrane and electrolyte types, we will begin to systematize our understanding of the interactions leading to observed phenomena. In addition, we will outline our recent work in developing a description of the factors influencing uptake of ions into membranes.

Acknowledgements

We gratefully acknowledge the current support of this work by the Office of Naval Research and the U.S. Department of Energy, Office of Electricity Delivery and Energy Reliability (Dr. Imre Gyuk).

629874 Molecular Engineering of Anion-Conducting Polymers for Electrochemical Energy Conversion Technologies

Abstract:

Anion exchange membranes (AEMs) based on hydroxide-conducting polymers are a key component for anion-based electrochemical energy technology such as fuel cells, electrolyzers, and advanced batteries. Although these alkaline electrochemical applications offer a promising alternative to acidic proton exchange membrane electrochemical devices, access to alkaline-stable and high-performing polymer electrolyte materials has remained elusive until now. Despite vigorous research of AEM polymer design, examples of high-performance polymers with good alkaline stability at an elevated temperature are uncommon. Traditional aromatic polymers used in AEM applications contain a heteroatomic backbone linkage which is prone to degradation via nucleophilic attack by hydroxide ion.

In this presentation, I will highlight recent progress at the Bae group of Rensselaer Polytechnic Institute in the development of advanced hydroxide-conducting polymers and technology transfer effort to commercialize materials for AEM applications. We have developed new synthetic methods that produce polymer design made of all C–C bond backbone and a flexible chain-tethered quaternary ammonium group and that provide an effective solution to the problem of alkaline stability. The advantage of good solvent processability, synthetic versatility, and convenient scalability of the reaction process has generated considerable interest of these polymers, and they are considered leading candidates for commercial standard AEM. AEM fuel cells, electrolyzer, and vanadium redox flow battery tests of some of the developed polymer membranes showed excellent performance, suggesting that this new class of AEMs open a new avenue to electrochemical devices with real-world applications.

631783 New Membranes for Solar Fuels Systems

Abstract:

Anion exchange membranes (AEMs) play an essential role in artificial photosynthesis devices, which (photo)electrochemically convert CO2 and water into useful products. AEMs allow the transport of charge carriers between electrodes while minimizing the transport of CO2 reduction products (e.g., ethanol). Fundamental transport studies in AEMs relevant to artificial photosynthesis are uncommon. Herein, we describe the preparation of an imidazolium-functionalized poly(phenylene oxide) membrane. Membrane transport properties were controlled by systematic variation of the degree of imidazolium functionalization, which induced changes in the membrane water volume fraction. Ethanol permeability and ionic conductivity increased with the membrane water volume fraction. Consequently, the membranes with a relatively high ionic conductivity exhibited a relatively high ethanol permeability, presenting a trade-off in the transport properties desirable for artificial photosynthesis applications. This work seeks to enable the optimization of AEMs for artificial photosynthesis through the systematic study of membrane structure (water volume fraction) and its relevance to alcohol transport and electrolyte ion conductivity in prototype electrochemical CO2 conversion devices.

626042 CO_2–CO Energy Conversion Cycle Enabled By a CO_2-Selective Membrane

Abstract:

This presentation describes a method of CO_2 utilization for the production of CO and electricity via a fuel cell system that is enabled by a CO_2-selective membrane. We have synthesized highly CO_2-selective facilitated transport membranes, showing an extraordinarily high CO_2/CO selectivity of >1200 at 100–120°C (see the left figure below). The CO_2-selective membranes comprised polyvinylamine as the fixed-site carrier and the salt of potassium 2-aminoisobutyrate (AIBA-K) or 2-(1-piperazinyl)ethylamine 2-aminoisobutyrate (AIBA-PZEA) as the mobile carrier to facilitate the CO_2 transport.

One approach to utilize CO_2 for electricity production is a CO_2–CO energy conversion cycle as shown in the right figure below. CO_2 enters the cathode side of the first solid oxide fuel cell (SOFC) where it is partially reduced to CO while using H_2 as the fuel to the anode side of this fuel cell. The CO_2 and CO mixture is fed to the membrane, which separates it into a CO_2-rich permeate and a CO-rich retentate. The enriched CO_2 is recycled back to the H_2–CO_2 SOFC. The CO-rich retentate, for electricity generation, enters the anode side of the second SOFC, where air is used as the oxidant and the anode exhaust is also recycled to the first SOFC. Preliminary techno-economic analysis indicates that the developed facilitated transport membrane can render a retentate of >80% CO with a <1% CO loss. Owing to the high CO content, the CO–air fuel cell can produce a power density of ca. 350 mW/cm^2, which is 75% higher than the scenario where the CO_2 is not removed by the membrane. Experimentally, we have also demonstrated that the H_2–CO_2 SOFC can provide a power density of 0.11–0.37 mW/cm^2 at partial CO_2 conversion. Overall, the electricity generation can be achieved with zero carbon emission.

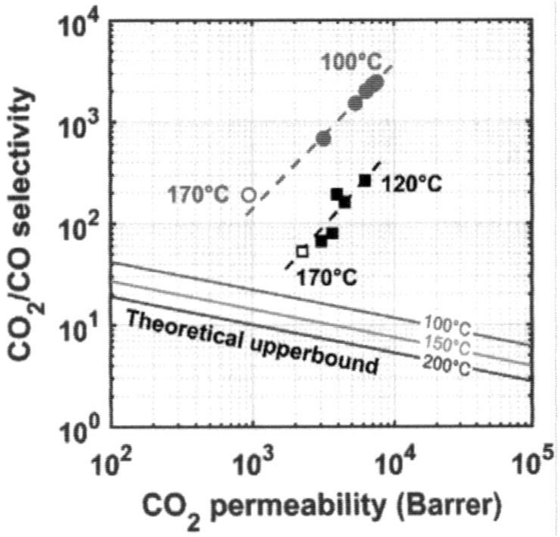

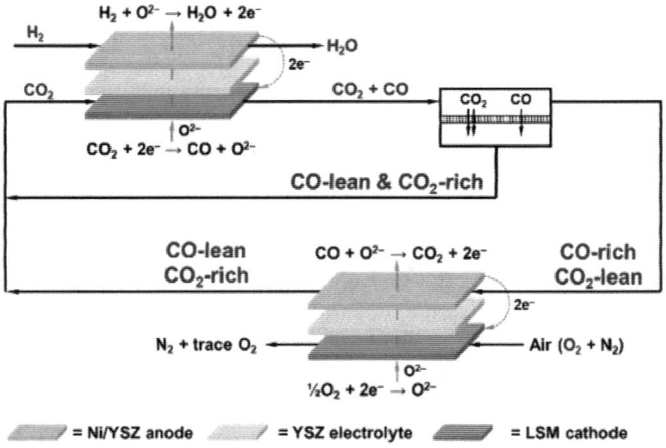

625584 Model-Based Investigation of Upstream CHO Cell Culture Process for Production of Monoclonal Antibodies with Desired N-Linked Glycosylation

Abstract:

N-linked glycosylation is post-translational protein modification that takes place during synthesis of monoclonal antibodies (mAbs) [1]. N-linked glycan composition or glycan index, including total afucosylation, high mannosylation, fucosylation, and/or galactosylation indexes, are often critical quality attributes (CQAs) of mAb based therapeutic proteins [2]. Researchers have found controlling the cell culture media components such as uridine, galactose, manganese, and ammonia concentration can tweak the glycosylation composition with the desired CQA levels [3-5]. During bioreactor operations, many process parameters or operating conditions can also affect protein glycosylation such as temperature, pH and dissolved oxygen [6-8]. However, the effect of process parameters on mAb glycosylation can vary significantly between different cell lines. To better understand the protein glycosylation process and further optimize the bioreactor operations, process modeling can be used to support quality by design (QbD) by correlating critical process parameters with critical quality attributes and further defining a design space for process operations.

We have investigated the effect of temperature and pH on protein glycosylation for a model Herceptin biosimilar (or Trastuzumab) production using a CHO cell culture cultivation in an upstream fed-batch bioreactor. Mechanistic kinetic model was developed based on the experimental data obtained under different temperature and pH combinations. The mechanistic model contains two parts, including an unstructured cell culture model and a structured intracellular model. For our unstructured cell culture model, viable cell density, glucose, ammonia, and protein concentrations are captured by empirical equations. In the structured model, the intracellular kinetics of biochemical reactions cascade of the N-linked glycosylation process are simulated. The kinetic model is used to understand the effect of temperature and pH on the bioprocess. Finally, feasibility analysis with surrogate-based adaptive sampling is applied to determine a bioreactor operation design space. In this combined experimental and computational study, we have determined how adjusting the bioreactor operating conditions can achieve critical quality attributes of the targeted product. Ongoing work will implement integration with a control system to achieve real-time process control for mAb glycosylation.

References

1. Akmal, M. A.; Rasool, N.; Khan, Y. D. J. P. o., Prediction of N-linked glycosylation sites using position relative features and statistical moments. **2017,** *12* (8), e0181966.
2. Szekrenyes, A.; Szigeti, M.; Dvorakova, V.; Jarvas, G.; Guttman, A., Quantitative comparison of the N-glycosylation of therapeutic glycoproteins using the Glycosimilarity Index. A tutorial. *TrAC Trends in Analytical Chemistry* **2020,** *122*, 115728.
3. Gramer, M. J.; Eckblad, J. J.; Donahue, R.; Brown, J.; Shultz, C.; Vickerman, K.; Priem, P.; van den Bremer, E. T.; Gerritsen, J.; van Berkel, P. H., Modulation of antibody galactosylation through feeding of uridine, manganese chloride, and galactose. *Biotechnol Bioeng* **2011,** *108* (7), 1591-602.
4. Kotidis, P.; Jedrzejewski, P.; Sou, S. N.; Sellick, C.; Polizzi, K.; del Val, I. J.; Kontoravdi, C., Model-based optimization of antibody galactosylation in CHO cell culture. **2019,** *116* (7), 1612-1626.

5. Ehret, J.; Zimmermann, M.; Eichhorn, T.; Zimmer, A., Impact of cell culture media additives on IgG glycosylation produced in Chinese hamster ovary cells. *Biotechnol Bioeng* **2019,** *116* (4), 816-830.
6. Villiger, T. K.; Scibona, E.; Stettler, M.; Broly, H.; Morbidelli, M.; Soos, M., Controlling the time evolution of mAb N-linked glycosylation - Part II: Model-based predictions. **2016,** *32* (5), 1135-1148.
7. Sou, S. N.; Sellick, C.; Lee, K.; Mason, A.; Kyriakopoulos, S.; Polizzi, K. M.; Kontoravdi, C., How does mild hypothermia affect monoclonal antibody glycosylation? *Biotechnol Bioeng* **2015,** *112* (6), 1165-76.
8. del Val, I. J.; Kontoravdi, C.; Nagy, J. M., Towards the implementation of quality by design to the production of therapeutic monoclonal antibodies with desired glycosylation patterns. *Biotechnol Prog* **2010,** *26* (6), 1505-27.

624781 Next-Generation Vaccines and Therapeutics: Towards Resilient Pharmaceutical Supply Chains

Abstract:

The promising clinical outcomes of genetic engineering have paved the way to curing life-threatening diseases, including cancers. Several innovative gene therapy drugs have already been launched over the last few years and this rapid growth is forecasted to continue with the U.S. Food and Drug Administration (FDA) anticipating the approval of 10-20 cell and gene therapy products per year by 2025 [1]. Approximately 70% of the cell and gene therapies in the clinical trials utilise viral vectors for gene transfer [2]. The versatility of viral vectors has also emerged in preventive healthcare, with their recent approval as vaccine platforms (e.g. Vaxzevria (AstraZeneca) and the Janssen COVID-19 vaccine) [3][4].

Given the high demand and limited manufacturing capacity, viral vector manufacturers are asked to make decisions on how to best utilise and expand the available manufacturing resources in the presence of numerous demand uncertainties. Some of the challenges are related to production and capacity planning under uncertain clinical trial outcomes and dosage requirements. Furthermore, manufacturers of viral vector-based COVID-19 vaccines tackle the hurdles of rapid and in-risk manufacturing, transitioning to large-scale deployment of novel vaccine products in record times. In this case, SARS-Cov-2 has clearly demonstrated the need for sophisticated tools that can assist responsive decision-making and support planning under uncertainty.

This work presents a novel mixed-integer linear programming (MILP) model for the design and optimisation of viral vector supply chains (Figure 1). An array of applications for viral vectors and their respective demand scales are considered, thereby capturing the emerging paradigm shift from *one-size-fits-all* manufacturing and distribution to targeted healthcare [5]. The framework relies on techno-economic analyses of well-established viral vector manufacturing protocols, which are modelled and simulated in SuperPro Designer. The formulated optimisation model identifies good candidate supply chain structures, capacities and production plans for multi-product multi-suite facilities, distribution plans and transportation flows. For this, we consider (i) a set of products and demand scenarios, (ii) capital and operating manufacturing expenditures for a range of production scales, as well as storage and transport costs for sensitive formulated products, (iii) candidate locations for network nodes, namely primary, secondary manufacturing and distribution centres. The resilience of the strategic decisions is then assessed under demand uncertainty through comparisons of network structures with respect to (i) cost, (ii), scalability and (iii) flow time.

References

[1] **U.S. Food and Drug Administration,** "Statement from FDA Commissioner Scott Gottlieb, M.D. and Marks, M.D., Ph.D., Director of the Centre for Biologics Evaluation and Research on new policies to advance development of safe and effective cell and gene therapies", https://www.fda.gov/news-events/press-announcements/statement-fda-commissioner-scott-gottlieb-md-and-peter-marks-md-phd-director-center-biologics, (2019).

[2] **Ginn, S., Amaya, A. K., Alexander, I. E., Edelstein, M., Abedi, M.R.,** "Gene therapy clinical trials worldwide to 2017: An update", *The Journal of Gene Medicine,* **20**(5), e3015, (2018).

[3] **European Medicines Agency, "**Vaxzevria (previously COVID-19 Vaccine AstraZeneca)", https://www.ema.europa.eu/en/medicines/human/EPAR/vaxzevria-previously-covid-19-vaccine-astrazeneca#overview-section, (2021).

[4] **U.S. Food and Drug Administration**, "Janssen COVID-19 Vaccine", https://www.fda.gov/emergency-preparedness-and-response/coronavirus-disease-2019-covid-19/janssen-

covid-19-vaccine, (2021).

[5] **Sarkis, M., Bernardi, A., Shah, N., Papathanasiou, M.,** "Emerging Challenges and Opportunities in Pharmaceutical Manufacturing and Distribution", *Processes*, **9**(3), 457, (2021).

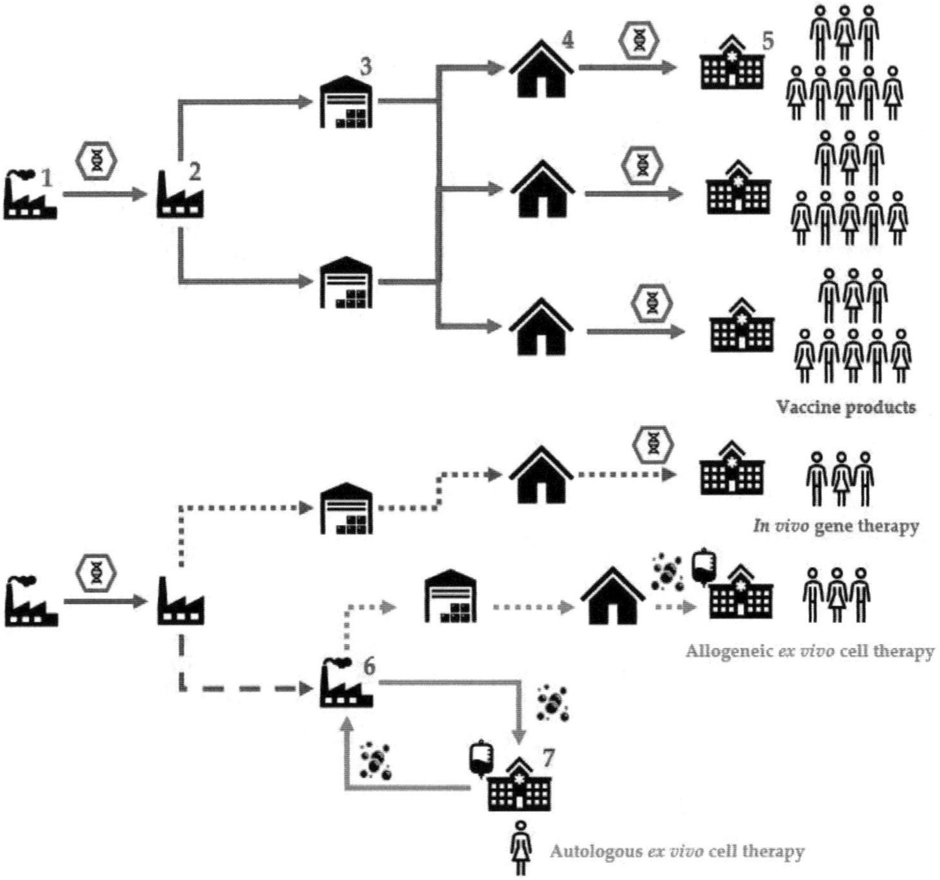

Figure 1: General schematic of supply chains for viral vector applications, comprising (1) primary manufacturing, (2) secondary manufacturing/fill-to-finish, (3) regional storage, (4) district storage, (5) hospitals/administration points as well as (7) specialised leukapheresis sites, (6) cell therapy manufacturing facilities.

624019 Systematic Decomposition & Evaluation of a Process Design Space for Monoclonal Antibody (mAb) Manufacturing

Abstract:

Three of the major trials within drug development and commercialization are phase 1 to 3 clinical trials. As these trials proceed product-process understanding increases and these are utilized via a feedback loop to the different development areas. The complex feedback loop consists of knowledge from input and output material properties (product related) and unit operations performance (process related). Monoclonal antibodies (mAbs) are of importance because they are proteins that are purposely engineered to treat various severe diseases for example cancer or hemophilia (Wolf Pérez et al., 2019). A typical pharmaceutical process for a specific active pharmaceutical ingredient (API) is batch-based because batch processes provide the flexibility of quality control at different intermediate steps and batch tracing inside the process boundaries and outside, that is, the product supply chain. Of interest, whether a continuous or batch process, is the evaluation of technologies that have a positive impact on process performance. Therefore, it is of interest to systematically evaluate relevant technologies for improving process performance and sustainability beyond high level "what-if analyses" to provide the foundation for advanced process understanding, design and optimization (Yang et al., 2020).

The mAb platform process to be investigated is pre-selected to be a batch process and therefore, any pre-selected flowsheet model for representation of the process is dynamic. Therefore, in analyzing upscale data for the platform process (lab and/or pilot), a systematic approach where multiple ideas associated with evaluating new technologies with respect to mass throughput and time execution is of importance. For the platform process, based on the large number of alternative placements of new technologies (integer decisions) and estimation of the input/outputs (continuous decisions) a systematic approach consisting of two layers of decisions is proposed. The decomposition layer addresses the integer decision problem, that is, placement of the new technology and the evaluation layer; having fixed the placement; addresses the continuous decision, that is, the expected gain in process performance, capacity and sustainability. This is performed using a simulation-based approach of material flow analysis with embedded non-linear process behavior.

The objective of this presentation is to present the systematic method for mAb process design space generation and screening. First, the problem and method will be presented. Second, the concept of discrete-event simulation will be introduced and how this can be used within the method for decomposition and evaluation of the design space. Finally, the method will be exemplified through an example of a promising technology for mAb manufacturing.

References

Wolf Pérez A. M., Sormanni P., Andersen J. S., Sakhnini L. I., Rodriguez-Leon I., Bjelke J. R., Gajhede A. J., De Maria L., Otzen D. E., Vendruscolo M. and Lorenzen N. 2014. In vitro and in silico assessment of the developability of a designed monoclonal antibody library. mAbs, 11(2), 388–400.

Yang, O., Qadan, M. & Ierapetritou, M. Economic Analysis of Batch and Continuous Biopharmaceutical Antibody Production: a Review. 2019. Journal of pharmaceutical innovation, 15, 182–200.

621647 Keynote Talk: Integrated Quality By Design in (Bio)Pharmaceutical Manufacturing

Abstract:

This presentation describes a Quality-by-Design approach for an accelerated process development workflow for the design of integrated (bio)pharmaceutical manufacturing systems from drug substance to drug product. This approach draws R&D and management into closer alignment and execution while defining product critical quality attributes, the operational design space, the manufacturing process, and the control strategy by integrated use of the ICH quality quartet – namely Q8(R2), Q9, Q10, and Q11. The workflow employs process intensification, micro-scale technologies, modular unit operations with integrated process control and monitoring systems, systems integration, and full plant automation. The presentation describes how to best develop and transfer knowledge between steps in the workflow via mechanistic models, data analytics, and machine learning. The approach is illustrated in more than a half dozen small- and large-molecule applications, including where process equipment was first designed digitally using first-principles models, then constructed and implemented experimentally, with the experimental results confirming model predictions. These applications include a wide variety of upstream, downstream, and formulation unit operations for small-molecule and biological drug manufacturing, as well as the design and operation of fully automated end-to-end manufacturing systems.

624003 Keynote Talk: An *in silico* approach for Monoclonal Antibody (mAb) Process Research & Early Development

Abstract:

Monoclonal antibodies (mAbs) are a type of biopharmaceutical that can be used to treat various diseases, for example, hemophilia and other serious diseases (Wolf Pérez et al., 2019). Improvements for the development of mAbs can be categorized as follows, product related improvements, for example, expression, solubility etc. and process related improvements, for example, novel flowsheets, new technologies (unit operations) etc. The synergistic effect of product-process decisions is also of importance because it addresses the integrated problem, that is, how can process design decisions be accounted for in the product design phase. Consider the example, given a specified mAb, purify the mAb using a chromatography column and selected eluent at a predetermined concentration. If the resulting aggregated pool volume for a finite time horizon for a pilot and/or industrial process requires a non-standard tank size that increases capital investment, how can this be a priori mitigated? If product and process design is performed sequentially then the design decision is unchanged however, if product-process design is performed integrated then, additional technologies can be evaluated for up-concentration. Therefore, the process is sized to the required scale and not oversized which has both capital investment and time-to-market implications.

Production planning-scheduling (PPS) is a field within operations research that solves the dynamic problem as to what to produce when (Harjunkoski et al., 2014). It can be performed using optimization-based methods and simulation-based methods, for example, discrete event simulation. The former differs from the latter because it provides an optimal solution for a set of decisions however, the latter provides the opportunity to deep dive into the details of the plan-schedule. Therefore, the symbiosis between the two approaches can be explained as follows. The former generates the optimal plan-schedule and the latter simulates how this is executed. In other words, the latter acts as a digital version of the process, a so-called digital twin. Therefore, we can use discrete event simulation to also evaluate the impact of different technologies and various design decisions on pharmaceutical processes, evaluate economic and sustainability impacts and improve decision making a priori to final investment (Yang et la., 2020).

The objective of this presentation is to present an *in silico*, systematic framework for multiscale process design and idea evaluation for biopharmaceutical products. A case study applied to mAbs will be presented. The method will be presented, and its application exemplified through idea evaluation related to new technologies for mAb production. Finally, the screening of the evaluation space using a combination of economic and sustainability metrics will be presented.

References

Harjunkoski I, Maravelias CT, Bongers P, Castro PM, Engell S, Grossmann IE, Hooker J, Méndez C, Sand G and Wassick J. 2014. Scope for industrial applications of production scheduling models and solution methods. Computers and Chemical Engineering, 62, 161–193.

Wolf Pérez A. M., Sormanni P., Andersen J. S., Sakhnini L. I., Rodriguez-Leon I., Bjelke J. R., Gajhede A. J., De Maria L., Otzen D. E., Vendruscolo M. and Lorenzen N. 2014. In vitro and in silico assessment of the developability of a designed monoclonal antibody library. mAbs, 11(2), 388–400.

Yang, O., Qadan, M. & Ierapetritou, M. 2019. Economic Analysis of Batch and Continuous Biopharmaceutical Antibody Production: a Review. Journal of pharmaceutical innovation, 15, 182–200.

627576 Optimization Framework for the Electroreduction of CO_2 into Chemicals

Abstract:

The past few years have seen rapid advances in the experimental design of the electroreduction of carbon dioxide (ECO2R) for the synthesis of chemicals and fuels, leading to its foreseeable industrial implementation. A transition to carbon-free chemicals could have a considerable impact on the future of the chemical industry but requires holistic methods to assess the technological, economic, and environmental performance of ECO2R processes and optimize process network designs. In this work, we propose a methodology to assess the large-scale implementation of ECO2R based on the analysis of market data, traditional process synthesis, techno-economic assessment, and supply chain optimization.

We build upon previous studies that have dealt with the modeling and assessment of ECO2R processes (Jouny et al., 2018; Orella et al., 2019; Roh et al., 2020), hybrid fossil- and CO_2-based routes (Ioannou et al., 2020) and supply chain optimization of carbon utilization processes (Leonzio et al., 2020; Zhang et al., 2020). Our method consists of a multi-level approach to network synthesis with detailed models of each one of the technologies involved. The network is modeled as a superstructure that includes different options of carbon capture, electrolysis, and separation.

To illustrate the capabilities of the method, we define a case study on the substitution of fossil-based ethylene, ethanol, and formic acid by their CO_2-sourced alternatives. Different scenarios based on technology adoption and efficiency, electricity prices, and CO_2 source are considered to target the break-even conditions necessary to adopt the technology. The results obtained show that the model is a useful tool to identify the most promising routes to convert CO_2 into valuable chemicals. The methodology is valuable for the multiple actors involved in the waste revaluation sector, from private companies to policymakers.

References

Ioannou, I., D'Angelo, S.C., Martín, A.J., Pérez-Ramírez, J., Guillén-Gosálbez, G., 2020. Hybridization of Fossil- and CO 2 -Based Routes for Ethylene Production using Renewable Energy. ChemSusChem cssc.202001312. https://doi.org/10.1002/cssc.202001312

Jouny, M., Luc, W., Jiao, F., 2018. General Techno-Economic Analysis of CO 2 Electrolysis Systems. Ind. Eng. Chem. Res. 57, 2165–2177. https://doi.org/10.1021/acs.iecr.7b03514

Leonzio, G., Foscolo, P.U., Zondervan, E., 2020. Optimization of CCUS Supply Chains for Some European Countries under the Uncertainty. Processes 8, 960. https://doi.org/10.3390/pr8080960

Orella, M.J., Brown, S.M., Leonard, M.E., Román-Leshkov, Y., Brushett, F.R., 2019. A General Technoeconomic Model for Evaluating Emerging Electrolytic Processes. Energy Technol. 1900994. https://doi.org/10.1002/ente.201900994

Roh, K., Bardow, A., Bongartz, D., Burre, J., Chung, W., Deutz, S., Han, D., Heßelmann, M., Kohlhaas, Y., König, A., Lee, J.S., Meys, R., Völker, S., Wessling, M., Lee, J.H., Mitsos, A., 2020. Early-stage evaluation of emerging CO 2 utilization technologies at low technology readiness levels.

Green Chem. 22, 3842–3859. https://doi.org/10.1039/C9GC04440J

Somoza-Tornos, A., Giraldo-Carvajal, V., Espuña, A., Graells, M., 2019. Targeting material exchanges in industrial symbiosis networks, Computer Aided Chemical Engineering. https://doi.org/10.1016/B978-0-12-818634-3.50255-1

Zhang, S., Zhuang, Y., Liu, L., Zhang, L., Du, J., 2020. Optimization-based approach for CO2 utilization in carbon capture, utilization and storage supply chain. Comput. Chem. Eng. 139, 106885. https://doi.org/10.1016/j.compchemeng.2020.106885

627651 Material Property Targets for Emerging Adsorptive Water Treatment and Resource Recovery Systems

Abstract:

There is a critical need to develop sustainable separations technologies to meet current and emerging challenges, for example, i) the removal of trace metals such as lead from water to ensure acceptable contaminant levels[1] and ii) the recovery of lithium from brines to keep up with the raw material demand due to vehicle electrification[2]. Yet, the synthesis of numerous advanced materials in the past three decades have failed to produce a corresponding revolution in separation applications[3]. In this talk, we present a mathematical modeling framework for material property targets in general adsorption separation systems; using the framework, we benchmark existing materials and identify theoretical limits for the lead removal and lithium recovery applications [4].

In the paradigm of process targeting[5], feasible high-level goals such as solute recovery targets are fixed before detailed process optimization. Once targets are fixed, fundamental thermodynamics and material and energy balances are used to find alternate, often non-trivial pathways to achieve the targets. Process targeting is commonly applied to flowsheet optimization wherein the individual unit operations are optimized after setting a feasible overall production (or utility consumption) target[6]. Traditionally, design is carried out using sequential techniques in which unit operations are optimized in a logical order which is most often the sequence based on the process flowsheet. Sequential design techniques often place a huge demand on the downstream processes which makes them very difficult to design efficiently. Process targeting overcomes the limitations of sequential design by explicitly accounting for inter-unit interactions early in the design process.

We extend process targeting to study adsorptive separation systems from a multiscale perspective: we set process targets at the device-scale to study the feasibility of using existing adsorbent materials as the separating medium. We use material balances and isotherms to set process targets. At the materials scale, we use structure-property relationships to define the theoretical limits of adsorptive membranes and packed-bed adsorbers. We demonstrate that while there is still scope for improvement, the sorbents for next-generations separations such as the lead removal from water and lithium extraction from brine already exist and are limited by engineering at the device-scales or higher. In other words, our framework helps illuminate if specific applications are process- or materials-limited. We highlight our belief that adsorption-based devices will play a key role in realizing the fit-for-purpose paradigm of water treatment which places emphasis on treating water to meet the varied specifications of its end users[7].

References

[1] Hanna-Attisha, M., LaChance, J., Sadler, R., & Schnepp, A. (2016). Elevated Blood Lead Levels in Children Associated With the Flint Drinking Water Crisis: A Spatial Analysis of Risk and Public Health Response. *American Journal of Public Health*, 106(2), 283–290. https://doi.org/10.2105/AJPH.2015.303003

[2] Li, H., Eksteen, J., & Kuang, G. (2019). Recovery of lithium from mineral resources: State-of-the-art and perspectives – A review. *Hydrometallurgy*, 189. https://doi.org/10.1016/j.hydromet.2019.105129

[3] Luo, J., & Crittenden, J. (2019). Nanomaterial Adsorbent Design: From Bench Scale Tests to Engineering Design. *Environmental Science & Technology,* 53(18), 10537–10538. https://doi.org/10.1021/acs.est.9b04371

[4] Eugene, E. A., Phillip, W. A., Dowling, A. W., (2020). Material Property Targets for Emerging Nanomaterials to Enable Point-of-Use and Point-of-Entry Water Treatment Systems. *Preprint submitted to ACS ES & T Engineering*
https://chemrxiv.org/articles/preprint/Material_Property_Targets_for_Emerging_Nanomaterials_to_Enable_Point-of-Use_and_Point-of-Entry_Water_Treatment_Systems/12526190/1

[5] Fox, J. A., & Stacey, N. T. (2019). Process targeting: An energy based comparison of waste plastic processing technologies. *Energy (Oxford)*, 170, 273–283. https://doi.org/10.1016/j.energy.2018.12.160

[6] Hildebrandt, D., Glasser, D., Patel, B., Sempuga, B. C., & Fox, J. A. (2015). Making processes work. *Computers & Chemical Engineering*, 81, 22–31. https://doi.org/10.1016/j.compchemeng.2015.03.022

[7] Zodrow, K., Li, Q., Buono, R., Chen, W., Daigger, G., Dueñas-Osorio, L., Elimelech, M., Huang, X., Jiang, G., Kim, J.-H., Logan, B., Sedlak, D., Westerhoff, P., & Alvarez, P. (2017). Advanced Materials, Technologies, and Complex Systems Analyses: Emerging Opportunities to Enhance Urban Water Security. *Environmental Science & Technology,* 51(18), 10274–10281. https://doi.org/10.1021/acs.est.7b01679

628097 Integrated Waste-to-Energy and Energy Management Platform: A Demonstration in an Eco-Industrial Park

Abstract:

The rapid expansion of urban populations leads to the increases in the generation of municipal solid waste (MSW) and energy consumption, causing great pressure on the waste treatment and energy systems. Waste-to Energy (WtE) technologies, which processes waste into electricity, heat, and biogas, are popular waste treatment alternatives to landfill that address the problem of rising energy consumption simultaneously. Nowadays, both the conventional centralized MSW and energy management systems are evolving toward a more decentralized form with smaller capacity. This shift would requires an adapted decision-support method to assist decision-makers in analyzing and optimizing the WtE systems. In this work, an integrated waste-to-energy and energy management platform was developed and it is comprised of two parts: 1) waste-to-energy decision-support system (WTEDSS) and 2) energy management decision support system (EMDSS).

The WTEDSS addresses the process of electricity generation from waste. Firstly, the historical amount of waste generation is analysed to projects future waste generation for different waste streams. Then, mixed-integer linear programming (MILP) optimization is used to determine the optimum capacities of waste treatment facilities (centralized vs decentralized, anaerobic digestion (AD) vs incineration) and the allocation of waste intakes from different waste generation sites for different facilities for each year by maximizing the net present value (NPV) of the waste management system.

The EMDSS deals with the management of electricity generated by the WtE facilities using a game-theory-based demand-side management (DSM) incorporating energy storage components. The DSM is used to reduce the Peak-to-Average ratio smoothen the dips in load profile caused by supply constraints. A novel data-driven optimization framework is incorporated to select and size energy storage systems at suitable locations of the electricity network. The emerging blockchain technologies are also introduced to guarantee the seamless and secure implementation of the proposed energy management scheme, illustrating how a decentralized DSM approach could be implemented in practice.

As a demonstration, the integrated platform is applied to Jurong Island, an eco-industrial park in Singapore. The platform help decision-makers with examining the design and operation of the waste-to-energy and energy management systems under different scenarios. A case study showed that decentralized waste-to-energy systems were preferred if the current capital cost of the decentralized AD system could be lowered. The incorporation of DSM and energy storage systems were able to maintain the supply and demand balance, reduce utility bills of consumers, and reduce stress on grid in daily operation.

630716 Design and Integration of Thermal Energy Storage Systems for Power Plants

Abstract:

Integration of intermittently available renewable energy with electricity grids leads to increased variability and uncertainty in power generation, thereby pushing many conventional power plants to speedily ramp their loads to maintain grid reliability. Considerable loss of power plant energy is incurred due to rapid cycling and time delay during energy transmission. Frequent cycling can also induce low overall efficiency and high failure rates of high-temperature units in the long run [1]. Energy storage can resolve these challenges. Among the several existing energy storage alternatives, thermal energy storage is particularly promising because of cost-effective design, operational flexibility, high efficiency, stability and long storage lifetime. Thermal storage has been widely employed in the cooling and heating system of buildings, chemical processes as well as solar plants in the past [2-5]. However, the optimal design and integration of thermal energy storage with large-scale fossil-fueled power plants have not been investigated rigorously. There are two types of thermal energy storage, namely high-temperature thermal storage (HTTS) and cryogenic energy storage (CES). HTTS employs molten salt or phase change materials (PCM) as storage medium. The operating temperature of molten salt ranges from 70 to 800 °C. Conversely, the operating temperatures of PCM are between 130-580 °C. Depending on the storage medium, process configuration and operation, the costs range from $0.5–60 per kg of the material used. CES provides a highly efficient mechanism to store excess energy by liquefying air to obtain cryogen [6-8]. The stored cryogen is further used to generate energy during high energy demand scenarios. To increase efficiency of the CES process, cryogenic turbines and internal heat transfer loops with intermediary thermic fluids are used for energy retention. There also exists a large range of options for thermal fluids, storage pressures, trains for compressing and cooling, and expansion cycles for energy generation, which affect the capital and operational cost of a CES system.

We propose a model-based process design and synthesis approach for thermal energy storage systems that store energy from high-/medium-/low-pressure steam extracted from the power plant and discharge energy on demand by regenerating steam from stored energy which is sent back to the power plant. The designs are complicated as the heat transfer involves both sensible and latent heat, different types of storage media, various techno-economic and energetic trade-offs, and numerous possible configurations and operational stages for heat integration. To this end, we have developed a superstructure configuration of the storage systems, thereby allowing the various alternative configurations to adapt different charging, discharging, and integration strategies effectively. The superstructure optimization problem is formulated as a mixed-integer nonlinear program (MINLP), which is then solved to identify the optimal process configurations, number of cycles, charging/discharging strategies, and storage medium. The objective function and the constraints consider various trade-offs between ramp rates, load balances, lifespan, and capital costs, and leading to optimal energy storage operations. Our results indicate that the performance and the cost of energy storage depend strongly on charging steam conditions and the qualities of both charging and discharging steam. Using realistic power plant conditions and charging/discharging demands, we illustrate how the MINLP-based synthesis of storage configurations provides the target process configurations, capacities, and operational decisions with an efficiency of 85%-90%, thereby reducing the overall cost of energy storage for a conventional power plant. The insights obtained from the process synthesis study is further incorporated in developing surrogate models for dynamic integration of storage technologies for distributed energy storage for different scenarios. The development of these technology models will pave way for solving an overall storage downselection problem for reducing the cycling of conventional power plants in face of intermittent renewable energy integration.

References:

[1]Kumar, N., Besuner, P., Lefton, S., Agan, D., & Hilleman, D. (2012). Power Plant Cycling Costs. doi: 10.2172/1046269

[2]Edwards, J., Bindra, H., & Sabharwall, P. (2016). Exergy analysis of thermal energy storage options with nuclear power plants. *Annals of Nuclear Energy*, *96*, 104–111. doi: 10.1016/j.anucene.2016.06.005

[3]Richter, M., Möllenbruck, F., Starinsk, A., Oeljeklaus, G., & Görner, K. (2015). Flexibilization of Coal-fired Power Plants by Dynamic Simulation. *Proceedings of the 11th International Modelica Conference, Versailles, France, September 21-23, 2015*. doi: 10.3384/ecp15118715

[4]Miró, L., Gasia, J., & Cabeza, L. F. (2016). Thermal energy storage (TES) for industrial waste heat (IWH) recovery: A review. *Applied Energy*, *179*, 284–301. doi: 10.1016/j.apenergy.2016.06.14

[5]Zhu, Y., Zhai, R., Peng, H., & Yang, Y. (2016). Exergy destruction analysis of solar tower aided coal-fired power generation system using exergy and advanced exergetic methods. Applied Thermal Engineering, 108, 339–346. doi: 10.1016/j.applthermaleng.2016.07.116

[6]Zhang, Q., Grossmann, I. E., Heuberger, C. F., Sundaramoorthy, A., & Pinto, J. M. (2015) Air separation with cryogenic energy storage: optimal scheduling considering electric energy and reserve markets. AIChE Journal. doi: 10.1002/aic.14730

[7]Li, Y., Cao, H., Wang, S., Jin, Y., Li, D., Wang, X., & Ding, Y. (2014) Load shifting of nuclear power plants using cryogenic energy storage technology. Applied energy. doi: 10.1016/j.apenergy.2013.08.077

[8] Abdo, R. F., Pedro, H. T., Koury, R. N., Machado, L., Coimbra, C. F., & Porto, M. P. (2015). Performance evaluation of various cryogenic energy storage systems. Energy. doi: 10.1016/j.energy.2015.08.008

632270 A Multiscale Electro-Chemical Model for Simulating Dendrite Formation in Lithium-Ion Batteries

Abstract:

The growing market demand for Lithium-ion batteries (LIBs) with a high energy density has given an impetus to a plethora of studies investigating the utilization of Lithium anode, which is considered as the 'Holy-Grail' anode due to its high energy density (~ 3860 mAh/g) as compared to traditional graphite anodes (~ 372 mAh/g) [1]. However, commercial application of Li metal anode has not been possible because it is plagued by irregular Li deposition, which causes dendrite formation and results in poor cycling efficiency, capacity fade, and in extreme cases, short-circuiting of batteries [2]. Previous studies show that dendrite formation is majorly affected by microscopic battery parameters [1] (i.e., mechano-chemical properties of the SEI species), and macroscopic operating conditions (i.e., cell voltage, current density, and Li-ion concentration) [3]. However, the complex interaction between these two types of factors (i.e., macroscopic and microscopic) has not been adequately addressed in the literature.

Inspired by this, we have constructed a multiscale model that combines a high-fidelity kMC simulation with an electrochemical continuum model, which in this case is a modified form of the single-particle model (SPM) [4]. Specifically, the kMC model employs a series of microscopic events like the desolvation of Li-ions from the electrolyte, their diffusion in the SEI,[1][6] and finally the electrodeposition of Li-ions on the anode to simulate the microscopic dendrite formation phenomenon [5],[6]. The electrochemical model considers Fickian diffusion of Li-ions in the electrolyte region to trace the spatio-temporal evolution of , and then couples this with various equations to determine cell voltage and the state of charge (SOC). The proposed multiscale model is simulated for varying current densities at different values of initial Li-ion concentration to investigate its effect on microscopic dendrite formation (quantified using roughness), and the temporal evolution of various macroscopic variables. Furthermore, the simulation results for dendrite growth and SOC (which is a measure of the charging status of LIBs) are successfully validated with experimental results from the literature.

Overall, the proposed multiscale electrochemical model can (a) generate the evolution of macroscopic variables such as current density, Li-ion concentration, cell voltage, and SOC, (b) explain their effect on the microscopic spatio-temporal evolution of dendrites (kMC model), and (c) be extended to simulate multiple charge-discharge cycles to help devise effective dendrite mitigation strategies in future studies.

Literature Cited:

1. Qin X, Shao M, Balbuena PB. Elucidating mechanisms of Li plating on Li anodes of lithium-based batteries. Electrochemica Acta. 2018 Sep 10; 284:485-94.
2. Wen J, Yu Y, Chen C. A review on lithium-ion batteries safety issues: existing problems and possible solutions. Materials Express. 2012 Sep 1;2(3):197-212.
3. Hao F, Verma A, Mukherjee PP. Mechanistic insight into dendrite–SEI interactions for lithium metal electrodes. Journal of Materials Chemistry A. 2018;6(40):19664-71.
4. Zhang D, Popov BN, White RE. Modeling lithium intercalation of a single spinel particle under potentiodynamic control. Journal of the Electrochemical Society. 2000 Mar 1;147(3):831.
5. Sitapure N, Lee H, Ospina-Acevedo F, Balbuena PB, Hwang S, Kwon JS. A computational approach to characterize formation of a passivation layer in lithium metal anodes. AIChE

Journal. 2021 Jan; 67(1): e17073.
6. Angarita-Gomez S, Balbuena PB. Insights into lithium ion deposition on lithium metal surfaces. Physical Chemistry Chemical Physics. 2020;22(37):21369-82.

624138 Computationally Efficient Distillation Energy Targeting Model for Superstructure-Based Process Synthesis

Abstract:

Separation processes account for a significant fraction of the total energy consumption in chemicals and fuels facilities[1]. Thus, to design energy efficient systems, especially during the early design phase, it is important to develop methods that allow us to easily estimate separations-related energy requirements. Accordingly, given the importance of distillation in the process industries, the goal of this work is to develop an improved method for the calculation of energy demand of complex distillation networks. Shortcut methods, which do not rely on rigorous mass-energy balances, are particularly useful for this purpose[2,3]. In particular, the Underwood equations[2] have been adopted in many works because of their simplicity, that is, the computationally cheap calculation of minimum vapor flow rates to achieve the separation task and from the flow the calculation of the minimum energy requirement. Notably, to solve the Underwood equations, components that are present in the stream to be separated need to be known a priori so that all relevant roots of the Underwood equations can be calculated.

However, when synthesizing an integrated reactor-separation system, component flow rates of the stream to be separated can vary, and some components can have even zero flow rates, depending on decisions in the upstream reactor subsystem(s). Thus, components in the stream to be separated may not be known a priori. Such streams are referred to as *undetermined*[4]. The presence of undetermined streams complicates the use of the Underwood equations; for instance, existing approaches disaggregate the stream so that each disaggregated stream can have a known set of components. Then, each disaggregated stream is connected to a separate distillation network/column. This approach can easily lead to a large optimization problem and therefore prevent the use of Underwood equations (and other shortcut models) when multiple undetermined streams are present.

To address this limitation, we propose a novel reformulation of the Underwood equations, which allows the automatic calculation of all relevant roots efficiently even for undetermined streams. We introduce binary variables to detect components in the stream to be separated, and these binary variables are then used to constrain the Underwood roots accordingly. Also, we introduce strong valid constraints which significantly enhance computational efficiency. Then, utilizing this reformulation, we propose a distillation energy targeting model inspired by the Fully-Thermally-Coupled (FTC) configuration[5], which has the minimum energy demand among all configurations for the distillation of a zeotropic mixture. The proposed model can be used to calculate the energy target to separate a stream, potentially undetermined, into pure components without finding detailed distillation network configurations. Thus, it is particularly useful for preliminary process synthesis with multiple candidate reactions, where many undetermined streams can naturally appear. Also, we introduce some extensions to consider cases where some components do not need to be separated from each other. For example, for fuel production, some components do not need to be separated from each other if all property constraints can be met.

We present a number of case studies. First, we present the preliminary synthesis of a multi-stage itaconic acid production process. Then, we present the preliminary synthesis of ethanol upgrading bio-refinery, which has more complexity. These case studies illustrate how the proposed energy targeting model can be used to formulate and solve interesting and challenging process synthesis problems.

References

1. Sholl DS, Lively RP. Seven chemical separations to change the world. *Nature News*. 2016;532(7600):435
2. Bausa J, Watzdorf RV, Marquardt W. Shortcut methods for nonideal multicomponent distillation: I. Simple columns. *AIChE journal*. 1988;44(10), 2181-2198.
3. Underwood AJ V. Fractional distillation of multicomponent mixtures. *Ind Eng Chem*. 1949;41(12):2844-2847.
4. Ryu J, Maravelias CT. Computationally efficient optimization models for preliminary distillation column design and separation energy targeting. *Computers & Chemical Engineering*. 2020;143:107072.
5. Halvorsen IJ, Skogestad S. Minimum energy consumption in multicomponent distillation. 3. More than three products and generalized Petlyuk arrangements. *Ind Eng Chem Res*. 2003;42(3):616-629.

624709 Dynamic Simulation of an Ammonia Synthesis Plant Fed By Stranded Natural Gas in Aspen Hysys

Abstract:

The Haber Bosch process is one of the most important industrial processes to modern society. In 2015, 160 million tons of ammonia was produced mainly from the Haber Bosch process, with a global value of $250 billion [1,2]. The process has benefited from over 100 years of improvements and modifications. Today, the Haber Bosch process is being modified to operate on alternative feedstocks in an attempt to preserve the world's natural gas reserves [2]. Modifying the Haber Bosch process to use Stranded Natural Gas (SNG) as a feedstock, can allow for the monetization of an unexploited resource [3]. SNG, is natural gas that is in a location that is uneconomical to recover due to its quantity, composition or both. SNG sources are intermittent and the quality of natural gas from the source can fluctuate dramatically. Previously a Haber Bosch process fed by SNG was studied at steady state [3], which does not accurately capture the transient nature of the feed source. Dynamic simulation of a Haber Bosch process fed by SNG will allow for accurate: operational, economic and environmental assessments of the modification.

This presentation will communicate the development, performance and validation of a dynamic model of a 100 tNH_3/day plant in Aspen Hysys [4]. The process of Fig. 1 uses feed streams of natural gas, water, and air to synthesize ammonia in a three-stage fixed bed reactor. Steam reforming of methane is performed to generate hydrogen from a variable stranded natural gas feed modeled after natural gas wells in the Bakken Basin of North Dakota. A water gas shift reactor is employed to convert CO from the steam reformer to CO_2. The CO_2 is removed ahead of the ammonia synthesis reactor using a Monoethanolamine stripping unit. Ammonia is synthesized at 200 atm and 450^oC, and is then refrigerated to -20^oC before being stored. The model includes heat integration of wastewater streams to reduce water consumption and pressure relief systems to prevent over-pressurization of system components. The model also includes controllers programmed to control system temperature, pressure, and flows. The steady state results of the model are compared and agree with the performance of previous works [4]. The following case studies are presented in this paper: startup under variable natural gas composition and flow, and normal operation under variable natural gas composition and flow. The model is found in good agreement with prior work and, therefore, it can be used as a benchmark virtual testbed for evaluation of intensified schemes of ammonia synthesis such as those presented in [4].

Acknowledgements

This work was partially sponsored by the United Technologies Corporation Institute for Advanced Systems Engineering (UTC-IASE) of the University of Connecticut. Any opinions expressed herein are those of the authors and do not represent those of the sponsor.

References

[1] Fao, *World fertilizer trends and outlook to 2018;* Annual Report 14; Food and Agriculture Organization of the United Nations (FAO): Rome, Italy, 2015; ISBN 978-92-5-108692-6.

[2] L. Wang *et al.*, "Greening Ammonia toward the Solar Ammonia Refinery," *Joule*, vol. 2, no. 6, pp. 1055–1074, 2018.

[3] Z. Gou, Monetization of Stranded Gas through Ammonia and Urea Production, Texas A&M University 2017

[4]. L. Burrows *et al.*, (In Review) "Dynamic Simulation of an Ammonia Synthesis Plant fed by Stranded Natural Gas in Aspen Hysys^{â}," *Chem. Eng. J.*

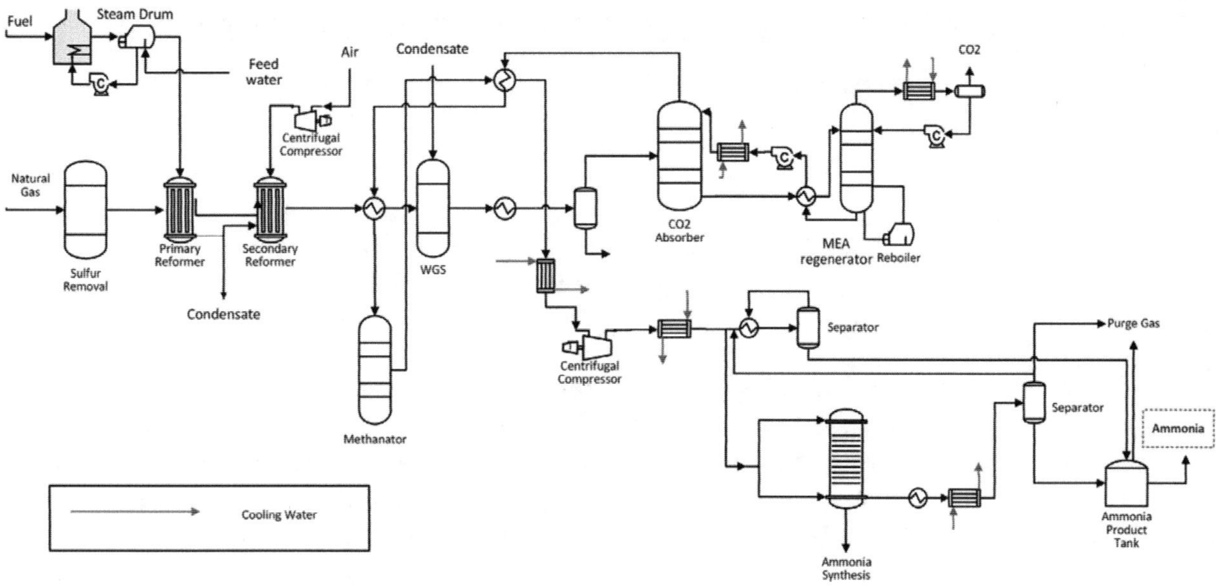
Fuel
Steam Drum
Feed water
Air
Condensate
CO2
Centrifugal Compressor
Natural Gas
Sulfur Removal
Primary Reformer
Secondary Reformer
Condensate
WGS
CO2 Absorber
MEA regenerator
Reboiler
Methanator
Centrifugal Compressor
Separator
Purge Gas
Separator
Ammonia
Ammonia Product Tank
Ammonia Synthesis
Cooling Water

623000 Sequence Modulates Polypeptoid Hydration Water Structure and Dynamics

Abstract:

Polypeptoids have become an important class of polymers, capable of precisely defined sequences while remaining gram-synthesizable—properties that have driven a rapidly expanding set of applications and also make polypeptoids ideal model heteropolymers for insight into sequence-structure-function relationships. However, such relationships have been challenging to establish for structurally disordered polypeptoids, which adopt a wide range of conformations in solution and are thus difficult to characterize. Here, we first present a computational workflow to characterize the conformational landscape of disordered polypeptoids using enhanced-sampling molecular dynamics. We validate this workflow using experimental measurements of the full end-to-end distance distributions from Double Electron-Electron Resonance (DEER) spectroscopy, which show excellent agreement with the simulations. We then leverage the workflow to probe shifts in polypeptoid hydration behavior with sequence, which is central to many key polypeptoid applications, such as antifouling coatings. We show that water near polypeptoids with hydrophobic monomers is less dense, more tetrahedral, and slower than water near polypeptoids with hydrophilic monomers. This effect is independent of sequence-induced changes in polypeptoid conformation, suggesting that changes in chemistry alone are sufficient to induce shifts in water structure and dynamics at molecular scales. Furthermore, these shifts in water structure and dynamics are correlated, indicating that experimental probes of local water dynamics may also indirectly provide insight into structural shifts.

630496 Role of Charge Patterning and Hydrophobicity in Peptide-Based Complex Coacervates

Abstract:

Complex coacervates have been used as a versatile materials platform to explore the material microenvironment with a goal of facilitating a plethora of biochemical and biomedical applications ranging from biosensors and catalysts to refrigeration free vaccines. Complex coacervates are materials that are formed by the associative phase separation of two oppositely charged polymers (polycation and polyanion), which is driven by an electrostatic attraction and an entropically driven complexation. Here, we design and synthesize various patterned peptides as model polymers to study the role of charge patterning and hydrophobicity in our coacervate formation. Although there has been extensive effort at accurately describing this phase behavior phenomenon, there is still a need to understand how it is affected by the specific chemistry of the polymers involved. Here, we use a combination of a systematic practical approach and a validated innovative theoretical approach to incorporate sequence effects and hydrophobicity into our physical understanding of our system and thereby broadening the range of chemical functionalities in our polymers.

624803 Impact of Collagen-like-Peptide (CLP) Triple Helix Design on CLP Melting Transition and Assembly: A Coarse-Grained Molecular Dynamics Simulation Study

Abstract:

Recent advances in materials design, synthesis, and simulation have allowed the creation of biomimetic materials with responsive yet controllable physicochemical properties. Materials that self-assemble into desired morphologies such as fibrils and supramolecular networks are of particular interest, where their ability to self-assemble can be tuned by applying external stimuli such as heat, light, pH, and salt for a range of applications including in drug delivery and tissue engineering. In this talk, I will present our recent coarse-grained (CG) molecular dynamics (MD) simulation studies on the melting transitions, fibrillar assembly, and gelation of collagen-like peptides (CLPs) as a guide for designing self-assembling peptide-based biomaterials. CLPs are thermoresponsive biopolymers in which each CLP chain is made up of repeat units of amino acid triplets, (X-Y-G), where X and Y are usually proline (P) and hydroxyproline (O), respectively. Like native collagen, CLPs have been shown to assemble to form triple helices, fibrils, and gels in aqueous solutions thus exhibiting self-assembly at multiple length scales. In this work, we extend our CG CLP model to simulate CLP heterotrimers in which the length of each of the three CLP strands forming the triple helix can be different. Inspired by the heterotrimeric nature of natural collagens, we investigate CLP heterotrimers with sticky ends which self-assemble to form fibrils and fibrillar networks driven by inter-chain and inter-helix hydrogen bonding. We explore how design parameters including the length and number of sticky ends in each CLP triple helix impact its thermal stability at low CLP triple helix concentration and assembly at varying CLP triple helix concentration. Overall, our work highlights the predictive capabilities of CG MD simulations in guiding experiments, as these complex peptide systems with unique molecular insights can inform new system designs and streamline the discovery of new, biomimetic platforms.

631473 Controlled Alignment of Collagen and its Influence on the Proliferation of Human Schwann Cells

Abstract:

Collagen is a key component of the extracellular matrix and well-oriented domains of collagen are relevant for mimicking the local cell environment in vitro. While there has been significant attention directed towards the alignment of collagen, formation of large-scale oriented domains remains a key challenge. Type I collagen self-assembles to form liquid crystalline (LC) mesophases in acidic conditions at concentrations above 50 mg/ml. The LC mesophase provides an efficient platform for large-scale alignment and patterning of collagen coated substrates. However, several questions remain regarding the self-assembly of collagen which results in the formation of the LC phase. In addition, there exist challenges related to solubilizing and processing of collagen at high concentrations.

In this contribution we report on centimeter-scale alignment in collagen-coated glass substrates using solutions that are well below the LC forming concentrations. We explain the formation of these uniform domains via shear-induced liquid crystallinity of the collagen solutions. We characterize the proliferation of human Schwann cells along the direction of alignment of the collagen-coated substrates.

632566 Opto-Chemical Characterization and Determination of Nanostructural Organization in Complex Leafhopper Brochosome Protein Assemblies

Abstract:

Biological systems rely on complex and robust assembly of biomolecular building blocks, yielding hierarchical macromolecular structures with unique functions and characteristics. Brochosomes are a fascinating example of highly ordered macromolecular structures exhibiting hollow and pitted Buckyball shaped nanostructures. Brochosomes are produced in the hindgut of leafhoppers (Hemiptera; Cicadellidae) and are composed of a unique class of intricately arranged and disulfide bond crosslinked secretory proteins. Despite recent progress, we lack a full understanding of structural features of these materials, including overall spatial arrangement of brochosome building blocks. In this talk, I will discuss our recent work in understanding the short- and long-range structural order of brochosomes using opto-chemical atomic force microscopy and x-ray scattering techniques. Results have indicated the presence of optical activity consistent with secondary protein structures along brochosome surfaces, while small-angle x-ray scattering has indicated that some specific structural orientation may be present. Furthermore, single and clustered brochosomes have shown regions of high optical absorbance across several IR-wavelengths, which may be related to their potential as a material for camouflage applications. Overall, these findings will inform next-generation synthetic methods for producing complexed self-assembled hierarchical structures.

621975 Mucus: Cactus-like Conformations of Associative Polymers

Abstract:

The unique fractal structure of macromolecules in human bronchial epithelial mucus is determined by small-angle neutron scattering. Hybrid Monte Carlo/molecular dynamics simulations of a minimal coarse-grained model of polymers with associative groups distributed evenly along the chain contour is found to reproduce the experimental form factor. A unique "cactus" model resulting from loops of different orders is proposed. Predictions with number and spacing between associative groups are given for the scaling of the polymer radius of gyration, contact probability, and pairwise interchain interactions.

625592 Barrier Coatings Derived from Cellulose and Chitin

Abstract:

There is a strong drive to discover and develop alternatives to conventional plastics that offer the ability to be manufactured and used in a circular manner. In a circular economy, as opposed to a linear one, materials are derived from renewable resources or recycled content, and at the end of life, they are able to be circulated back into production via a chemical, physical or biological pathway. A critical need exists to develop such materials for plastic packaging, which represents the largest contributor to unrecyclable or difficult-to-recycle plastic waste. This talk describes innovations in production of barrier films and coatings suitable for food or pharmaceutical packaging, based on combinations of cellulose- and chitin-based nanomaterials. Cellulose nanocrystals, CNCs, and chitin nanofibers or nanocrystals (ChNFs, ChNCs), are oppositely-charged, water-dispersible fibers that could form the basis of a platform of renewable, high-performance materials. In particular this talk will highlight work in creating multilayer coatings with excellent oxygen barrier properties on a variety of substrates. Challenges to industrial implementation and approaches to overcome these will also be analyzed.

632460 Engineering Electrostatic Interactions between Proteins and Biopolymers for Intracellular Phase Separation

Abstract:

Protein and nucleic acid phase separation is key to several biological processes, from cell signaling to RNA processing. In addition to specific biological interactions between these biomolecules, electrostatic interactions between positively charged protein domains and negatively charged nucleic acids have been shown to play a role in regulating the formation of these intracellular condensed phases. To directly probe the importance of electrostatic interactions in intracellular phase separation, we have engineered a panel of both cationic and anionic proteins and evaluated their phase behavior in simulated *in vivo* conditions as well as directly inside living cells. *In vitro*, phase separation between proteins and RNA was evaluated by turbidity measurements and optical microscopy. We also monitored intracellular phase separation in *E. coli* by fluorescence microscopy. Both *in vitro* and in bacteria, we have found that supercationic proteins form complex coacervates with RNA and enrich a small fraction of endogenous proteins in the condensed phase. Additionally, we have identified short cationic sequences that are sufficient to drive intracellular complex coacervation of globular proteins with modest net charge. Phase separation can be predicted by the overall net protein charge, regardless of distribution between the globular domain and the short peptide sequence. Building on our findings with model proteins, we have also engineered intracellular condensates that contain enzymes and demonstrated that they maintain activity in the condensed phase.

633304 Kinetics of Soy Protein Adsorption at the Fluid Interface: Interfacial Rheology

Abstract:

In this work, we used interfacial rheology to probe the soy protein globulins adsorption at the water/oil interface. We determined different energy barriers against the protein adsorption: diffusion, surface pressure, and electrostatic repulsion. It was quite clear that at early stages of adsorption, the kinetics were diffusion-controlled, with interface concentration~t^0.5. However, as the protein continued to travel to the fluid interface, the energy barriers arising from surface pressure and electrostatic repulsion appeared to affect the process. The DLVO theory was used to explain the effect of pH and ionic strength on the adsorption kinetics. Also, by adjusting the centrifuge process, protein particles with four different average particle diameters (0.1-10.0 um) were prepared to examine the effect of the particle size on these energy barriers.

623812 Cephalexin and Amoxicillin Crystal Shape Modification By Manipulating the Supersaturation and Wet Milling

Abstract:

Cephalexin and amoxicillin belong to the family of beta-lactam antibiotics and are among the most widely used small-molecule drugs. Both compounds are sparingly soluble in water, the solvent in their enzymatic synthesis process, and form hydrated crystals under typical synthesis conditions [1]. Unfortunately, both cephalexin and amoxicillin crystallization processes are reported to produce needle-like crystals with high aspect ratios, which can lead to significant agglomeration, long filtration times, and poor downstream processability [2]. There have been several efforts to address the above issue and to modify the crystal habit and shape in these systems. For example, it has been suggested that seeding and controlling the supersaturation generation through controlling either the cooling rate [3], synthesis reaction rate [4], or feeding rate [5] can be used to improve the shape of cephalexin monohydrate crystals.

In the present work, we examine the impact of wet milling on the crystal shape, aspect ratio, and size distribution in cephalexin and amoxicillin crystallization processes. Wet milling is commonly used to reduce the crystal size in different processes and to tighten the size distribution. Our results suggest that wet milling not only reduces the mean size of cephalexin monohydrate crystals, but it also leads to further improvements in crystal aspect ratio during the enzymatic synthesis and crystallization process. Based on a systematic study of different milling parameters, we will discuss the impact of milling on crystals of both cephalexin and amoxicillin. Furthermore, we investigate potential improvements in the aspect ratio of the final crystals by combining the milling with a series of pH manipulation steps where small, needle-like crystals are dissolved by periodically increasing the system pH (leading to higher solubility), giving a way to produce larger crystals with improved aspect ratios (Figs. 1A and 1B). The impact of the improved aspect ratio on the filtration rate of both crystals will also be discussed (Fig. 1C).

Fig. 1. Example of improvement in the aspect ratio of cephalexin monohydrate crystals by applying a series of pH swings (A), or periodic milling (B). Both cases lead to the formation of more platelike crystals compared to initially needle-like crystals. Example of the impact of aspect ratio (or size) on filtration rate (C).

Cited literature

[1] McDonald, Matthew A., Grant D. Marshall, Andreas S. Bommarius, Martha A. Grover, and Ronald W. Rousseau. "Crystallization kinetics of cephalexin monohydrate in the presence of cephalexin precursors." *Crystal Growth & Design* 19, no. 9 (2019): 5065-5074.

[2] Perini, Giulio, Carlos Avendaño, William Hicks, Anna R. Parsons, and Thomas Vetter. "Predicting filtration of needle-like crystals: A Monte Carlo simulation study of polydisperse packings of spherocylinders." *Chemical Engineering Science* 230 (2021): 116151.

[3] Ostergaard, Iben, Heidi Lopez de Diego, and Haiyan Qu. "Crystallization of cephradine polymorphs and hydrates from mixed solvents of methanol and water." *Chemical Engineering & Technology* 42, no. 7 (2019): 1512-1518.

[4] McDonald, Matthew A., Grant D. Marshall, Salami, Hossein, Andreas S. Bommarius, Martha A. Grover, and Ronald W. Rousseau. "Crystallization Optimization By Enzymatic Control of a Reactive Crystallization". AIChE annual meeting, San Francisco (2020)

[5] Li, Mingchen, Zeren Shang, and Baohong Hou. "Optimizing the Aspect Ratio of Cephalexin in Reactive Crystallization by Controlling Supersaturation and Seeding Policy." *Transactions of Tianjin University* 25, no. 4 (2019): 348-356.

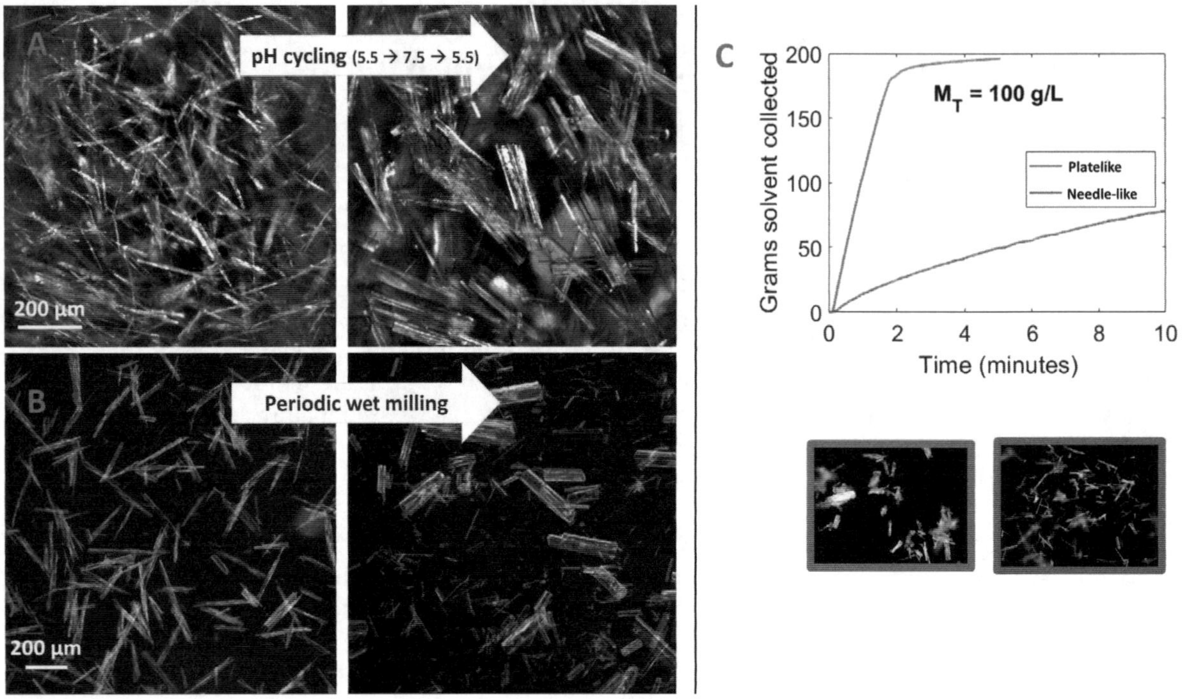
A
pH cycling (5.5 → 7.5 → 5.5)
200 µm
B
Periodic wet milling
200 µm
C
M_T = 100 g/L
Platelike
Needle-like
Grams solvent collected
Time (minutes)

625129 Accelerated Early-Stage Enabling API Crystallization Process Development and Scale-up

Abstract:

Development of crystallization, filtration/washing, and drying processes is an essential cornerstone of delivering quality Active Pharmaceutical Ingredient (API) in the pharmaceutical industry. During early-stage development, rapid timelines and the need to generate high quality API to enable GLP toxicity studies and Phase I clinical trials necessitate rapid and targeted data rich experimentation to develop robust and scalable processes. While early-stage crystallization development often falls upon the chemists, early intervention with an engineering and scalability-focused mindset can significantly reduce the development burden later in the pipeline. Such development efforts on Compound A, and the enabling data-rich workflow leveraged to yield a robust crystallization, filtration, and drying process is detailed in this work. A rapid 6-week development period, followed by a successful scale up from gram to kilo scale over a subsequent two-week period enabled a GLP toxicity study to proceed for the early-stage Compound A.

Swift crystallization solvent selection was aided by a combination of in-silico modeling using COSMO quantum chemistry software in conjunction with experimental UHPLC based solubility measurements. Selection of a suitable solvent-antisolvent system was further informed using DynoChem®, which was leveraged to generate full solubility curves over a range of temperatures and compositions with limited experimental data. Leveraging these tools is often essential in early-stage process development due to limited material availability. A favorable solubility profile for a given composition was identified, and a simple cooling crystallization was designed to afford high yield and purity. Material lean experiments using a variety of process analytical technologies (PAT) were designed to determine the metastable zone width and explore seeding conditions (seed load and supersaturation ratio) to ensure consistent crystal form control while avoiding oiling out. A complex form landscape during filtration and drying was mapped with DSC/TGA/MS and PXRD and navigated through to ensure the physical properties of the API were preserved during filtration, washing, and drying. Carefully tailored displacement washes were developed and leveraged to avoid crystal form change. A terminal wet milling and thermocycling process was also developed to control crystal surface area in conjunction with appropriate drying process understanding to achieve target specifications in the lowest cycle time. The process was successfully scaled-up to kilo scale to enable the GLP toxicity study.

Disclosures:

All authors are employees of AbbVie and may own AbbVie stock. AbbVie sponsored and funded the study; contributed to the design; participated in the collection, analysis, and interpretation of data, and in writing, reviewing, and approval of the final publication.

630526 Development and Scale-up of a Crystallization Process for a Kinetically-Unfavorable Polymorph

Abstract:

ABSTRACT: This work describes the development and scale-up of a crystallization process to isolate a thermodynamically stable but kinetically unfavorable polymorph of a small organic agrochemical prone to oiling. The process involves simultaneous addition of solution and antisolvent into a slurry of the desired polymorph, similar to semi-batch, reverse addition antisolvent crystallization.[1] For this system, the simultaneous addition process is more robust to feed variation than a traditional antisolvent process and affords suitable polymorph control, recovery, cycle time, and purity. However, the process produces a thick, but shear-thinning slurry of fine, needle-like particles that is challenging to mix well. Mixing analysis tools, including both the Fluid Mixing Processes (FMP) Excel Design Guide and computational fluid dynamics simulations were utilized to enable successful scale-up of this mixing-sensitive process from lab to pilot plant.

[1] R. Pena, C.L. Burcham, D.J. Jarmer, D. Ramkrishna, Z.K. Nagy, "Modeling and optimization of spherical agglomeration in suspension through a coupled population balance model," *Chemical Engineering Science,* 167 pp. 66-77, 2017.

632628 Acid-Catalyzed Esterification Governs the Chain Elongation and the Oriented Attachment in Cof-5 Synthesis

Abstract:

Covalent-Organic Frameworks (COFs) have found application in a wide range of fields such as catalysis, membrane separations, adsorption, and drug delivery due to their high symmetry, porosity, crystallinity, and large surface area. However, long timescales of the synthesis, difficulties in characterization, formation of amorphous intermediate, and the limited understanding of the underlying reaction network have prevented the controlled synthesis of COF crystals. Furthermore, the existing solvothermal synthesis techniques are difficult to reproduce. To overcome these limitations, improve reproducibility, and obtain quantitative rate contributions of different polymerization phases, we employ experimental studies of a wide range of operating conditions to synthesize COF-5 crystals. The yield and grain sizes of the COF crystals were characterized using reliable and reproducible ex-situ techniques such as Fourier Transformed Infrared Spectroscopy (FTIR) and powdered X-ray diffraction (P-XRD). Optimization of the temperature, reactant concentrations, type of solvents, and catalysts have provided significant mechanistic and resulted in low reaction times, high yields, and larger particle sizes. The insights gained from the comprehensive experimental studies allowed the development of a microkinetic model which simulates millions of nodes in the COF reaction network. The error-minimized parameters in the microkinetic models reproduce the experimental results of COF-5 yield and grain sizes in a predictive manner. Furthermore, the microkinetic approach also validates the proposed reaction mechanism obtained from the experimental studies. Results suggest that the acid-catalyzed esterification governs the chain elongation and the oriented attachment of COF-5 crystals.

621619 Enabling a Selective Dissolution Scheme for the Removal of Fines in Crystallization with a Hydrocyclone: Modeling and Experimental Validation

Abstract:

Effective methods to tune the particle size distribution (PSD) are of major interest in the field of crystallization, with the aim of facilitating the downstream processing, improving the mechanical properties of the powder, and, in the case of crystalline pharmaceuticals, modulating the absorption rate of the active ingredient in the human body.

Fines, i.e., crystals of a few tenths of micrometers in size, are often an undesired portion of the PSD, as they are responsible for extended filtration times. For their removal, temperature cycling is usually applied, consisting of one or more repetitions of a partial dissolution step followed by a growth step [1,2]. Temperature cycling has the advantage of being a very simple procedure; however, depending on the PSD and on the compound-specific behavior of crystallization, its efficiency may be limited, and the outcome may not be well predictable. Two factors limit the efficiency and predictability of temperature cycling. First, the partial dissolution step is non-selective, as it acts on the whole suspension and also affects larger crystals, thus decreasing the productivity; second, the evolution of the PSD strictly depends on the kinetics of dissolution and growth, which are generally unknown, and require tedious lab-work to be evaluated.

To address both shortcomings of temperature cycling, we have proposed and analyzed in detail through modeling an alternate process featuring a hydrocyclone [3], referred to as a selective dissolution process and outlined in Figure 1a. Hydrocyclones are common and inexpensive devices employed in several fields of chemical engineering, where they can act as classifiers to separate particles based on their size [4]. In the selective dissolution process, first the hydrocyclone separates finer from coarser crystals, then the suspension containing the fines is heated, so that only the particles in the targeted size range are dissolved. Further, the separation in the hydrocyclone depends on its design and operating conditions, and on the physical properties of the suspension, thus making the process less reliant on the kinetics of crystallization.

In this contribution we present both the main findings of the above-cited modeling study and the results of novel validation experiments, highlighting the trends and trade-offs that characterize the operation of a hydrocyclone and their effect on the overall performance of the crystallization process. Particular attention is devoted to process design, and a pathway for the selection of a suitable hydrocyclone design for a given target of fines removal is outlined.

The experimental campaign has been performed with β L-glutamic acid in water. 3D printing has been exploited to manufacture the hydrocyclones in-house (shown in Figure 1b), so as to enable superior flexibility in the design of units with different geometries. The experimental results demonstrate the advantage of the selective dissolution process over standard temperature cycling, and validate the modeled trends. Finally, the experimental data has been used to fine-tune the model predicting the separation performance of the hydrocyclone, which can be therefore used reliably for process design.

[1] Nagy, Z. K.; Aamir, E.; Rielly, C. D. Internal Fines Removal Using Population Balance Model Based Control of Crystal Size Distribution under Dissolution, Growth and Nucleation Mechanisms. *Growth Des.* **2011**, *11*, 2205–2219.

[2] Salvatori, F.; Mazzotti, M. Manipulation of Particle Morphology by Crystallization, Milling, and Heating Cycles: Experimental Characterization. *Eng. Chem. Res.* **2018**, *57* (45), 15522–15533.

[3] Binel, P.; Mazzotti, M. A Selective Dissolution Process Featuring a Classification Device for the Removal of Fines in Crystallization. *Eng. Chem. Res.* **2021**, *60* (1), 614–628.

[4] Svarovsky, L. Hydrocyclones. In *Solid-Liquid Separation*; Svarovsky, L., Ed.; Butterworth-Heinemann: Oxford, U.K., 2001; pp 191–245.

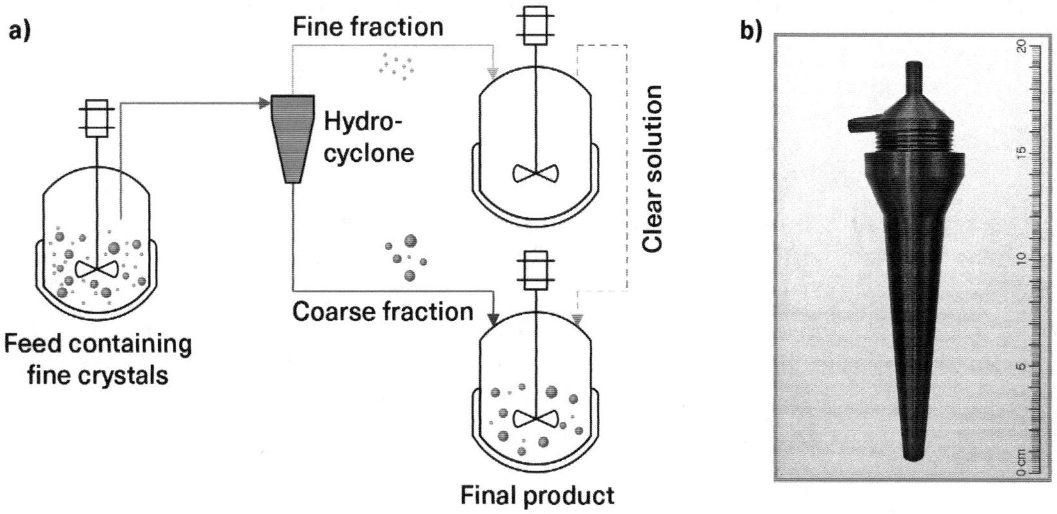

Figure 1. (a) Scheme of the selective dissolution process. The feed suspension is pumped through the hydrocyclone, which splits it into two streams; one stream contains most fine crystals while the other contains most coarse crystals. The stream containing the fines is collected in a vessel where complete dissolution occurs. The clear solution is then combined with the suspension containing the coarse crystals, and the supersaturation is depleted until equilibrium, i.e., the end of the process. (b) Photograph of a 3D printed hydrocyclone. For improved printing performance and flexibility, it is composed of a top cylindrical part and a bottom conical part. The top part is screwed into the bottom part, ensuring a tight fit and a smooth transition between the two.

627313 Crystallization Modeling of a Pharmaceutical Compound for Digital Twin Based in-Silico Optimization with Experimental Validation

Abstract:

The design of crystallization processes to achieve robust critical quality attributes (QCAs) is a practical challenge in the pharmaceutical industry. The current industrial standard still highly depends on Quality-by-Design (QbD), which requires intensive experimentation using traditional design of experiments (DoE) to find the operation conditions in the experimental space that satisfy the QCAs. With the current emerging Industry 4.0 practices the emerging paradigm has become the introduction of digital twins, which provide a predictive tool to be used in the optimization of the crystallization processes, or for in-silico DoE studies.[1]

In this work, we present the development and validation of a digital twin for the crystallization process of a pharmaceutical API (Compound A), from Takeda Pharmaceuticals International Co., which forms needle-like crystals. Several model training and validation experiments were performed with temperature cycling to promote the growth of the crystals for kinetic studies. For data acquisition and system monitoring during these experiments, process analytical technology tools (PAT) were used, including Mettler Toledo's ParticleTrack G400 for in-line chord length distribution (CLD), for detecting the crystallization events such as nucleation and dissolution, as well as for the crystal count measurement, the Zeiss MCS621 ATR-UV/Vis spectrophotometer for concentration measurement, and the Mettler Toledo's ParticleView v19 for in-situ microscope images. Additionally, Malvern's Mastersizer 3000 Hydro MV was used as an off-line characterization tool to measure crystal size distribution (CSD) and the data from the PAT tools were directly used in the parameter estimation step.

Two novel features were introduced during the parameter estimation for digital twin development. First, to make the model parameter estimation more robust, the focused beam reflectance measurement (FBRM) data was directly used in the parameter estimation without further transformations. Secondly, a novel size dependent growth (SDG) rate expression was used in the model to capture the CSD dynamics considerably better than the standard SDG rate models. The developed digital twin includes secondary nucleation, SDG, and dissolution mechanisms, which were identified by the designed training experiments and the objective that required thermocycling for the desired CQAs. The aim was to grow the crystals as large as possible without exceeding a batch time of 40 h while having product crystals with bulk density values larger than 0.25 g/mL. For this purpose and from previous experience, the training and validation experiments were designed to have temperature cycles.[2,3] The developed model was used for in silico DoE and process optimization and the simulation results were validated experimentally, demonstrating the benefits of model-based digital design for crystallization process development. In addition, the benefits of having a digital twin for in-silico optimization were demonstrated by comparing the findings from an experimental DoE work on Compound A to the ones from the optimized profiles, which were obtained by using the digital twin. The results demonstrated that the developed model was accurate and predictive enough to validate the general conclusions related to the required key features of the optimal temperature profiles identified by the experimental DoE. These results indicated that in order to obtain larger crystals with narrower CSD, multiple temperature cycles should be implemented at higher temperature ranges (50-65 °C). It was also shown that having a higher range for cycling with multiple cycles has a larger effect on product crystals size and bulk density than changing the seed type and loading.

The conclusions of the in-silico DoE and digital design were validated by experiments, demonstrating that the model was able to identify the key features of the optimal operating conditions for the crystallization of Compound A, when the aim was to produce large crystals by implementing temperature cycles. The orders of magnitude difference in the time required to reach this conclusion by in-silico and experimental DoE also demonstrated the benefits of having a digital twin of the process that can enable rapid crystallization process development by using a digital design framework.

Declaration of Competing Interest

The authors declare that they have no known competing financial interests or personal relationships that could have appeared to influence the work reported.

References:

1. Su, Q. *et al.* A perspective on Quality-by-Control (QbC) in pharmaceutical continuous manufacturing. *Comput. Chem. Eng.* **125**, 216–231 (2019).
2. Jiang, M. *et al.* Modification of crystal shape through deep temperature cycling. *Ind. Eng. Chem. Res.* **53**, 5325–5336 (2014).
3. Simone, E., Klapwijk, A. R., Wilson, C. C. & Nagy, Z. K. Investigation of the Evolution of Crystal Size and Shape during Temperature Cycling and in the Presence of a Polymeric Additive Using Combined Process Analytical Technologies. *Cryst. Growth Des.* **17**, 1695–1706 (2017).
4. Mydlarz, J. & Jones, A. G. On the estimation of size-dependent crystal growth rate functions in MSMPR crystallizers. *Chem. Eng. J.* **53**, 125–135 (1993).

627585 Superstructure Optimization Enabled Design Heuristics and Material Property Targets for Continuous Diafiltration Membrane Cascades

Abstract:

In a recent consensus report, the US National Academies highlight opportunities to transform Separations Science to better address global environmental challenges with sustainable technologies. [1] For example, several forecasts predict that vehicle electrification will lead to a massive demand for lithium which will exceed its current production capacity within a decade [2]. , While recycling lithium ion batteries (LIBs) has many environmental benefits including reduced environmental impact from conventional lithium mining techniques and diverting waste and pollution caused by disposing valuable elemental resources (Li, Co, Ni, etc.), it remains uneconomical compared to Li harvesting from primary sources (e.g., mining). [3-5] This example highlights how multidisciplinary innovations across scales are often required to solve global engineering challenges. For LIB recycling, process systems engineering can help accelerate technology development by i) systematically optimizing process flowsheets tailored to site specific criteria and ii) setting quantitative materials, device, and unit operation performance targets to guide research at smaller length scales. Inspired by the above mentioned National Academies report, we recently proposed a molecules-to-systems engineering design framework that integrates data-science, domain expertise, and experimental measurements to accelerate innovations in separations and beyond. [6] This talk highlights opportunities for superstructure optimization to enable molecular-to-systems engineering of novel membrane separations.

In diafiltration systems, dialysate is strategically fed to the retentate side of a membrane to mitigate concentration polarization which enables more efficient batch or continuous staged membrane cascades. Although diafiltration membrane systems have been studied for over 20 years, their industrial use is limited to high-value niche separations including buffer exchange and protein purification. One possible reason for less widespread use of diaflitration is the lack of systems engineering research related to diafiltration system design, operation, and control.

In this talk, we present a novel superstructure optimization framework to design diafiltration membrane cascades [7]; we show optimally designed diafiltration systems are efficient Li/Co fractionation steps for LIB recycling processes and quantify the process-scale benefits of next-generation membrane materials. Specifically, our framework encodes all realistic diafiltration cascade configurations into a superstructure; through continuous optimization, we systematically search over these configurations (with complex recycle and feed injection strategies) while considering membrane areas, flowrates, and concentrations of all streams as decision variables. The optimal system designs show how dialysate buffer is used to offset concentration effects in membrane filtration to recover highly concentrated metal rich solutions. Using an epsilon-constrained multiobjective optimization, we identify Pareto-optimal configurations that balance permeate (Li) and retentate (Co) product recoveries and explore tradeoffs in staging complexity versus membrane area. We then distill thousands of optimal diafiltration configurations into six design heuristics. We justify each design heuristic using governing transport and thermodynamic phenomena and draw analogies to design short-cut methods for other staged separations (e.g., absorption, distillation). Finally, we highlight how our framework enables top-down design by studying the effects of alternate membrane characteristics on the performance of the cascade. To our knowledge, this is the first optimization-based analysis of diafiltration cascade systems. We conclude by sharing opportunities for process systems engineering to advance diafiltration technologies and more broadly membrane science.

References:

[1] National Academies of Sciences, Engineering, and Medicine. *A Research Agenda for Transforming Separation Science.* (2019).

[2] Sonoc, A., & Jeswiet, J. (2014). A Review of Lithium Supply and Demand and a Preliminary Investigation of a Room Temperature Method to Recycle Lithium Ion Batteries to Recover Lithium and Other Materials. *Procedia CIRP*, 15(C), 289–293. https://doi.org/10.1016/j.procir.2014.06.006

[3] Li, L., Deshmane, V.G., Paranthaman, M.P., Bhave, R.R., Moyer, B. A., Harrison, S., (2018). Lithium Recovery from Aqueous Resources and Batteries: A Brief Review. *Johnson Matthey Technology Review*, 62(2), 161–176. https://doi.org/10.1595/205651317X696676

[4] Swain, B. (2017). Recovery and recycling of lithium: A review. *Separation and Purification Technology*, 172, 388–403. https://doi.org/10.1016/j.seppur.2016.08.031

[5] Flexer, V., Baspineiro, C. F., & Galli, C. I. (2018). Lithium recovery from brines: A vital raw material for green energies with a potential environmental impact in its mining and processing. *The Science of the Total Environment*, 639, 1188–1204. https://doi.org/10.1016/j.scitotenv.2018.05.223

[6] Eugene, E., Phillip, W., Dowling, A. (2019). Data Science-Enabled Molecular-to-Systems Engineering for Sustainable Water Treatment, *Current Opinion in Chemical Engineering,* 26, 122–130.

[7] Eugene, E., Phillip, W., Dowling, A. (2019). Material Property Goals to Enable Continuous Diafiltration Membrane Cascades for Lithium-Ion Battery Recycling, *Proceedings of the 9th International Conference on Computer Aided Process Design (FOCAPD),* 469-474.

627094 Optimal Integration of Process Design and Dynamic Transitions for Catalytic Distillation Columns: A Discrete-Steepest Descent Framework

Abstract:

Catalytic Distillation (CD) is a typical Process Intensification (PI) application that combines both separation and heterogeneous reaction into a single unit. This process has gained attention given that optimal design and dynamic optimization strategies requires the implementation of efficient Mixed-Integer Nonlinear Programming (MINLP) optimization strategies. Recent studies has shown that it is possible to use CD to produce multiple products [1]–[4]. In particular, transition policies and steady-state process designs have been successfully optimized as independent tasks for multigrade reactive distillation units; however, the simultaneous integration of both tasks have not been considered. Usually, the design of the equipment and the optimization of dynamic transitions are performed sequentially; however, there is no guarantee that a given design can comply with the desired restrictions during operation, e.g., hydrodynamic or product specification constraints. An approach to simultaneously optimize the design and the dynamic transitions of a multigrade CD unit is critical to avoid suboptimal or dynamically infeasible designs at the conceptual design stage.

The Discrete-Steepest Descent Algorithm (D-SDA) is a recent optimization framework that has been applied to the design and synthesis of CD columns with rigorous equilibrium and non-equilibrium models [5], [6]. The D-SDA handles these problems by fixing discrete variables, solving the resulting Nonlinear Programming (NLP) subproblems, and comparing the solution obtained within a neighborhood of the discrete variables at each iteration. If there is no better solution to the NLP subproblems within the neighborhood, then the algorithm stops; otherwise, the algorithm takes the steepest-descent direction to continue the search by performing another iteration. The D-SDA incorporates a reformulation step that allows to decrease the combinatorial difficulties of the problem associated with the presence of discrete design variables, e.g., number of stages and the location of feeds and catalytic stages. In addition, this algorithm applies a rigorous re-initialization strategy at each iteration to handle model nonlinearities, e.g., non-ideal vapor and liquid mixtures, reaction kinetic models, pressure drop calculations and hydrodynamic constraints. The D-SDA has proven to be a useful tool to solve PI problems involving optimal process synthesis and design at steady-state. Nevertheless, this framework is not yet suitable to deal with problems involving process dynamics. This method relies on a partial discrete enumeration based on neighborhoods exploration; accounting for process dynamics may lead to prohibitive computational times since each NLP subproblem becomes a dynamic optimization problem. Also, these subproblems would require custom and near feasible initializations to converge due to the model nonlinearities. Still, no other determinist MINLP solver has shown better performance than the D-SDA when solving CD steady-state design problems [5], [6]. This is because the D-SDA was formulated as a specialized algorithm that takes advantage of the distinctive properties that emerge in CD distillation superstructures, e.g., the presence of binary variables that represent the location of catalytic stages, reflux stream, boil-up stream and feed streams; the presence of binary-binary and binary-continuous products, and the absence of phases at not existing trays that lead to zero flows. Thus, the D-SDA is a promising framework to handle CD superstructures that consider the transient operation of this process.

The aim of this work is to present the optimal integration of process design and dynamic transitions of a multigrade CD process using a new version of the D-SDA algorithm. The problem incorporates dynamic and steady-state operation requirements into a single large-scale Mixed-Integer Dynamic Optimization (MIDO) problem. The objective is to minimize investment and production costs, intensify production throughput and optimize the dynamic operation of a multigrade CD system. An improved version of the D-SDA that returns attractive solutions in reasonable computational times is used to tackle this challenging problem. The new D-SDA algorithm solves the integrated MIDO problem as a multi-scenario problem, where each scenario corresponds to one of the dynamic transitions involved in the process. Moreover, the new D-SDA incorporates a refined initialization strategy and an iterative procedure. At the initialization step, a feasible initialization is found using the conventional D-SDA applied to a steady-state version of the multi-scenario model. This results in an initial CD design that can accommodate multiple grades of a product. Iterations over discrete design variables is performed next. An iteration comprises the solution of the dynamic NLP subproblem for the discrete point being evaluated, and a collection of candidate neighbors that may improve the objective function. The key feature that differentiates the proposed iterative procedure from the conventional D-SDA's methodology is the implementation of a new systematic procedure that eliminates feasible neighbors for which the overall objective function is not expected to improve, i.e., neighbors are discarded without the need to solve their corresponding dynamic NLP subproblems. For the remaining neighbors, the dynamic NLP subproblems are solved and used to search for an optimal solution using the steepest descent direction method. The improvements made in the new D-SDA framework avoid the solution of multiple NLP dynamic optimization subproblems thus making the original MINLP problem computationally tractable. Another advantage of incorporating this solution procedure is the refinement of the D-SDA's re-initialization strategy, by using steady-state solutions to progressively generate educated initializations for dynamic NLP subproblems. As a result, the new D-SDA approach overcomes the limitations of the conventional D-SDA when applied to simultaneous process design and dynamic transition problems involving MIDO formulations.

The improved D-SDA framework was applied to the optimal design and optimal dynamic transitions of a multigrade CD unit, with Ethyl Tert-Butyl Ether (ETBE) as the desired product. ETBE columns are typically designed to produce nearly pure ETBE; nevertheless, previous studies have shown the potential benefits of producing ETBE at lower compositions in terms of cost reduction and efficiency [7], [8]. On the other hand, having ethanol impurities in ETBE is not always admissible, e.g., when the aim is to produce non-ethanol gasoline [9], [10]. Therefore, it is necessary to investigate the optimal design of a multigrade CD unit and verify the economic and dynamic viability of this novel operating mode for ETBE production. The results show that the new D-SDA is suitable to generate a design and a dynamic operation scheme for a CD column that produces ETBE at 63%, 83% and 95% molar composition. Also, it was found that the revised D-SDA methodology allows to obtain a design that is different and more affordable than that obtained with a sequential design and dynamic transition strategy. Moreover, savings in computational time were recorded; this is because the assumptions made allowed us to decrease the number of neighbors to be verified at each iteration, which resulted in a decrease on the number of dynamic optimization subproblems that need to be solved. Overall, the application of the new D-SDA framework allowed us to find attractive solutions to the rigorous ETBE case study in acceptable computational times. Future work includes testing this strategy with more challenging case studies in the field of PI; account for products' due dates, production demands and optimal sequencing constraints in the formulation, and investigate the performance of the new D-SDA framework under closed-loop operation.

References

[1] D. Haßkerl, C. Lindscheid, S. Subramanian, P. Diewald, A. Tatulea-Codrean, and S. Engell, "Economics optimizing control of a multi-product reactive distillation process under model uncertainty," *Computers & Chemical Engineering*, vol. 118, pp. 25–48, Oct. 2018, doi: 10.1016/j.compchemeng.2018.07.003.

[2] R. L.-N. de la Fuente and A. F. Tlacuahuac, “Optimal Start-Up and Product Transition Policies of a Reactive Distillation Column,” *Ind. Eng. Chem. Res.*, vol. 46, no. 7, pp. 2092–2111, Mar. 2007, doi: 10.1021/ie061312z.

[3] N. Medina-Herrera, S. Tututi-Avila, A. Jiménez-Gutiérrez, and J. G. Segovia-Hernández, “Optimal design of a multi-product reactive distillation system for silanes production,” *Computers & Chemical Engineering*, vol. 105, pp. 132–141, Oct. 2017, doi: 10.1016/j.compchemeng.2017.01.014.

[4] N. Medina-Herrera, A. Jiménez-Gutiérrez, L. A. Ricardez-Sandoval, and S. Tututi-Avila, “An approach for dynamic transitions in multiproduct reactive distillation columns,” *Chemical Engineering and Processing - Process Intensification*, vol. 153, p. 107967, Jul. 2020, doi: 10.1016/j.cep.2020.107967.

[5] D. A. Liñán, D. E. Bernal, L. A. Ricardez-Sandoval, and J. M. Gómez, “Optimal design of superstructures for placing units and streams with multiple and ordered available locations. Part II: Rigorous design of catalytic distillation columns,” *Computers & Chemical Engineering*, vol. 139, p. 106845, Aug. 2020, doi: 10.1016/j.compchemeng.2020.106845.

[6] D. A. Liñán, D. E. Bernal, J. M. Gómez, and L. A. Ricardez-Sandoval, “Optimal synthesis and design of catalytic distillation columns: A rate-based modeling approach,” *Chemical Engineering Science*, vol. 231, p. 116294, Feb. 2021, doi: 10.1016/j.ces.2020.116294.

[7] E. W. de Menezes, R. Cataluña, D. Samios, and R. da Silva, “Addition of an azeotropic ETBE/ethanol mixture in eurosuper-type gasolines,” *Fuel*, vol. 85, no. 17, pp. 2567–2577, Dec. 2006, doi: 10.1016/j.fuel.2006.04.014.

[8] E. W. de Menezes and R. Cataluña, “Optimization of the ETBE (ethyl tert-butyl ether) production process,” *Fuel Processing Technology*, vol. 89, no. 11, pp. 1148–1152, Nov. 2008, doi: 10.1016/j.fuproc.2008.05.006.

[9] “D7618 Standard Specification for Ethyl Tertiary-Butyl Ether (ETBE) for Blending with Aviation Spark-Ignition Engine Fuel.” 2017, doi: 10.1520/D7618-13R17.

[10] “D8235 Standard Specification for Ethyl Tertiary-Butyl Ether (ETBE) for Blending with Gasolines for Use as Automotive Spark-Ignition Engine Fuel.” 2018, doi: 10.1520/D8235-18.

625334 Novel Module-Based Design and Optimization Approach for Intensified Membrane Reactor Systems

Abstract:

In recent years, manufacturing of smaller modular and intensified chemical plants as opposed to large refineries has been motivated by the potential increased efficiency of smaller process units and the cost savings through mass production of these units. However, there is a tradeoff when intensifying these processes. As their sizes are reduced and unit operations are combined, there is a loss in the degrees of freedom for controlling the process. Thus, although the intensified unit may be more efficient than the traditional unit operations approach at the designed nominal conditions, it is more difficult to reject disturbances, scale the process up or down, or handle different feedstocks. This work aims to address the design and control challenges caused by the integration of phenomena and the loss of degrees of freedom (DOF) that occur in the intensification of membrane reactor units.

The challenge of reduced DOF in intensified processes is one that has been identified in literature as early as 2003 [1] and has continued to be studied and discussed to this day [2]–[6]. To understand this body of research, it is important to categorize the DOF into two main categories. The first are the design DOF that include all the equipment design parameters such as length, diameter, membrane thickness, mass of catalyst, etc. The others are the operational DOF which encompass the operating conditions and manipulated variables used to control the process. If on one end of the design spectrum, there are modular, highly efficient, but potentially poorly controlled units and on the other end of the spectrum there are large, less efficient, but easier to control processes, then intuition suggests that a hybrid design approach considering both cases may provide the desired balance between efficiency and controllability.

First, a novel approach to designing membrane reactor units is proposed. This approach consists of designing smaller modules based on specific phenomena such as heat exchange, reactions, and mass transport and combining them in series to produce the final modular membrane-based unit. A membrane model is developed using the AVEVA Process Simulation platform that allows for the modeling of this novel hybrid design approach for membrane reactor systems [7]. This approach to designing membrane reactors is then assessed using a process operability analysis [8]–[10] to maximize the operability index, as a way of quantifying the operational performance of intensified processes. This is done by converting the mixed-integer programming problem posed that considers phenomena placing and optimization of separate blocks to a gradient-based, non-linear programming problem with only continuous variables that connect the individual blocks. An optimization algorithm developed in Python is then used to optimize the modular membrane reactor system to find the best arrangement of modules to maximize the operability index, subject to both design and economic constraints.

Previous work [6] has demonstrated that the operability index can be improved as much as 37% using this novel design approach for membrane reactor systems, translating to an improvement in achievability for a potential control structure implementation. Through the development of this novel design and optimization approach and by applying economic and design constraints for the first time, this work will identify the upper limit for operability improvement of modular and intensified membrane reactor systems.

References

[1] G. Schembecker and S. Tlatlik, “Process synthesis for reactive separations,” *Chem. Eng. Process. Process Intensif.*, vol. 42, no. 3, pp. 179–189, 2003, doi: 10.1016/S0255-2701(02)00087-9.

[2] M. Baldea, “From process integration to process intensification,” *Comput. Chem. Eng.*, no. 81, pp. 104–114, Mar. 2015, doi: 10.1016/j.compchemeng.2015.03.011.

[3] US DOE, “Advanced Manufacturing Office: Process Intensification Workshop,” Department of Energy, Alexandria, VA, Report of conclusions from a meeting sponsored by the Department of Energy., 2015. Accessed: Jun. 24, 2019. [Online]. Available: https://www.energy.gov/sites/prod/files/2016/03/f30/2015%20DOE%20AMO%20Process%20Intensification%20Workshop%20Report_2.pdf.

[4] D. Vlachos, M. Ierapertritou, P. Dauenhauer, and A. Hock, “Modular Manufacturing Workshop,” National Science Foundation and the Department of Energy, Arlinton, VA, Report of conclusions from a meeting sponsored by the National Science Foundation and the Department of Energy, 2017. Accessed: Jun. 24, 2019. [Online]. Available: https://cpb-us-w2.wpmucdn.com/sites.udel.edu/dist/d/1086/files/2018/07/MMW_Report-2o9jwun.pdf.

[5] N. M. Nikačević, A. E. M. Huesman, P. M. J. Van den Hof, and A. I. Stankiewicz, “Opportunities and challenges for process control in process intensification,” *Chem. Eng. Process. Process Intensif.*, vol. 52, pp. 1–15, Feb. 2012, doi: 10.1016/j.cep.2011.11.006.

[6] B. A. Bishop and F. V. Lima, “Novel Module-Based Membrane Reactor Design Approach for Improved Operability Performance,” *Membranes*, vol. 11, no. 2, Art. no. 2, Feb. 2021, doi: 10.3390/membranes11020157.

[7] B. A. Bishop and F. V. Lima, “Modeling, Simulation, and Operability Analysis of a Nonisothermal, Countercurrent, Polymer Membrane Reactor,” *Processes*, vol. 8, no. 1, Art. no. 1, Jan. 2020, doi: 10.3390/pr8010078.

[8] V. Gazzaneo and F. V. Lima, “Multilayer Operability Framework for Process Design, Intensification, and Modularization of Nonlinear Energy Systems,” *Ind. Eng. Chem. Res.*, vol. 58, no. 15, pp. 6069–6079, Apr. 2019, doi: 10.1021/acs.iecr.8b05482.

[9] J. C. Carrasco and F. V. Lima, “Novel operability-based approach for process design and intensification: Application to a membrane reactor for direct methane aromatization,” *AIChE J.*, vol. 63, no. 3, pp. 975–983, 2017, doi: 10.1002/aic.15439.

[10] J. C. Carrasco and F. V. Lima, “Bilevel and parallel programing-based operability approaches for process intensification and modularity,” *AIChE J.*, vol. 64, no. 8, pp. 3042–3054, 2018, doi: 10.1002/aic.16113.

630062 Techno-Economic Analysis of a Dynamic Packed Reactive Distillation Column for Renewable Biosurfactant Production

Abstract:

The rapidly growing global market of surfactants is expected to grow from 14.3 billion to 20.9 billion by the year 2026 [1]. Increasing greenhouse gas emissions and environmental concerns are driving policymakers to impose stricter regulations on synthetic surfactant production [2]. Because biosurfactants have lower carbon footprints, they have been gaining interest both in industry and academia [3]. However, the key challenge for the biosurfactant market remains economic feasibility, given that the current price of synthetic surfactant is approximately $2/kg [4]. Traditional surfactant processes use distinct process equipment for reaction and separation; however, combining the reactive and distillation processes into a single reactive distillation column (RDC) lowers capital investment [5]. In this study, we have designed and modeled a novel packed bed reactive distillation column for producing renewable biosurfactants using Lauric acid and Methyl furan as feed. The reactive section produces furan compounds and water; the distillation section separates the furan compound and water. By introducing a control scheme in the RDC model to maintain the distillate purity of water after separation in the condenser, we were able to demonstrate in simulation the technical advantages of reactive distillation columns over traditional reactive separation, especially energy integration in the combined system and prevention of reverse reaction by product removal. Our economic analysis shows that producing biosurfactants using reactive distillation can potentially reduce the biosurfactant plant's overall operational capital cost by 12% and 20%, respectively, compared to the traditional process. Furthermore, the economic study estimated the market selling price (MSP) of biosurfactant to be $1.58/kg, which is 20% lower than the corresponding current value for synthetic surfactant. These results show that biosurfactant produced using RDC technology has the potential to achieve significantly lower carbon footprint with attractive economics.

Reference:

[1] "Natural Surfactants Market Size 2021, Sales, Key Raw Materials Price Trend, Description, Business Overview, Top Countries Records, Future Growth by 2026 - MarketWatch." https://www.marketwatch.com/press-release/natural-surfactants-market-size-2021-sales-key-raw-materials-price-trend-description-business-overview-top-countries-records-future-growth-by-2026-2021-03-01 (accessed Apr. 01, 2021).

[2] K. Rashid, "DESIGN, ECONOMICS, AND REAL-TIME OPTIMIZATION OF A SOLAR/NATURAL GAS HYBRID POWER PLANT," 2019.

[3] Q. Shu, H. Lou, T. Wei, X. Liu, Q. Chen, and G. Fricker, "pharmaceutics Contributions of Glycolipid Biosurfactants and Glycolipid-Modified Materials to Antimicrobial Strategy: A Review Academic Editors: Barbara Luppi and," *Pharmaceutics*, vol. 13, p. 227, 2021, doi: 10.3390/pharmaceutics13020227.

[4] P. Singh, Y. Patil, and V. Rale, "Biosurfactant production: emerging trends and promising strategies," *Journal of Applied Microbiology*, vol. 126, no. 1. 2019, doi: 10.1111/jam.14057.

[5] J. G. Segovia-Hernández, S. Hernández, and A. Bonilla Petriciolet, "Reactive distillation: A

review of optimal design using deterministic and stochastic techniques," *Chem. Eng. Process. Process Intensif.*, vol. 97, pp. 134–143, Nov. 2015, doi: 10.1016/J.CEP.2015.09.004.

630961 Intensified Process Design Alternatives for the Diethyl and Ethyl-Methyl Carbonate Production

Abstract:

Reactive distillation allows enhancing selectivity in multi-reaction equilibrium systems, where more than one component is valuable. The production of diethyl carbonate (DEC) is particularly attractive for its wide usage as a fuel additive and lithium battery applications. It can be produced by transesterification of dimethyl carbonate (DMC), which consists of two reactions, having ethyl methyl carbonate (EMC) as a valuable reaction intermediate. This work aims to present three different reactive distillation systems (RC), one to favour the production of DEC, another for the production of EMC, and one to produce both compounds simultaneously (DEC-EMC). Such processes were optimized considering the total annual cost (TAC) to compare them, which includes capital and operating plant costs. Also, a techno-economic analysis was carried out considering the TAC, the price of the products, and the raw materials costs. The distillation columns were designed and simulated in the process simulator Aspen Plus and optimized through Python via particle swarm optimization (PSO) technique. Product purities of 0.99 and 0.98-mole fraction for the DEC and EMC were considered, respectively.

The results showed that producing only EMC provides 50% more utilities compared to the column that produces only DEC, even though the TAC of the EMC design is 36% higher than the DEC design. The TAC of the column that generates both components provides 47% more utilities over the DEC production process. Therefore, the design that showed the best economic performance was the EMC production process, considering the current product market costs.

632074 A Process Intensification Synthesis Framework for the Design of Divided Wall Columns

Abstract:

Novel separation concepts based on process intensification (PI) principles have received burgeoning interest from the chemical industry to address the dual challenge of energy efficiency and environmental sustainability [1,2]. As one of the commercialized PI technologies, divided wall column (DWC) features a fully thermally coupled and single-shell distillation column for the separation of multi-component mixtures. The task-integrated design scheme, with improved thermodynamic efficiency, can lead to approximately 30% savings in capital expenditure, space, and energy [3]. For the computer-aided design of divided wall columns, extensive efforts have been made leveraging process synthesis and superstructure optimization to systematically generate optimal DWC configurations and simultaneously evaluate their performance metrics with other distillation alternatives (e.g., distillation sequences, thermally coupled distillation) [4-6]. Recent advances in process intensification synthesis [7-9], in which phenomenological building blocks are used to intensify the chemical systems towards ultimate performance limits, further opens up the opportunity to innovate the multi-component separation process from process fundamental perspective and to exploit the potential by divided wall columns.

In this work, we present a process intensification synthesis approach for the design of divided wall columns based on recent extensions of the Generalized Modular Representation Framework [10-11]. A superstructure-based representation leveraging modular phenomenological building blocks is utilized to intensify the fundamental chemical performance (e.g., mass transfer, heat transfer) and to systematically generate novel process structures without pre-postulation of equipment design (including but not limited to divided wall columns). To accurately describe the possible liquid-vapor and liquid-liquid phase behaviors of the multi-component mixture, rigorous thermodynamic models (e.g., UNIQUAC) are explicitly incorporated in the synthesis model. The synthesis problem is formulated as a single mixed-integer nonlinear optimization problem. The applicability and versatility of the proposed framework will be showcased via an industrial case study on methyl methacrylate purification by Dow Global Technologies [12]. Two new divided wall column designs are obtained, both of which can achieve equipment size reduction and substantial energy savings (Design 1: 18%, Design 2: 37%) compared to the original patent design. A two-column design is also generated which provides promising energy savings while needs to overcome the process bottleneck on product purity by incorporating membrane-assisted separation.

Reference

[1] Kiss, A. A. (2014). Distillation technology – Still young and full of breakthrough opportunities. Journal of Chemical Technology & Biotechnology, 89(4), 479-498.

[2] Tian, Y., Demirel, S. E., Hasan, M. M. F., & Pistikopoulos, E. N (2018). An overview of process systems engineering approaches for process intensification: State of the art. Chemical Engineering and Processing: Process Intensification, 133, 160-210.

[3] Dejanović, I., Matijašević, L., & Olujić, Ž. (2010). Dividing wall column – A breakthrough towards sustainable distilling. Chemical Engineering and Processing: Process Intensification, 49(6), 559-580.

[4] Dünnebier, G., & Pantelides, C. C. (1999). Optimal design of thermally coupled distillation columns. Industrial & Engineering Chemistry Research, 38(1), 162-176.

[5] Caballero, J. A., & Grossmann, I. E. (2014). Optimal synthesis of thermally coupled distillation sequences using a novel MILP approach. Computers & chemical engineering, 61, 118-135.

[6] Ramirez-Corona, N., Jiménez-Gutiérrez, A., Castro-Agüero, A., & Rico-Ramírez, V. (2010). Optimum design of Petlyuk and divided-wall distillation systems using a shortcut model. Chemical Engineering Research and Design, 88(10), 1405-1418.

[7] Tula, A. K., Eden, M. R., & Gani, R. (2019). Computer–aided process intensification: Challenges, trends and opportunities. AIChE Journal, 66(1).

[8] Demirel, S. E., Li, J., El-Halwagi, M., & Hasan, M. F. (2020). Sustainable Process Intensification Using Building Blocks. ACS Sustainable Chemistry & Engineering, 8(48), 17664-17679.

[9] Pichardo, P., & Manousiouthakis, V. I. (2017). Infinite DimEnsionAl State-space as a systematic process intensification tool: Energetic intensification of hydrogen production. Chemical Engineering Research and Design, 120, 372-395.

[10] Proios, P., & Pistikopoulos, E. N. (2005). Generalized modular framework for the representation and synthesis of complex distillation column sequences. Industrial & engineering chemistry research, 44(13), 4656-4675.

[11] Tian, Y., & Pistikopoulos, E. N. (2018). Synthesis of operable process intensification systems – Steady-state design with safety and operability considerations. Industrial & Engineering Chemistry Research, 58(15), 6049-6068.

[12] Jewell, D. W., Pendergast, J. G., & Worley, W. G. (2019). *U.S. Patent No. 10,392,337*. Washington, DC: U.S. Patent and Trademark Office.

623480 Optimization Modeling for Advanced Syngas to Olefin Reactive Systems Under Parameter Uncertainty

Abstract:

Reactor designs with mixed catalysts play an important role transforming a multiple reactor system to single-shot reactors. In addition to savings in capital and ease of implementation, single-shot reactors are useful to break equilibrium limitations, therefore increasing the yield and selectivity of desired product as shown in [1, 2]. However, the nonlinear and highly exothermic nature of mixed-catalyst systems makes it difficult for commercial process simulation and optimization tools to optimize these systems. This talk describes the development and application of optimization strategies for mixed-catalyst, single-shot reactors for syngas to olefin (STO) processes.

Finding the optimal catalyst distribution is challenging and requires advanced solution strategies for singular optimal control problems, which are poorly conditioned and often lead to flat response surfaces. The graded bed approach is applicable to these problems for preliminary results, but more sophisticated solution approaches are needed in order to find the exact optimal profiles. To determine these profiles, we develop a partial moving finite element approach, which decreases the complexity and solution times significantly compared to previous singular optimal control optimization strategies [3].

The resulting optimal control strategy is applied to a comprehensive STO reaction mechanism with detailed rate expressions. However, the experimentally obtained kinetic parameters come with some level of uncertainty due to measurement and estimation errors. If not considered, the uncertainty in the parameters may lead to violations in hard constraints, like thermal runaways or bad product quality. Nominal control of the STO problem cannot account for the parameter uncertainty and thus, robust optimization algorithms are needed. A two-stage multi-scenario approach is implemented to the STO problem [4]. In the first stage, the optimal control profile is found given a certain number of scenarios. With the fixed control profile, the largest sum of constraint violations is solved in the second stage. The scenario corresponding to the largest violation is added to the problem in the first stage, and the algorithm stops when no constraints are violated in the second stage.

The overall goal of this RAPID project is to find the optimal catalyst mixing ratio along the bed for the mixed catalyst STO reactor, respecting safety constraints, like avoiding thermal runaways. The two optimization approaches (graded bed and partial moving grid) are applied to the STO reaction model. Moreover, uncertainties in the form of confidence regions for the model parameters will be considered in the optimization problem for a robust reactor design.

[1] D. L. S. Nieskens, A. Ciftci, P. E. Groenendijk, M. F. Wielemaker, A. Malek, "Production of Light Hydrocarbons from Syngas Using a Hybrid Catalyst" I&EC Research, 2017

[2] S. K. Mazidi, M. T. Sadeghi, M. A. Marvast, "Optimization of Fischer-Tropsch Process in a Fixed-Bed Reactor Using Non-Uniform Catalysts" Chem. Eng. Technol., 2013

[3] W-F. Chen, L. T. Biegler, "A Simultaneous Approach for Singular Optimal Control Based on Partial Moving Grid" AIChE Journal, 2019

[4] Y. Wang, L. T. Biegler, M. Patel, J. Wassick, "Robust Optimization of Solid-Liquid Batch Reactors under Parameter Uncertainty" Chemical Engineering Science, 2020

628253 A Study of Thermal Management in Commercial Face Masks

Abstract:

The indisputable and widespread Coronavirus disease (COVID-19) has caused tremendous social life changes. COVID-19 has expeditiously taken the lives of innocent individuals. Combined with social distancing and quintessential hygiene habits, facemasks can slow the coronavirus spread and protect wearers and the people around them. This project aims to study the thermal management of commercially available face masks using FLIR One thermal camera. The goal is to enhance the user's thermal comfort as heat generated from prolonged use causes fatigue, skin rash, and discomfort to an individual. In this experiment, a healthy young man volunteered to wear the masks and conduct the study on himself. Three commercially used masks, KN95, disposable masks, and reusable cloth masks were utilized in this experiment. The volunteer wore each mask for 15 minutes, and the thermal images were continuously captured using the FLIR One camera five feet apart. The breathing temperature in the masks was recorded as a function of time. The ambient temperature and humidity were also recorded at the time of the experiments. The Heat Flow Rate equation was used to find the heat transfer through each mask's membranes. A gradual increase in temperature on the mask's surface was observed. The thermal images show that the KN95 has the best thermal management overtime. The morphology and chemical structure of the different layers of the mask were investigated using the Dino-lite premier and the Attenuated Total Reflection (ATR) respectively. The rate of heat flow of the Non-surgical disposable masks is higher than its counterparts in comparison. Face masks with better thermal management will be helpful for prolonged use and enhance comfort. The future goal of this work is to study the effects of Stimuli-responsive polymer to advance thermal management of the ideal mask.

Keywords: Thermal management, thermal image, face mask, coronavirus.

633277 Synthetic biology platforms for the sustainable production of superior flavor and fragrance ingredients

Abstract:

In the wake of the looming climate crisis and the rapid transformation in both economy and society caused by the coronavirus pandemic, corporations see an opportunity to redouble their efforts towards a new, more sustainable development model for the years to come. In the past decade, biotechnology and synthetic biology have made great advances in the sustainable biomanufacturing of precision chemicals that are clean, safe, natural, and have a low carbon footprint. The flavors and fragrances market is estimated at over $20 billion and is one of the industries that is experiencing a more rapid transition from chemical to natural. Moreover, the industry's susceptibility to supply chain disruptions has prompted the major F&F companies to invest heavily in biotechnology, as it does allow a very rapid development and commercialization of new ingredients, with delocalized production independent of environmental and geopolitical instability, yielding products of reliable and reproducible quality. In this context, Conagen has developed synthetic biology platforms that leverage both bioconversion and fermentation-based production of ingredients with improved sustainability, cost, and quality to support the F&F industry and others. Conagen's Terpenoid, Lipid & Lactone, and Phenolic & Flavonoid platforms will be introduced. Besides low-calorie, high-intensity sweeteners, sweetener enhancers and bitter blockers, Conagen's product portfolio includes several fruit and vanilla flavors and floral, fruity, citrusy, creamy and woody fragrances. Some of these are existing ingredients with improved sustainability, cost, and/or quality attributes while others are novel ingredients with improved performance or that have been developed to fill a market gap. Overall, Conagen offers one of the largest and most diverse portfolios of SynBio-derived F&F ingredients in the industry.

624866 On the Quantitative Investigation of the Antimicrobial Efficacy of Grape Seed Extracts Against Food-Related Bacterial Pathogens.

Abstract:

INTRODUCTION

Over the last years, the trends in food consumption have changed significantly. Consumers are demanding products that are minimally processed, without the use of chemical preservatives or antibiotics. In addition, there is a growing concern about the role of antimicrobial resistance (AMR) in disease outbreaks. Antimicrobial resistance (AMR) is driven by the extensive-use of antibiotics and preventive measures need to be taken against their excessive use[1]. Moreover, due to climate change, environmental awareness is more present as compared to the past, urging the food industry follow a more sustainable approach on the production design. Traditional processing methods such as pasteurization or sterilization cannot meet all the above demands and novel processing control strategies should, therefore, be developed[2].

The possibility to use natural antimicrobials to replace chemical preservatives and the extensive-use of antibiotics has been one of the major goals of the food industry. Fruit and vegetable by-products are particularly interesting as sources of natural antimicrobials. Valorising waste of a food processing line which otherwise would have been disposed is an added value for researching their antimicrobial activity, as it contributes to the reduction of food waste[3]. Grape by-products are derived from grape processing to produce wine and juice. They represent 20% of the total weight of the fruit of the grape, making their disposal a challenging problem for wineries and for the grape juice industry. Grape by-products are composed by the skins, the seeds and in some cases the stems of the fruit. However, the seeds contain higher concentrations of antioxidant and antimicrobial compounds, such as polyphenols, in comparison with other grape by-products. Most studies conducted to date, are focused on the extraction of polyphenols from locally produced grape pomaces and test whether these extracts have a microbial inhibitory effect, without specific systematic monitoring of the microbial dynamics/kinetics[4]. The limited studies that have quantified the inhibition following treatment with extracts from grape by-products used specific food products, deeming the results relevant only for those specific food products[5].

Overall, in order to use grape by-products effectively as a natural antimicrobial strategy in a larger scale, it is important to study the detailed microbial dynamics of, Gram-negative and Gram-positive, foodborne pathogens robustly controlled systems (2D or 3D food models) instead of food products, as different physicochemical and rheological properties can be better monitored. Such fundamental studies will enable a better understanding of the exact concentration ranges at which grape pomace extracts are bacteriostatic and bactericidal, their mode of action and how they could be utilized, individually or in combination with other novel technologies (ultrasound, cold atmospheric plasma, pulsed electric fields) to attain sustainable food processing. *The aim of our work is therefore to perform fundamental systematic studies of the impact of grape seed extracts on the microbial dynamics of food related pathogens in model systems.*

METHODS

Grape seed extracts (GSE) were purchased from Bulk powders. The inactivation effect of GSE was tested against *L. monocytogenes and E. coli* inoculated in TSBYE supplemented with 1% and 4% GSE. Additionally, two levels of initial microbial population were tested, i.e., 10^5 and 10^8 CFU/ml. The effect of 1% GSE against mid-exponential or early stationary phase *L. monocytogenes* was also studied, to enable an understanding of the impact of the growth phase on the potential antimicrobial resistance. After the inoculation of *L. monocytogenes* in the nutrient broth, microbial dynamics were monitored for 24h. Furthermore, potential morphological changes of the cells treated with GSE, were observed using scanning electron microscopy (SEM). To identify the percentage of sub-lethally injured cells and dead cells, flow cytometry methods were carried out, staining with propidium iodide (PI) and bis-(1,3-dibutylbarbituric acid) trimethaine oxanol (BOX), respectively.

RESULTS

Treatment with GSE showed great antimicrobial activity against Gram-positive *L. monocytogenes* and Gram-negative *E. coli,* especially for lower initial inoculum levels (10^5 CFU/ml) and for mid-exponential cultures phase cells. The microbial inactivation was substantially enhanced for higher GSE concentrations ,i.e., 4%. Analysis of the treated samples using flow cytometry showed significant differences on the ratio of dead and sub-lethally injured cells when the initial inoculum concentration, the GSE concentration and/or growth phase were changed, in alignment with the microbial inactivation kinetics.

CONCLUSIONS

This work shows that GSE have good potential as a natural antimicrobial strategy against food related bacterial pathogens. This insight to the microbial dynamics is essential for the development of novel and robust green/sustainable processing control strategies, i.e., application of GSE individually or coupled with other novel technologies in a hurdle approach.

Acknowledgements:

This work was supported by the Department of Chemical and Process Engineering of the University of Surrey, the Doctoral College of the University of Surrey and R-processing Limited. E.V. is grateful to the Royal Academy of Engineering, for an industrial Fellowship.

References:

1. European Food Safety Authority. The European union summary report on antimicrobial resistance in zoonotic and indicator bacteria from humans, animals and food in 2017. *EFSA J.* 2019;17(2). doi:10.2903/j.efsa.2019.5598
2. Teixeira JA. Grand Challenges in Sustainable Food Processing. *Front Sustain Food Syst.* 2018;2(June):2017-2019. doi:10.3389/fsufs.2018.00019
3. Villalobos-Delgado LH, Nevárez-Moorillon GV, Caro I, Quinto EJ, Mateo J. *Natural Antimicrobial Agents to Improve Foods Shelf Life*. Elsevier Inc.; 2019. doi:10.1016/b978-0-12-817190-5.00004-5
4. Xu Y, Burton S, Kim C, Sismour E. Phenolic compounds, antioxidant, and antibacterial properties of pomace extracts from four virginia-grown grape varieties. *Food Sci Nutr.* 2016;4(1):125-133. doi:10.1002/fsn3.264
5. Sagdic O, Ozturk I, Yilmaz MT, Yetim H. Effect of Grape Pomace Extracts Obtained from Different Grape Varieties on Microbial Quality of Beef Patty. *J Food Sci.* 2011;76(7). doi:10.1111/j.1750-3841.2011.02323.x

627237 Systematic Screening of Materials for Consumer Products Using Semantically Enabled LCA

Abstract:

In the European Union, the manufacture and use of chemical substances as well as produced and imported goods must comply with a comprehensive legislative framework (e.g. REACH, RoHS). The main focus of this legislation is the protection of human health and the environment that can be posed by chemicals. Thus, substances can be restricted or banned if they are characterized as hazardous. The manufacture of consumer products, such as home appliances, is a complicated process involving a wide range of materials and chemicals. Health and environmental impacts of such processes should be evaluated, monitored and adjusted to the lowest achievable level taking into account the legislative framework and the competitiveness of the industry.

The LIFE MATHER Platform, developed in the context of the LIFE MATHER Project, constitutes a new chemical monitoring data platform that integrates corporate data (materials, chemicals, manufacturing process) with health and environmental data in order to evaluate possible alternative materials according to existing and/or upcoming legislation. The principles of LCA (Life Cycle Analysis) are applied, thus building a rich inventory regarding consumer products (home appliances) with detailed data flows: items of the appliance (Bill of Materials, BoM), construction of materials (Full Material Disclosure, FMD) and manufacturing data (energy, electricity etc.). The inventory is enriched with open and commercial LCA databases (e.g. Ecoinvent).

The LIFE MATHER Platform conceptualizes health and environmental aspects using knowledge models, enables reasoning by employing semantics and ontologies, builds properties and its automated technology enables high-throughput screening and analysis. It is prepared to manage large volumes of data streams as may be available from customers or external providers in general. The substitution of chemicals in existing products leads to “new products” (green products) ahead of design and new product features emerge. It is noteworthy that the approach is generalized, thus enabling the incorporation of other classes of appliances, while the challenge lies with the inclusion of processes beyond appliances.

The authors acknowledge financial support by the H2020-EU.1.3.3. Grant RENESENG II (H2020-778332).

625233 Innovating a Bendable Concrete Railroad Tie with Enhanced Fatigue Durability Via CO_2 Utilization

Abstract:

The concrete industry is a significant CO_2 emitter, with one of its primary components, Portland cement, accounting for 8% of anthropogenic CO_2 emissions through its production. As the global construction market is continuously expanding, effective strategies for mitigating concrete CO_2 footprint are much needed. In this study, we propose a novel approach that utilizes and sequesters CO_2 to create a highly durable bendable concrete (technically known as *engineered cementitious composites*, ECC) through precast carbonation curing of ECC rail ties. Our goal was to examine the potential for this production strategy to enhance rail tie service life and reduce overall lifecycle CO_2 emissions in railroad tie applications. To do this, we verified fatigue resistance of the CO_2-sequestered ECC with respect to conventional concrete, as this is a critical criterion for railroad tie service life prediction, and developed a comprehensive lifecycle assessment for the CO_2-cured ECC ties.

Preliminary results suggest that CO_2-sequestered ECC can achieve >30% CO_2 uptake per cement mass after 24-hour carbonation curing. Carbonation curing increased ECC's flexural strength by 20% and promoted crack width control capability with maximum post-fatigue crack width being reduced from 128μm to 79μm. The flexural strength of carbonated ECC reached 12.03 MPa, exceeding the American Railway Engineering and Maintenance-of-Way Association (AREMA) requirement of 5.89 MPa for rail seat positive, negative, and center bending tests. Moreover, carbonation curing significantly improved the fatigue life of non-reinforced ECC to pass the standard-required 3 million cycles under flexural loading, with comparable performance to traditional prestressed railroad ties. This eliminates the need for the costly and time-consuming prestressing process for traditional tie production. Carbonation curing can also increase the fatigue life by 20% (under 0.5 stress level) and 12% (under 0.4 stress level). The enhanced fatigue life through CO_2 utilization is anticipated to improve the railroad safety, resiliency, and sustainability, while mitigating CO_2 emissions both at concrete manufacture and through infrastructure lifetime.

626393 Managing Model and Data Alternatives within the Design of Ionic Liquid Enabled Separations of High Global Warming Potential Hydrofluorocarbon Refrigerants

Abstract:

Mandated by the 1987 Montreal Protocol, chlorofluorocarbon (CFC) refrigerants have been gradually replaced by HFCs to prevent ozone depletion. Many of these second-generation HFC mixtures, however, have a high global warming potential (GWP) and the 2016 Kigali agreement ordered their gradual phase-out [1]. Due to the often azeotropic compositions of these HFC refrigerants, existing separation methods for removing the high GWP components are currently infeasible or not practical, but it is wasteful to incinerate HFC mixtures as some HFC components have low GWPs and can be recycled. We hypothesize that custom ILs can be designed to remove low GWP HFC components from specific HFC mixtures [2,3]. Moreover, thermodynamic and process modeling can accelerate screening and design of optimal ILs for azeotropic HFC separation [4,5]. To date, many thermodynamic model families including activity coefficient models (Margules, Wilson, and nonrandom two liquid), cubic equations of state (van der Waals, Peng-Robinson, Soave-Redlich-Kwong), and soft-SAFT [6-11] have been proposed to predict HFC solubilities in ILs.

In this talk, we apply uncertainty quantification and statistical model selection to thermodynamic modeling of HFC-IL mixtures. Our ultimate goal is to answer several open modeling questions including (a) what level of model complexity is justified by experimental solubility data? (b) how can model overparameterization be avoided? (c) what is the impact of model choice and uncertainty on process design and operations decisions? (d) and, what type of data are most valuable to discriminate between models and reduce uncertainty in process scale decisions? We explore these questions in the context of separating R-410a, a mixture of difluoromethane (HFC-32) and pentafluoroethane (HFC-125). We determined that a van der Waals EoS with five fitted parameters is able to capture solubilities of HFC-32 and HFC-125 in six ILs, but uncertainty analysis showed four of the five parameters were correlated, indicating over-parameterization, and that another parameter was sloppy, indicating the model was insensitive to its value [5]. Additionally, we investigated the use of the Margules and nonrandom two liquid activity coefficient models to predict solubilities for these two HFCs in six ILs, and determined that these activity coefficient models provided poorer fits than the van der Waals EoS. We now explore and compare three cubic EoS (van der Waals, Peng-Robinson, Soave-Redlich-Kwong) combined with a variety of mixing rules to identify which accurate, but not over-complicated, model can best be used in a separation process model of R-410a. We compare frequentist and Bayesian thermodynamic model parameter estimation methods and implement Bayesian thermodynamic model selection for the models of interest. We perform identifiability analysis to determine the sensitivity of these models to the fitted parameters within the mixing rules studied. Additionally, we use uncertainty quantification and sensitivity analyses to explore the type of data (e.g. full or partial solubility curves, Henry's constants) and level of experimental accuracy which is necessary to generate reliable process models. This analysis offers quantitative insights into the appropriate data and models for designing and screening ILs for further study as HFC separating agents.

References

[1]United Nations Environment Programme. Ozone Secretariat. (2006). Handbook for the Montreal protocol on substances that deplete the ozone layer. UNEP/Earthprint.

[2]Plechkova, N. V., & Seddon, K. R. (2008). Applications of ionic liquids in the chemical industry. *Chemical Society Reviews*, *37*(1), 123-150.

[3]Chávez-Islas, L. M., Vasquez-Medrano, R., & Flores-Tlacuahuac, A. (2011). Optimal molecular design of ionic liquids for high-purity bioethanol production. *Industrial & Engineering Chemistry Research*, *50*(9), 5153-5168.

[4] SHIFLETT, M. B., & Yokozeki, A. (2006). Separation of difluoromethane and pentafluoroethane by extractive distillation using ionic liquid. *Chimica oggi*, *24*(2), 28-30.

[5] Morais, A. R. C., Harders, A. N., Baca, K. R., Olsen, G. M., Befort, B. J., Dowling, A. W., ... & Shiflett, M. B. (2020). Phase equilibria, diffusivities, and equation of state modeling of HFC-32 and HFC-125 in imidazolium-based ionic liquids for the separation of R-410A. *Industrial & Engineering Chemistry Research*, *59*(40), 18222-18235.

[6] M. Shiflett, M. Harmer, C. Junk, and A. Yokozeki. Solubility and diffusivity of 1, 1, 1, 2-tetrafluoroethane in room-temperature ionic liquids. *Fluid Phase Equilibria*, 242(2):220-232,2006.

[7] M. Shiflett and A. Yokozeki. Solubility and diffusivity of hydrofluorocarbons in room temperature ionic liquids. *AIChE Journal*, 52(3):1205-1219, 2006.

[8] M. Shiflett and A. Yokozeki. Solubilities and diffusivities of carbon dioxide in ionic liquids: [bmim][pf6] and [bmim][bf4]. *Industrial & Engineering Chemistry Research*, 44(12):4453-4464, 2005.

[9] S. Asensio-Delgado, D. Jovell, G. Zarca, A. Urtiaga, and F. Llovell. Thermodynamic and process modeling of the recovery of r410a compounds with ionic liquids. *International Journal of Refrigeration*, 118:365-375, 2020.

[10] A. Yokozeki. Solubility of refrigerants in various lubricants. *International Journal of Thermophysics*, 22(4):1057-1071, 2001.

[11] A. Yokozeki and M. Shiflett. Global phase behaviors of trifluoromethane in ionic liquid [bmim] [pf6]. *AIChE Journal*, 52(11):3952-3957, 2006.

626146 Data Analysis and Predictive Modeling for Wastewater Asset Management

Abstract:

Background

Water and wastewater distribution networks (WWDN) are the most valuable asset of a utility company and serves as a critical component of the urban infrastructure (Rahbaralam et al. 2020). The population growth and economic development have recently led to WWDN expansion, followed by the requirement of more sustainable and reliable performance of current water infrastructure (Bakri et al. 2015). In recent years, the American Society of Civil Engineers (ASCE) gave the water and wastewater infrastructure categories a D and D+ rating respectively, indicating a substandard system (ASCE 2017). Many factors contribute to these scores, the most important being asset deterioration and aging WWDNs (Mazumder et al. 2018). The infrastructure systems often experience a myriad of environmental conditions which have a negative impact on the integrity of the system (Mazumder et al. 2018). Besides, the last major investment into the water distribution network was done in the 1960s and 1970s, meaning that most systems are at least 50 years old (Malm et al. 2013). Most pipelines are built with a lifespan of about 80 to 100 years; therefore, many networks have used up over half of their expected lifespan without accounting for deterioration. As a result, many utility companies have been experiencing asset failures and malfunctions more frequently leading to large financial expenditures and negative environmental impacts. The current solution is to use a subjective and reactionary asset management approach, where a utility authority waits until a failure occurs to solve the malfunction (Baah et al. 2015; Fenner 2000). Figure 1 displays the overall process to developing an asset management plan as defined by the United States Environmental Protection Agency (US EPA) (Allbee and Rose 2017). The downfall of this process is the bias that is generated in the step “Assess Performance Failure Modes”. This bias leads to large amounts of uncertainty within the subjective asset management plan and can lead to capital losses and undesired expenses when an unforeseen event occurs. To that end, if we can find an objective-based predictive approach to strengthen this step in the asset management plan, then the negative economic and environmental impacts can be minimized.

Methodology

The asset management problem is solved using machine learning classification algorithms and Bayesian inference methods to generate a predictive deterioration model for each asset. A predictive model which can forecast the pipeline and manhole failures plays a vital role in the strategic rehabilitation planning of WWDN infrastructures (Rahbaralam et al. 2020). Therefore, these different analysis methods are employed to assess the accuracy of the predictive model. The objective is to minimize any potential failures, environmental hazards, and financial expenditures while providing a list of sound preventative measures as well as corrective actions by analyzing the historical data provided by the partner utility company. The analysis consists of evaluating the existing asset property data to develop correlations for risk factors and failure impacts. Using programming platforms such as Python and MATLAB, each asset is simultaneously analyzed, compared, and ranked. By comparing both analyses, the best method can be used in managing and predicting asset failures.

Results & Summary

The WWDN asset dataset provided by the partner company contains information about numerous variables such as the pipe diameter, pipe material type, pipe segment length, expected remaining life of pipe, and the type of flow within the pipe. After preprocessing the data, the different analysis methods are implemented to generate the best predictive deterioration model based on pipeline properties. By coupling the best analysis with accurate historical data, the model generates a failure score for each asset. This failure score allows for the prediction of asset malfunctions, allowing for each asset to be ranked on the basis of potential failures. The score is then investigated for each asset to identify the pipeline property that is the major contributor (bottleneck). Maintenance and corrective actions are then devised to bring the score to an acceptable level, which in turn minimizes the probability of asset malfunctions. The corrective actions then are compiled into a comprehensive maintenance schedule to provide an objective-based maintenance schedule. Thus, this objective-based predictive approach to asset management can minimize major economic expenditures as well as many negative environmental effects.

Acknowledgments

We thank the Atlantic County Utilities Authority, NJ for supporting and funding the research for this project. We also thank the Department of Chemical Engineering at Rowan University for their continued assistance with acquiring the tools required for this research.

References:

Allbee, Steve, and Duncan Rose. 2017. "Fundamentals of Asset Management Session 0 - Exectutive Overview." US EPA. https://www.epa.gov/sustainable-water-infrastructure/asset-management-workshops-training-slides.

ASCE. 2017. "2017 Infrastructure Report Card." ASCE. https://www.infrastructurereportcard.org/cat-item/wastewater/.

Baah, Kelly, Brajesh Dubey, Richard Harvey, and Edward McBean. 2015. "A Risk-Based Approach to Sanitary Sewer Pipe Asset Management." *Science of The Total Environment* 505 (February): 1011–17. https://doi.org/10.1016/j.scitotenv.2014.10.040.

Bakri, B., Y. Arai, T. Inakazu, A. Koizumi, H. Yoda, and S. Pallu. 2015. "Selection and Concentration of Pipeline Mains for Rehabilitation and Expansion of Water Distribution Network." *Procedia Environmental Sciences*, The 5th Sustainable Future for Human Security (SustaiN 2014), 28 (January): 732–42. https://doi.org/10.1016/j.proenv.2015.07.086.

Fenner, R. A. 2000. "Approaches to Sewer Maintenance: A Review." *Urban Water*, Sewer Systems and Processes, 2 (4): 343–56. https://doi.org/10.1016/S1462-0758(00)00065-0.

Malm, A., G. Svensson, H. Bäckman, and Gregory M. Morrison. 2013. "Prediction of Water and Wastewater Networks Rehabilitation Based Current Age and Material Distribution." *Water Supply* 13 (2): 227–37. https://doi.org/10.2166/ws.2013.011.

Mazumder, Ram K., Abdullahi M. Salman, Yue Li, and Xiong Yu. 2018. "Performance Evaluation of Water Distribution Systems and Asset Management." *ASCE Library* 24 (3): 24.

Rahbaralam, Maryam, David Modesto, Jaume Cardús, Amir Abdollahi, and Fernando M. Cucchietti. 2020. "Predictive Analytics for Water Asset Management: Machine Learning and Survival Analysis." *ArXiv:2007.03744 [Cs, Eess, Stat]*, July. http://arxiv.org/abs/2007.03744.

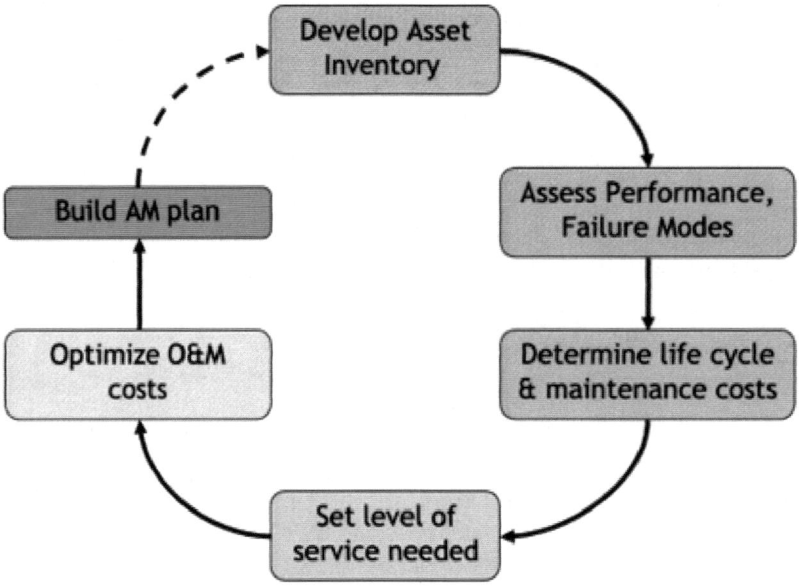

Figure 1. Development Process of an Asset Management Plan as Defined by the United States Environmental Protection Agency

623468 Regularized Bayesian Fusion for Toxin Concentration Estimation in an Industrial Wastewater Treatment Plant

Abstract:

The complex nature of biological and chemical processes taking place in a Waste Water Treatment Plant (WWTP) raises significant challenges to its management and optimal operation. In this work, we address the problem of estimating the concentration of a toxin present in an industrial effluent in order to timely take the necessary actions that secure it never exceeds the regulatory limit. However, the toxin concentration is measured at a low sampling rate (2-3 times per week), making its monitoring and control harder. To circumvent this limitation, models can be developed to estimate the concentration level more frequently. The biological diversity typically found in WWTP and the associated complex metabolic phenomena often make first principles modeling approaches either unfeasible or unreliable. Data-driven approaches provide an alternative solution, but laboratory measurements, as well as other off line instrumentation (such as imaging devices) are not sampled frequently, and only online sensor data from different points in the process are collected at higher rates. In order to make maximum use of the available data, a fusion methodology was developed that is able to handle simultaneously the multirate, asynchronous and heterogeneous (in data structure and quality/uncertainty) nature of the data collected from the WWTP. Data/information fusion methods offer flexible solutions to handle complex data [1], [2] but their application to process industries is still underexplored. The proposed fusion solution was developed under the InfoQ concept, whose purpose is to maximize the quality of information generated in an empirical study [3].

The proposed fusion scheme considers several single-source models (one source regards a particular origin of information in the process, usually a unit or an analytical device) that are flexibly combined and used depending on the availability of information. Their quality is also taken into account, and the smoothness of the successive estimates of the toxin level is controlled using a Bayesian regularization approach. Several regression methods from different corners of the data analytics landscape were considered for building the single-source models (namely, penalized, latent variable and tree-based ensemble regression methods). The models were developed and tuned using a nested double cross-validation strategy [5], [6] and the repeated prequential method [4] in order to handle the time series nature of data.

Our Regularized Bayesian Fusion strategy led to more frequent access to toxin concentration, based on the most updated information available, properly fused through Bayesian fusion, penalyzing excessive variation in the estimates due to unreliable/noisy estimates when less information is available. This methodology should facilitate the management and operation of WWTP and the ability to maintain the toxin concentration below the compliance level.

References

[1] F. Castanedo, «A Review of Data Fusion Techniques», *The Scientific World Journal*, vol. 2013, pp. 1–19, 2013, doi: 10/gb7x39.

[2] A. Diez-Olivan, J. Del Ser, D. Galar, e B. Sierra, «Data fusion and machine learning for industrial prognosis: Trends and perspectives towards Industry 4.0», *Information Fusion*, vol. 50, pp. 92–111, 2019, doi: 10/gf6wkf.

[3] M. S. Reis e R. Kenett, «Assessing the value of information of data-centric activities in the chemical processing industry 4.0», *AIChE J*, vol. 64, n. 11, pp. 3868–3881, 2018, doi: 10/gff327.

[4] V. Cerqueira, L. Torgo, e I. Mozetic, «Evaluating time series forecasting models: An empirical study on performance estimation methods», *arXiv:1905.11744 [cs, stat]*, 2019, Acedido: Mai. 06, 2020. [Em linha]. Disponível em: http://arxiv.org/abs/1905.11744.

[5] R. Rendall e M. S. Reis, «Which regression method to use? Making informed decisions in "data-rich/knowledge poor" scenarios – The Predictive Analytics Comparison framework (PAC)», *Chemometrics and Intelligent Laboratory Systems*, vol. 181, pp. 52–63, 2018, doi: 10/gfc6c8.

[6] T. J. Rato e M. S. Reis, «SS-DAC: A systematic framework for selecting the best modeling approach and pre-processing for spectroscopic data», *Computers & Chemical Engineering*, vol. 128, pp. 437–449, 2019, doi: 10/ghj9ks.

623884 Model-Based Design of Experiments in Pyomo and Its Application to Adsorptive CO_2 Capture Systems

Abstract:

Design of Experiments (DoE) methods optimize computational and physical experiments to maximize the information gain and minimize time and resource costs. Unlike the classical 'black-box' (a.k.a. factorial, response surface) DoE approach which decides the best design by the input-output relationship, model-based DoE (MBDoE) leverages high-fidelity models constructed from the underlying physical principles of the experimental system[1]. Taking advantage of the prior knowledge of the experimental system, MBDoE can discriminate between scientific hypotheses, posed as mathematical models, and facilitates optimization with efficient gradient-based methods. MBDoE has a rich history of success in chemical engineering including chemical kinetics[2], heat/mass transfer modeling[3], and biological modeling[4]. `Black-box' DoE is readily available in several software platforms including Design Expert, JMP, Modde, and several Python packages, such as pyDOE, pyDOE2, dexpy, and doepy. Yet, there are no popular general-purpose software platforms for MBDoE. Often instead a researcher implements their custom code in MATLAB, Python, GAMS, etc., which limits adoption in new application areas.

In this work, we present a general Python package for MBDoE using Pyomo models. Aiming at reducing the uncertainty in the system, MBDoE is conducted by minimizing the variance of the Fisher Information Matrix(FIM), which can be interpreted as minimizing the volume of the covariance ellipsoid. Classical alphabetic criteria such as D-optimality (determinant), A-optimality (trace), E-optimality (minimal eigenvalue), and a modified E-optimality (condition number) are considered to measure the size of the FIM [5]. Nonlinear programming sensitivity analysis, available through sIPOPT [6] or k_aug [7], is used to reduce the computational cost of assembling the FIM by over an order of magnitude.

The new capability is demonstrated in two case studies. Case study 1 considers a nonlinear reaction kinetic model with four highly-correlated kinetic parameters, which shows how MBDoE formalism quickly identifies model identifiability challenges. This case study acts as a tutorial for new users. Case study 2 considers CO_2 adsorption in a fixed-bed breakthrough experiment to characterize novel Metal-Organic Frameworks (MOFs) materials for CO_2 capture. The partial differential-algebraic equation (PDE) model couples mass and momentum transport phenomena with adsorption equilibria (isotherms) and kinetics; discretization in space (method of lines) and time (backward finite difference or collocation) results in over 20,000 sparse resulting in over 20,000 sparse algebraic constraints.

The goal of MBDoE is to infer the heat transfer coefficient (non-adiabatic operation) and a lumped kinetics transport parameter. Through MBDoE, we address the following questions: (1) Is the model identifiable with the current experimental configuration? (2) What is the value of modifying the experimental system (e.g., adding temperature sensors)? (3) How to conduct fixed bed experiments so that we can improve the accuracy of parameter estimation by DoE? The limited literature that considers DoE for a fixed bed adsorption system focuses on "black-box methods" such as factorial methods and response surfaces methods which do not offer insights into the fundamental physical process, in contrast to our proposed MBDoE approach.

In the future, validation mathematical models will be used for optimization and techno-economic analysis of novel CO_2 capture processes as part of the Carbon Capture Simulation for Industrial Impact (CCSI2) project.

Reference

[1]Franceschini, G., & Macchietto, S. (2008). Model-based design of experiments for parameter precision: State of the art. Chemical Engineering Science, 63(19), 4846-4872.

[2]Waldron, C., Pankajakshan, A., Quaglio, M., Cao, E., Galvanin, F. and Gavriilidis, A., 2020. Model-based design of transient flow experiments for the identification of kinetic parameters. Reaction Chemistry & Engineering, 5(1), pp.112-123.

[3]Balsa-Canto, E., Rodriguez-Fernandez, M. and Banga, J.R., 2007. Optimal design of dynamic experiments for improved estimation of kinetic parameters of thermal degradation. Journal of Food Engineering, 82(2), pp.178-188.

[4]Chakrabarty, A., Buzzard, G.T. and Rundell, A.E., 2013. Model-based design of experiments for cellular processes. Wiley Interdisciplinary Reviews: Systems Biology and Medicine, 5(2), pp.181-203.

[5]Box, G.E. and Lucas, H.L., 1959. Design of experiments in non-linear situations. Biometrika, 46(1/2), pp.77-90.

[6]Pirnay, H., López-Negrete, R. and Biegler, L.T., 2012. Optimal sensitivity based on IPOPT. Mathematical Programming Computation, 4(4), pp.307-331.

[7]Thierry, D. and Biegler, L.T., 2019. Dynamic real-time optimization for a CO2 capture process. AIChE Journal, 65(7), p.e16511.

[8]Nocedal, J., Wächter, A. and Waltz, R.A., 2009. Adaptive barrier update strategies for nonlinear interior methods. SIAM Journal on Optimization, 19(4), pp.1674-1693.

[9]Wächter, A., 2002. An interior point algorithm for large-scale nonlinear optimization with applications in process engineering (Doctoral dissertation, PhD thesis, Carnegie Mellon University).

[10]HSL, A., 2007. collection of Fortran codes for large-scale scientific computation. See http://www. hsl. rl. ac. Uk.

[11] Dowling, A.W., Vetukuri, S.R. and Biegler, L.T., 2012. Large-scale optimization strategies for pressure swing adsorption cycle synthesis. AIChE journal, 58(12), pp.3777-3791.

MACHINE LEARNING-BASED SOFT SENSORS FOR VACUUM DISTILLATION UNIT

Kamil Oster, Department of Mathematics, The University of Manchester, UK; Process Integration Limited, UK

Stefan Güttel, Department of Mathematics, The University of Manchester, UK

Lu Chen, Process Integration Limited, UK

Jonathan Shapiro, Department of Computer Science, The University of Manchester, UK

Megan Jobson, Department of Chemical Engineering and Analytical Science, UK

Introduction

Product quality assessment in the petroleum processing industry, such as crude distillation, can be difficult and time-consuming, *e.g.* due to a manual collection of liquid samples from the plant and the subsequent chemical laboratory analysis of the samples. The product quality is an important property that informs whether the products of the process are within the regulated specifications, such as ASTM Petroleum Standards. In particular, the delays caused by sample processing (collection, laboratory measurements, results analysis, reporting, *etc.*) can lead to detrimental economic effects. One of the strategies to deal with this problem is so-called soft sensors.

Soft sensors are a collection of models that can be used to predict and forecast some infrequently measured properties (such as laboratory measurements of petroleum products) based on more frequent measurements of quantities like temperature, pressure and flow rate provided by physical sensors [1]. Soft sensors short-cut the pathway to obtain relevant information about the product quality, often providing relevant measurements as frequently as every minute. One of the applications of soft sensors is for the real-time optimization of a chemical process by a targeted adaptation of operating parameters. Models used for soft sensors can have various forms, however, among the most common are those based on artificial neural networks (ANNs) [2].

While soft sensors can deal with some of the issues in the refinery processes, their development and deployment can pose other challenges that are addressed in this paper. Firstly, it is important to enhance the quality of both sets of data (laboratory measurements and physical sensors) in a so-called data pre-processing stage (as described in Methodology section) [3]. Secondly, once the data sets are pre-processed, different models need to be tested against prediction error and the model's interpretability. In this work, we present a framework for soft sensor development from raw data to ready-to-use models.

Methodology

The example process data used in this paper are taken from a vacuum distillation unit (VDU) of a refinery. All data are taken from an Asian refinery. The VDU, and all physical sensors and product lines (side draws), is shown in **Figure 1**. The first category of data is measurements of temperature (T), pressure (p) and flow rate (F) from physical sensors that make up the input data for the soft sensors. The data set consists of 40 physical sensors. 31 physical sensors were taken for further analysis, with the others rejected due to a high noise level and a large amount of missing values (sensors F20, F10, F30, F41), and coking causing problems with the sensors on the 6th side draw, V6SS (sensors T6, T61, F61, F62, F52).

The second category of data is ASTM-D2887 laboratory measurements of the boiling curve of a distillation product of the 3rd side draw vacuum distillation product, namely V3SS.

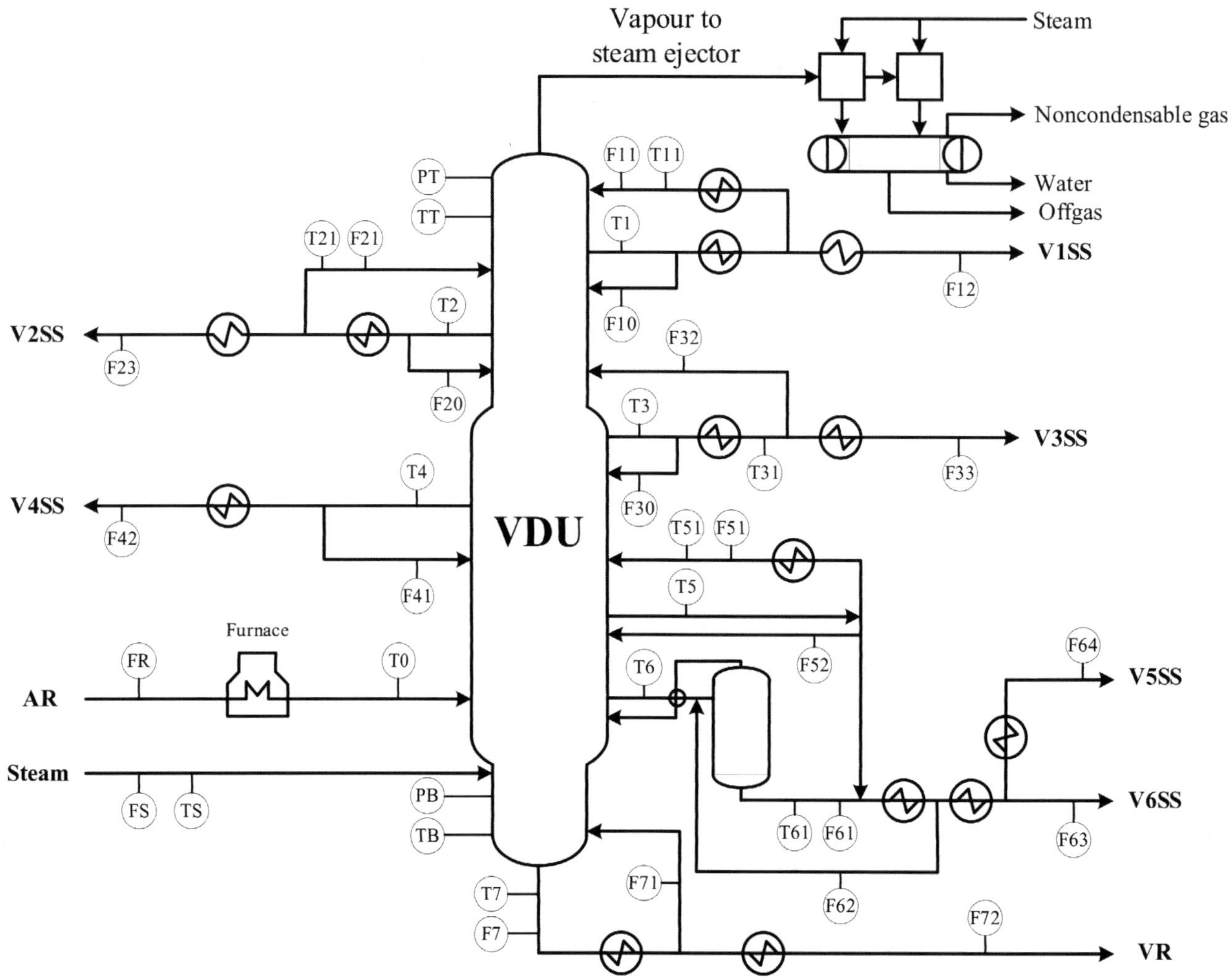

Figure 1. VDU diagram with all physical sensors (*T* - temperature, *p* - pressure, *F* - flow rate) and product lines (V1SS, V2SS, V3SS, V4SS, V5SS, V6SS, VR - vacuum residue), as well as inlets (AR - atmospheric residue).

The quality of the products of V1SS, V2SS, V4SS, V5SS, V6SS and VR is measured only once a week, which significantly limits the amount of data available to develop soft sensors. Hence, in this work, we only focused on V3SS that is measured daily. The developed soft sensors use measured data of physical sensors to infer points on the distillation curve (namely 2, 10, 30, 50, 70, 100 vol%), later called distillation points. The main aim of our framework is to ensure robustness, reliability, reproducibility and interpretability of the models. The overall framework (steps and data flow) is presented in **Figure 2.**

Firstly, the data from the laboratory measurements undergo pre-processing which includes removal of zero/missing/duplicated values and outliers. For the outlier detection, the interquartile method (IQR) was used [4]. For each distillation point, upper and lower bounds were determined and used to identify data points that exceeded the calculated IQR thresholds. The data for each distillation point were combined to establish data at which V3SS was outside the average behavior for at least one of the distillation points.

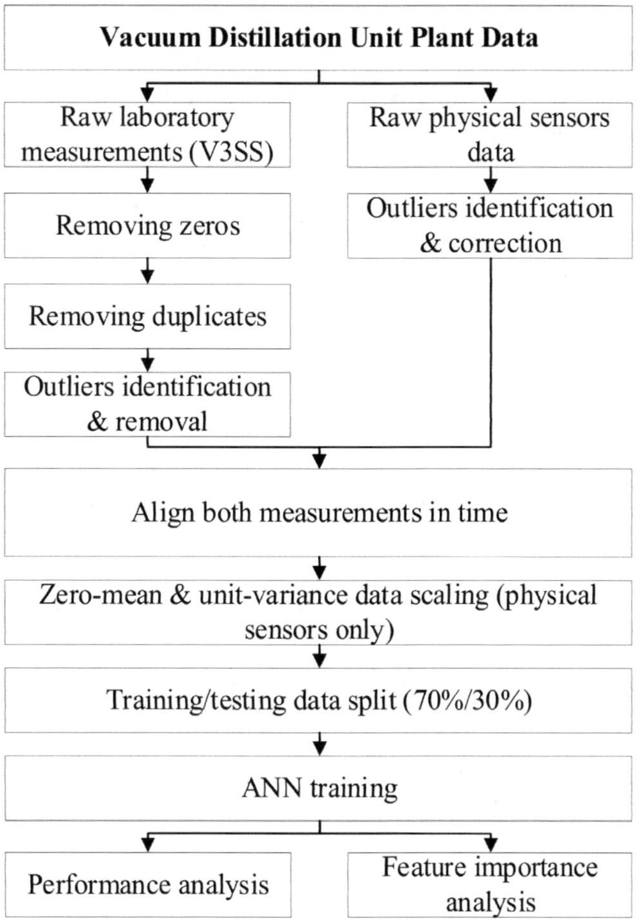

Figure 2. Steps of the framework and data flow.

A main issue in developing soft sensors is the assessment of the model's performance. For new models, it is important to understand how well the model can perform for the data used, specifically what prediction error threshold may be acceptable. One of the approaches to ascertain these thresholds is to use some established models. The V3SS measurements can be treated as a time series, therefore, methods for time series prediction can be employed [5].

We used one of the most commonly used time series models – Seasonal AutoRegressive Integrated Moving Average (SARIMA) [6,7]. The seasonality component in SARIMA is expected to originate from crude oil supply changes and work shift patterns. The *pmdarima* Python package was used to grid search the best SARIMA model for all V3SS distillation points. The algorithm aims to minimize the Akaike Information Criterion (AIC) in a stepwise manner to find the best model for the hyperparameters (using the Dickey-Fuller test of stationarity and the Canova-Hansen test for seasonality). The details of this approach can be found in the original documentation for the package [8]. Once the best SARIMA models for each distillation point were found, these models are then used to predict the distillation points data. Mean absolute errors between predicted (through SARIMA) and actual data for each distillation point are used as upper thresholds for the new ANN model's prediction errors.

In our previous work [9], we studied outlier detection in the plant sensor data (T, p and F). Generally, two types of outliers exist: short-term outliers (*i.e.* noise and erroneous data) and long-term outliers (*i.e.* malfunction for a longer period of time). The outlier detection is a two-step process: 1) identification and removal of short-term outliers; 2) identification of long-term outlier periods. Briefly, change points are identified across each of the signals and then used to partition the signals. The change point detection method is based on maximum log-likelihood estimation (assuming changes in the average response of the signal), ensuring that the partitioned signals comply with the central limit theorem, creating piecewise constant approximations of the mean and standard deviation. These piecewise approximated standard deviations are used to calculate upper and lower thresholds for the outlier detection (specifically triple standard deviation values - 3σ method), and piecewise 3σ upper and lower thresholds. The short-term outliers identified with this method were replaced with the piecewise constant means before proceeding to identify long-term outliers. As discussed in [9], principal component analysis (PCA) combined with Hotteling's T^2 statistics was used to identify the long-term outlier periods. The V3SS data during long-term outlier periods were discarded from further analysis.

The next step in the framework is the time alignment of both data sets, as the V3SS data were recorded daily, while plant data (p, T, F) were recorded every minute. To capture the state of the VDU process immediately prior to withdrawing a sample for laboratory measurement, a 1-hour average was calculated before each of the laboratory measurements. The pre-processed and time-aligned input and output data were scaled appropriately (with zero-mean and unit-variance). Afterwards, the data were split into training and testing sets (70% and 30%, respectively).

Our soft sensor model is based on a multilayer perceptron neural network. The input layer for the model receives 1-hour average plant data (with 31 features/sensors in total), while the output layer predicts the V3SS laboratory measurements (each record is an array of 7 distillation points, *i.e.* from 2 vol% to 100 vol%). The tested ANN model contains 2 hidden layers, each with 30 nodes. The input layer and all hidden layers use ReLu activation functions. The weights and biases of each hidden layer were initialized using the Xavier normal initializer, and zeros, respectively. The Keras/TensorFlow package was used to implement and train the neural networks. The neural network was optimized for a Huber loss function and with the Adam optimizer (both with default parameters from Keras/TensorFlow). Model checkpoint callback was used to prevent overfitting: after each epoch, the algorithm calls back to calculate the prediction error on the testing set to evaluate whether the model improved in relation to the testing set. In this approach, the algorithm only saves the model that showed improvement on the testing set, while updating the weights and biases on training set. The model training ran for up to 10,000 epochs.

To understand how all features correlate with the predicted output through the trained model, feature importance was analyzed. SHapley Additive exPlanations (SHAP) was used for this purpose [10]. Briefly, SHAP is based on calculating Shapley values as in game theory for each of the features (sensors). Each feature (sensor) value of the instance is a 'player' in a game, while the prediction is the 'payout'. Shapley values inform about fair distribution of the 'payout' amongst the features. Aside from minimizing the prediction error, it is important to ensure the model's interpretability, *e.g.* through feature importance analysis, to ensure that the impact of features - plant physical sensors on the predicted output is reflected in an overall understanding of the underlying physicochemical behavior of the VDU.

Results and Discussion

The laboratory measurements used in this paper follow the industrial standard ASTM D2887 *Standard Test Method for Distillation of Petroleum Products at Reduced Pressure* [11]. These measurements are carried out on a liquid sample withdrawn from the VDU to represent the product quality. In this paper, a product of 3rd side draw of VDU is studied, *i.e.* light-vacuum gas oil, namely V3SS.

The frequency of the measurements is 1 per day. Overall, the raw dataset consists of 6,037 data points. **Figure 3a** shows all 7 distillation points of the V3SS distillation curve - 2 vol%, 10 vol%, 30 vol%, 50 vol%, 70 vol%, 90 vol% and 100 vol%. Upon visual inspection of the raw data, it can be seen that a large number of data points are nulls (probable artifacts in the laboratory information management system). Moreover, duplicated records were found in the data. Overall, the raw data consisted of 4,391 nulls (72.7%

of all data) and 347 duplicates (5.7% of all data). After removing these artifacts, the remaining dataset was 1,299 data points (shown in **Figure 3b**). In the next step, outliers were detected and removed (in accordance with IQR method described in Methodology section), as shown in **Figure 3c.** In total, 91 outliers were detected and removed which leaves 1,208 data points. Interestingly, there were periods that would suggest issues with the functioning of the VDU system (35 outliers in the entire period between days 395 and 557). These time instances may correspond to crude oil supply changes that can lead to such deviations in product quality.

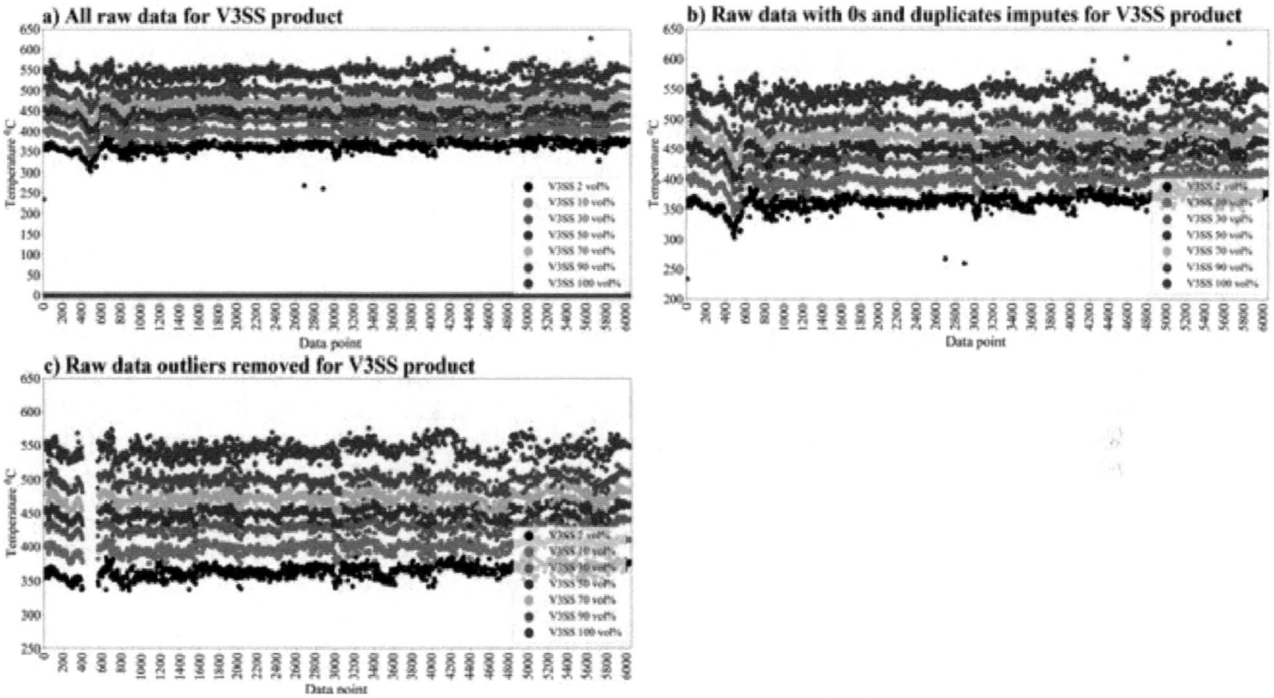

Figure 3. Output data (laboratory measurements of V3SS distillation product) pre-processing: **a)** original raw data; **b)** data after removing nulls and duplicated records; **c)** data after removing outliers.

As described in the Methodology section, SARIMA modelling was used to ascertain the V3SS accuracy, as a baseline for the predictive performance of the ANN. The results of the predicted data against the actual data for V3SS can be seen in **Figure 4a.** The values in the last column in the table (**Figure 4b**) will be used in a later part of this paper to assess the performance of the ANN, *i.e.* the maximum average error for the predictions for each distillation point. For example, it is expected that the ANN is able to predict V3SS 100 vol% with an error below 7.7°C.

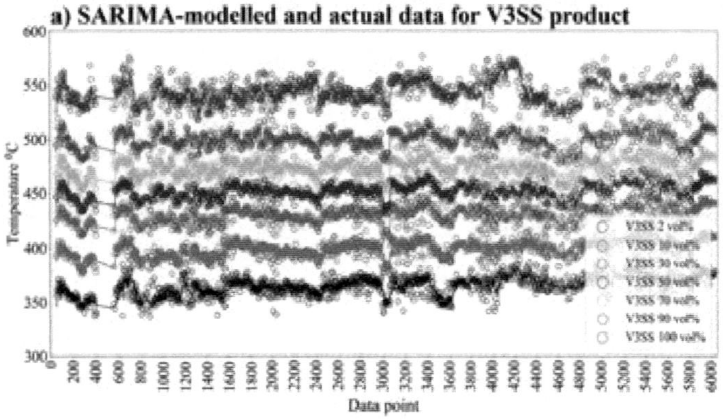

b) Summary of the results of SARIMA modelling

V3SS cut point	(p, d, q) - ARIMA	(P, D, Q, m) - SARIMA	MAE (°C)
2 vol%	(2, 1, 4)	(1, 1, 2, 10)	5.39
10 vol%	(3, 1, 6)	(4, 1, 1, 10)	4.63
30 vol%	(0, 1, 2)	(0, 1, 2, 10)	4.27
50 vol%	(0, 1, 2)	(0, 1, 2, 10)	4.45
70 vol%	(0, 1, 2)	(0, 1, 2, 10)	4.56
90 vol%	(0, 1, 1)	(0, 1, 2, 10)	4.94
100 vol%	(2, 1, 1)	(0, 1, 2, 10)	7.65

Figure 4. a) Comparison of SARIMA-modelled and actual data for V3SS distillation points; **b)** coefficients for ARIMA and SARIMA components with mean absolute error (MAE) of the deviations.

The complete results and discussion for both short- and long-term outliers detection were discussed in our previous paper [9]. Generally, each of the signals is treated separately to determine the short-term outliers (**Figure 5a**), while long-term outliers are determined using PCA, *i.e.* all signals are treated altogether; thus, the latter indicates malfunctioning in the chemical process (**Figure 5b**). Overall, the signals contained 0.78 – 7.31% of short-term outliers, with an average of 3.27%. In case of long-term outliers, two periods were identified (days 1—13 and 789—790), with 19,837 data points. Only 2 data points from laboratory measurements of V3SS were recorded during the identified long-term outliers. These laboratory measurements were removed from further analysis (total data points: 1,206).

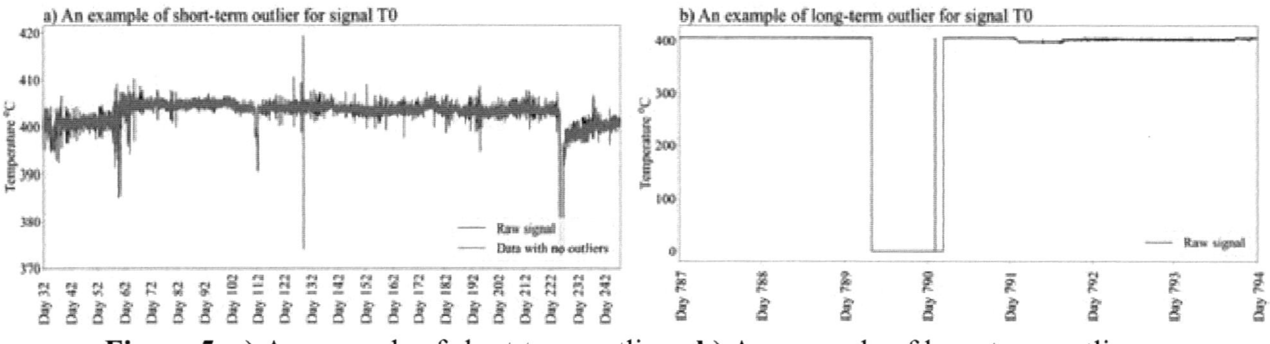

Figure 5. a) An example of short-term outliers; **b)** An example of long-term outliers.

At this stage, laboratory measurements and plant sensors are not aligned in time due to different frequencies. The time at which a liquid sample was withdrawn from VDU was used as the ending point to calculate 1-hour average before that measurement. These data were then scaled with zero-mean and unit-variance, and split into training and testing set (70% and 30% - 844 and 362 samples), before training the ANN. The size of the input and output matrices was 1,209 × 31 (samples × features/sensors) and 1,209 × 7 (samples × distillation points), respectively [12].

The minimum error for the testing set during training was achieved at 969 epochs (out of 10,000 epochs), MAE (°C) = 4.64 and 4.76 for the training and testing sets, respectively. The prediction errors (predicted—actual values for V3SS distillation point) are shown in the form of a histogram in **Figure 6a**.

The physicochemical properties of distillation products must lie within a certain range to comply with industrial requirements. Particularly, for some of them, the upper range is detrimental to the quality of the product. Hence, in the case of soft sensors, it is important to understand how many of the predictions, or to what extent, lie outside the specified limits. Nevertheless, specifications for VDU-derived products are difficult to quantify in terms of formal requirements. Instead, in this paper, we analyzed how many predictions lie outside a confidence interval appropriate for a confidence level of 95% (note that visually the errors appear to be normally distributed, as can be seen in **Figure 6a**) [13]. In the trained model, approximately 3% of data points in the testing set lie outside the confidence interval (grey columns in **Figure 6a**). The MAE of the predictions for each distillation point are collated in the table in **Figure 6b**. It can be seen that all values are below the upper SARIMA thresholds determined beforehand, which means that the model's performance is satisfactory for each of the distillation points.

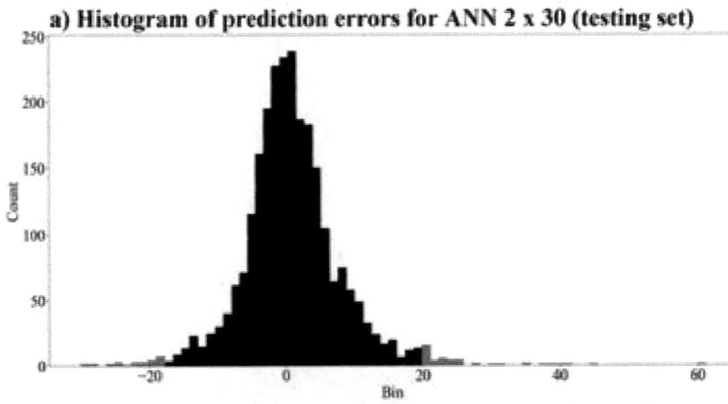

b) Prediction errors from SARIMA modelling and ANN

V3SS cut point	SARIMA threshold (°C)	ANN (2 x 30) MAE – testing set (°C)
2 vol%	5.39	5.35
10 vol%	4.63	4.27
30 vol%	4.27	4.04
50 vol%	4.45	4.13
70 vol%	4.56	4.17
90 vol%	4.94	4.38
100 vol%	7.65	6.98

Figure 6. a) Histogram of deviations in testing set (colors: black - bins inside 95% confidence level, grey - bins outside 95% confidence level); **b)** Comparison of SARIMA thresholds and MAE for the ANN.

The first point of measurements on the 3rd side draw (V3SS line) is the temperature sensor T3 (the VDU diagram was shown previously in **Figure 1)**. It is expected that this temperature has a significant impact on the model predictions. This can be explained by the fact that the sensor T3 measures the draw temperature of V3SS collected at the beginning of the 3rd side draw and distillation curves of the V3SS product (that make up the output of the model).

The feature importance analysis results are shown in **Figure 7** (in the decreasing importance order). It can be seen that T3 has the highest impact on the predictions for all distillation points. T0 has the second highest impact, which can be explained as it measures the feed temperature of the VDU columns (AR), which describes the vapor travelling up the column to liquify into different products, such as V3SS. The sensor T31 measures the return temperature of the reflux which cools down the heavy product in the V3SS stream. Hence, it is expected that sensor T31 and the product quality are correlated.

The VDU process operates under vacuum and so the control of the pressure in the process (PB and PT) is important. Hence, at least one of these pressure measurements should have high potency in the model predictions. As can be seen, sensor PT has a relatively high feature importance value.

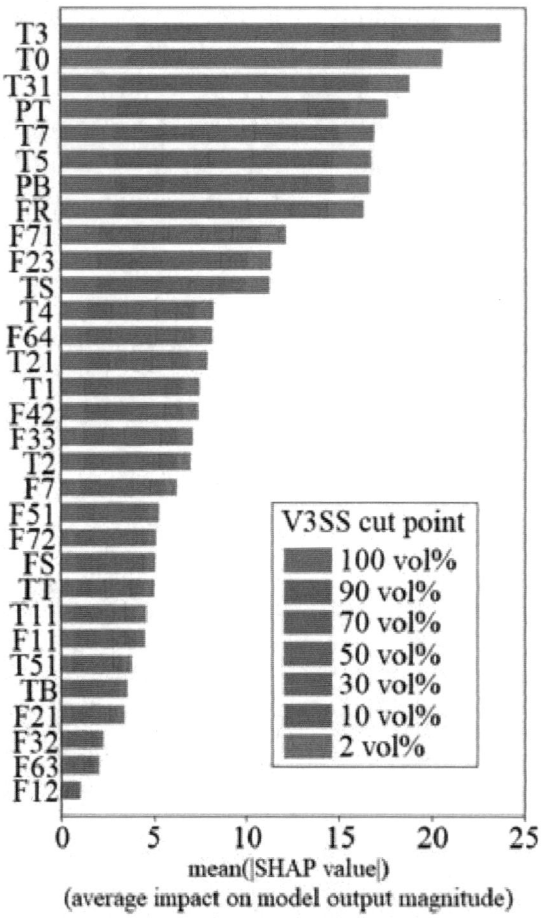

Figure 7. Feature importance analysis results (sorted in decreasing order).

Conclusions

We presented results of an ANN-based soft sensor model with 2 hidden layers and 30 hidden nodes in each hidden layer. Prior to training the model, a number of pre-processing steps was carried out on the plant sensors data (input for the ANN) and laboratory measurements (output for the ANN) to identify and correct artifacts, such as missing values and outliers.

The primary focus of the ANN was satisfying performance in relation to calculated SARIMA thresholds - *i.e.* the prediction errors of the model below the SARIMA errors. Secondly, we showcased an example of feature importance analysis. This was used to understand how all the plant sensors data (input) influence the prediction of the model which can indicate which sensors are more or less important for the model. This knowledge can support engineers onsite to control the process.

Acknowledgments

This work was supported by Innovate UK via a Knowledge Transfer Partnership (No. KTP10916) between the University of Manchester (UK) and Process Integration Limited (UK).

References

[1] A. Rogina, I. Šiško, I. Mohler, Ž. Ujević, N. Bolf, Soft sensor for continuous product quality estimation (in crude distillation unit), Chem. Eng. Res. Des. 89 (2011) 2070–2077.

[2] Y. Jiang, S. Yin, J. Dong, O. Kaynak, A review on soft sensors for monitoring, control and optimization of industrial processes, IEEE Sens. J. (2020).

[3] S. Sandbhor, N.B. Chaphalkar, Impact of outlier detection on neural networks based property value prediction, in: Inf. Syst. Des. Intell. Appl., Springer, 2019: pp. 481–495.

[4] S. Walfish, A review of statistical outlier methods, Pharm. Technol. 30 (2006) 82.

[5] R. Wadhvani, Review on various models for time series forecasting, in: 2017 Int. Conf. Inven. Comput. Informatics, IEEE, 2017: pp. 405–410.

[6] T.C. Nokeri, Forecasting Using ARIMA, SARIMA, and the Additive Model, in: Implement. Mach. Learn. Financ., Springer, 2021: pp. 21–50.

[7] J.-G. Wang, Z. Xie, Y. Yao, B.-H. Yang, S.-W. Ma, L.-L. Liu, Soft sensor development for improving economic efficiency of the coke dry quenching process, J. Process Control. 77 (2019) 20–28.

[8] T.G. Smith, pmdarima: ARIMA estimators for Python, Retrieved From. 309 (2017).

[9] K. Oster, S. Güttel, J.L. Shapiro, L. Chen, M. Jobson, Pre-treatment of outliers and anomalies in plant data: Methodology and case study of a Vacuum Distillation Unit, ArXiv Prepr. ArXiv2106.14641. (2021).

[10] S.M. Lundberg, S.-I. Lee, A unified approach to interpreting model predictions, in: Proc. 31st Int. Conf. Neural Inf. Process. Syst., 2017: pp. 4768–4777.

[11] A. D2887, Standard Test Method for Distillation of Petroleum Products at Reduced Pressure, 2019.

[12] I. Niño-Adan, I. Landa-Torres, D. Manjarres, E. Portillo, Soft-sensor design for vacuum distillation bottom product penetration classification, Appl. Soft Comput. 102 (2021) 107072.

[13] I. Mohler, Ž. Ujević Andrijić, N. Bolf, Soft sensors model optimization and application for the refinery real-time prediction of toluene content, Chem. Eng. Commun. 205 (2018) 411–421.

624022 Statistical Machine Learning for the DOW Data Challenge Problem

Abstract:

In this paper, we present a statistical machine learning approach to the DOW challenge dataset, which is obtained from a multi-column integrated process, as shown in Figure 1 (Braun et al., 2020). The process is composed of three key distillation columns, Primary Column, Feed Column, and Secondary Column. The main objective of the challenge is to identify key variables that affect impurity levels measured at the primary column outlet from more than 40 process variables. The challenge is to build a high-precision inferential sensor model to predict the impurity. To benchmark various solutions, a validation dataset is provided in addition to the training dataset, which contains data collected over a year of time. The validation dataset should not be used in any way for modeling, or to determine any hyperparameters in the modeling phase. It can only be used for showing the accuracy of the inferential sensor model.

Our proposed solution is a statistical machine learning approach which consists of i) process data exploratory analysis, ii) a method for variable selection, iii) a method to deal with non-negative physical property modeling using a soft-plus function; and iv) a method for real-time bias updating based on known data. We benchmark main algorithms among partial least squares (PLS), lasso, and the least angle regression solution (LARS). We demonstrate using the validation dataset that our method gives superior prediction results. The pros and cons of the statistical machine learning methods are given with practical implications for industrial users. We make use of and emphasize on the importance of domain knowledge in exploratory analysis and feature selections. We report the identification of mode-switching operation in the data that leads to proper data pre-processing and interpolations found in the impurity data. We provide a solution for irregularly sampled quality data modeling, which shows that it is unnecessary to interpolate the lab-test impurity data.

Figure 1. DOW Challenge process flowchart from which the datasets were collected.

References

1. Braun, I. Castillo, M. Joswiak, Y. Peng, R. Rendell, A. Schmidt, Z. Wang, L. Chiang, and B. Colegrove, Data Science Challenges in Chemical Manufacturing, IFAC World Congress, July 2020, Berlin, Germany.

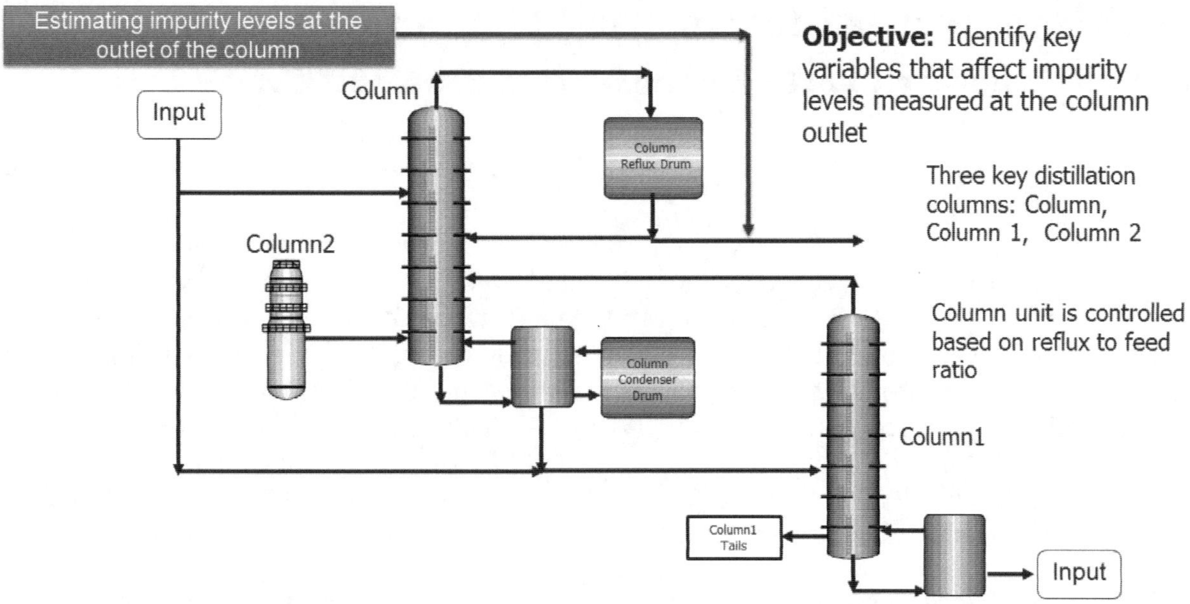

Estimating impurity levels at the outlet of the column
Objective: Identify key variables that affect impurity levels measured at the column outlet
Three key distillation columns: Column, Column 1, Column 2
Column unit is controlled based on reflux to feed ratio
Input
Column
Column Reflux Drum
Column2
Column Condenser Drum
Column1
Column1 Tails
Input

625619 Data-Driven Modeling and Optimization of an Industrial Scale Reverse Osmosis Desalination Plant

Abstract:

Population growth and globally depleting water sources contribute to immense stresses for water supply systems [1]. Water desalination technologies can lower these stresses, by utilizing saline water sources to generate water that can be used for various applications [2, 3]. Reverse osmosis (RO) superseded thermal separation technologies as the desalination industry leader due to its high energy efficiency, low space requirements, as well as process compactness [4], accounting for 68% of the globally produced desalinated water volume in 2019. Nevertheless, RO is still an energy and cost intensive process [5, 6], while optimizing the operation of RO systems can be computationally expensive due to the complexity of the interconnected membrane models. Data-driven surrogate models can help in overcoming this challenge, while still capturing the process behavior accurately.

Therefore, this work focuses on the optimal operation of desalination plants integrated with renewable energy sources through surrogate modeling. Firstly, a data-driven surrogate model for capturing the behavior of an industrial scale RO plant is developed. A neural network (NN) with rectified linear units (ReLU) is used to approximate collected data from the H2Oaks RO desalination plant in South-Central Texas. The data consists of process parameters per RO stage, as well as the energy consumption of the pumping system. Consequently, various possible surrogate model structures are investigated, and the performance compared. The developed NN is then transformed into a mixed-integer linear programming (MILP) formulation [7], and used for the derivation of minimal cost for the operation of the RO plant while analyzing energy-water trade-offs, thus enabling techno-economic feasibility analyses. The derived surrogate model can also be applied to the H2Oaks process control, as well as to investigate future investment decisions.

[1] R. Cory Allen, Yaling Nie, Styliani Avraamidou, Efstratios N. Pistikopoulos. Infrastructure Planning and Operational Scheduling for Power Generating Systems: An Energy-Water Nexus Approach, Computer Aided Chemical Engineering, 47, 233-238, 2019. https://doi.org/10.1016/B978-0-12-818597-1.50037-0

[2] Marcello Di Martino, Styliani Avraamidou, Julie Cook, Efstratios N. Pistikopoulos. An Optimization Framework for the Design of Reverse Osmosis Desalination Plants under Food-Energy-Water Nexus Considerations, Desalination, 503, 2021. https://doi.org/10.1016/j.desal.2021.114937

[3] Marcello Di Martino, Styliani Avraamidou, Efstratios N. Pistikopoulos. Superstructure Optimization for the Design of a Desalination Plant to Tackle the Water Scarcity in Texas (USA), Computer Aided Chemical Engineering, 48, 763-768, 2020. https://doi.org/10.1016/B978-0-12-823377-1.50128-2

[4] Muhammad Qasim, Mohamed Badrelzaman, Noora N. Darwish, Naif A. Darwish, Nidal Hilal. Reverse osmosis desalination: a state-of-the-art review, Desalination, 459, 59–104, 2019. https://doi.org/10.1016/j.desal.2019.02.008

[5] Edward Jones, Manzoor Qadir, Michelle T.H. van Vliet, Vladimir Smakhtin, Seong-mu Kang. The state of desalination and brine production: A global outlook, Science of the Total Environment, 657, 1343–1356, 2019. https://doi.org/10.1016/j.scitotenv.2018.12.076

[6] Jad R. Ziolkowska, Reuben Reyes. Prospects for Desalination in the United States - Experiences From California, Florida and Texas, Competition for Water Resources, 2017, pp. 298–316, https://doi.org/10.1016/B978-0-12-803237-4.00017-3.

[7] Bjarne Grimstad and Henrik Andersson. Relu networks as surrogate models in mixed-integer linear programs, Computers & Chemical Engineering, 131, 2019. https://doi.org/10.1016/j.compchemeng.2019.106580

Objective Assessment of Operator Training using Correspondence Analysis of Physiological and Behavioral Measures

Mohammed Aatif Shahab [a], Babji Srinivasan [a,b], Rajagopalan Srinivasan [b,c]

[a] Department of Applied Mechanics, Indian Institute of Technology Madras, Chennai, 600036, India

[b] Department of Chemical Engineering, Indian Institute of Technology Madras, Chennai, 600036, India

[c] American Express Lab for Data Analytics, Risk and Technology, Indian Institute of Technology Madras, Chennai, 600036, India

In a typical process industry, operators are responsible for monitoring the state of the plant and perform appropriate control actions during abnormal situations. The operator gathers the input from various sources (primarily human machine interface), interprets it based on their knowledge and expertise (using mental models of the process), makes a prognosis, and then reacts to the situation in the form of various actions (physical or verbal). Ironically, the complex automation and control techniques have made it challenging for the operators to develop an appropriate mental model of causal relationships related to plant behavior (Kluge et al., 2014). The likelihood of committing errors during abnormal situation is very high often because of the incorrect mental models of the process developed by the operators (Leveson and Stephanopoulos, 2014). Thus, it is imperative to assess the mental models of the process developed by the operator while they train, so as to ensure adequate preparedness.

Process industries have long resorted to the use of Operator Training Simulators (OTS) for imparting training to the operators. Several studies have been conducted to evaluate operator performance during these training sessions using OTS. Typically, these approaches measure the learning and performance in terms of outcomes (success/failure), response times, deviation of operator actions from predefined sequences (Patle et al., 2014), etc. Nevertheless, these studies do not explicitly focus on evolution of mental models of the process (operator's knowledge about the dynamics of the process) during the training sessions. This is an important aspect as it helps understand the operator's mental models and, therefore, can help develop effective individual-centric training programs. In addition, there are hardly few approaches that focus on understanding the cognitive behavior of operators, crucial to enhancing operators' skills and abilities (Kluge et al., 2014).. Therefore, it is necessary to account for cognitive aspects to assess the learning level of operators (Das et al., 2017) which can help improve training programs.

Recent development in sensor technology has enabled the researchers to get insights into the physiological behavior of humans (Manca et al., (2014). More specifically, eye tracking, a non-intrusive and non-invasive technology, has enabled researchers to get insights into humans' thought process (Srinivasan et al., 2019). Our previous research has shown interesting insights into the operator cognitive behavior (Bhavsar et al., 2017; Sharma et al., 2016; Kodapully et al., 2016). Our studies indicated that eye gaze analysis could identify the operator strategy in diagnosis and execution tasks during an experiment (Kodapully et al., 2016). Sharma et al.

(2016) reported that many operators, even though successful in dealing with the abnormalities, have eye gaze orientation like that of the unsuccessful operators, thus providing additional insights into the operator performance. Statistical studies revealed that gaze-based entropy measures could be used to understand the situational awareness of the operators while handling plant abnormalities (Bhavsar et al., 2017). A digraph-based approach was proposed by Das et al. 2017 to evaluate the mental models of the operators based on the gaze transitions among different areas of interest on HMI. However, these were limited to single task analysis and that the evolution of mental models during training was not addressed.

In this work, we adopted correspondence analysis, a multivariate statistical technique, to quantify the development of mental models of the process by the operator. Correspondence analysis provides the relation between two categorical variables, in our case, regions on HMI (called as areas of interest, AOIs) and the different abnormal scenarios pertaining to the disturbance in the plant. We used a simulated ethanol production plant as a testbed and conducted human subjects' experiments to evaluate the proposed approach. Ten participants were involved in the study. Each participant carried out repetitive tasks (repetition of a scenario) in different trials. A typical trial consists of six different disturbance scenarios. Overall, these ten participants performed 81 trials for each scenario leading to a total of 486 tasks. During these tasks, we recorded the process data, alarm information, operator actions data, and eye gaze data (using Tobi TX 300 eye tracker). To get the relation between the AOIs and different disturbance scenarios, we calculated the angle between these two on the asymmetric plot obtained from the correspondence analysis. A small angle between a scenario and an AOI represents the strong association between that scenario and AOI. Further, the distance to the origin of an AOI on the asymmetric plot indicates its importance in all scenarios. We observed signatures of learning with repetition of trials. For a typical operator in the last trial, the angle between scenario (corresponding to a particular disturbance) and relevant AOIs is found to be 0.90°, very close to zero compared to 57° in the first trial. This indicates that with repetitive tasks, the operator has learnt the root cause of the disturbance (coolant flow) and discovered the correct action to be taken in response to the disturbance. Further, in the last trial, the distance to the origin of trend panel is found to be 0.06 (compared to 1.72 in the first trial) implying that the operator has understood the importance of trend panel and is observing the response of his actions by monitoring the process variable trend in all scenarios of the trial. Statistical studies reveal that the proposed metrics (obtained from correspondence analysis) can quantify learning during operator training programs. In this paper, we will describe the proposed approach and report results from the human subject experiments conducted to validate it.

Key words: Mental Models, Operator Training, Eye Tracking, Correspondence Analysis

Acknowledgement- This work is partially funded by American Express Lab for Data Analytics, Risk and Technology, Indian Institute of Technology Madras

References

1. Bhavsar, P., Srinivasan, B., & Srinivasan, R. (2017). Quantifying situation awareness of control room operators using eye-gaze behavior. Computers & chemical engineering, 106, 191-201.

2. Das, L., Srinivasan, B., & Srinivasan, R. (2017). Cognitive behavior based framework for operator learning: knowledge and capability assessment through eye tracking. In Computer Aided Chemical Engineering (Vol. 40, pp. 2977-2982). Elsevier.
3. Kluge, A., Nazir, S., & Manca, D. (2014). Advanced applications in process control and training needs of field and control room operators. IIE Transactions on Occupational Ergonomics and Human Factors, 2(3-4), 121-136.
4. Kodappully, M., Srinivasan, B., & Srinivasan, R. (2016). Towards predicting human error: Eye gaze analysis for identification of cognitive steps performed by control room operators. Journal of Loss Prevention in the Process Industries, 42, 35-46.
5. Leveson, N.G. and Stephanopoulos, G. (2014), A system-theoretic, control-inspired view and approach to process safety. AIChE J., 60: 2-14. https://doi.org/10.1002/aic.14278
6. Manca, D., Nazir, S., Colombo, S., & Kluge, A. (2014). Procedure for automated assessment of industrial operators. Chemical Engineering Transactions, 36, 391-396.
7. Patle, D. S., Ahmad, Z., & Rangaiah, G. P. (2014). Operator training simulators in the chemical industry: review, issues, and future directions. Reviews in Chemical Engineering, 30(2), 199-216.
8. Puskás, J., Egedy, A., & Németh, S. (2018). Development of operator training simulator for isopropyl alcohol producing plant. Education for Chemical Engineers, 22, 35-43.
9. Sharma, C., Bhavsar, P., Srinivasan, B., & Srinivasan, R. (2016). Eye gaze movement studies of control room operators: A novel approach to improve process safety. Computers & Chemical Engineering, 85, 43-57.
10. Srinivasan, R., Srinivasan, B., Iqbal, M. U., Nemet, A., & Kravanja, Z. (2019). Recent developments towards enhancing process safety: Inherent safety and cognitive engineering. Computers & Chemical Engineering, 128, 364-383.

624024 Multi-Scale Optimization of a Fixed-Bed Multi-Tubular Reactor for CO_2 Methanation

Abstract:

The power-to-gas (PtG) technology has been known as one of the most viable solutions for storing intermittent renewable energy sources such as solar and wind to the existing gas grid. The chemical CO_2 methanation in a fixed-bed multi-tubular reactor was studied via a multi-scale optimization (MSO) approach covering an one-dimensional plug-flow reactor (1D PFR) model, a computational fluid dynamics (CFD) model, and a physical-informed neural network (PINN) model. The PINN model was composed of a feedforward neural network (NN) and physical-informed constraints where the mass and energy balances were introduced to the NN loss function using the automatic differentiation (AD) technique. The data for reaction kinetics, hydrodynamics, and heat transfer simulated by the CFD model were used for estimating the model parameter of 1D PFR via PINN. Taking the advantages of (1) 1D PFR on calculation speed, (2) CFD model on the rigorousness, and (3) PINN on the knowledge extraction from data, the MSO showed a good potential for fast and robust optimization of the fixed-bed multi-tubular CO_2 methanation process.

624869 Corning® Advanced-Flow™ Photo Reactors: A Light Touch to Continuously Improve Chemical Synthesis

Abstract:

Over the past 20 years, Corning has delivered its disruptive continuous flow process intensification platform, Corning® Advanced-Flow™ Reactors and their application technologies to the chemical processing industry to enable "fast" lab-scale flow process development as well as seamless scale-up of flow processes to commercial production.

Transferring chemical synthesis from traditional batch technology to continuous flow technology brings significant advantages in the reduction of cost, and in the management of safety – two factors that are typically associated with process scale-up. In the pharmaceutical, fine and specialty chemicals industries, product quality requirements have become crucial parameters for manufacturers to consider. With traditional batch technology, it is often not possible to maintain optimum product quality when scaling up a process in a short period of time. However, with Corning's continuous flow technology, seamless scale-up can be achieved via straight-forward methodology due to the consistent performance of Corning Advanced-Flow Reactors Technology, including 1000x improvement in heat transfer, 10-100x enhancements in multiphase mixing, 1000x reduction in chemical holdup compared to conventional stirred batch reactors.

Scaling up processes using Corning Advanced-Flow Reactors Technology can be faster than with traditional batch processing methods and time-to-market can be drastically reduced. In addition, this technology can help reduce development and production costs.

Recently, there is regained interest in the field of photochemistry that is driven by a need for synthetic alternative routes as well as new technological tools to perform photochemical reactions. Photochemistry can be a formidable tool to simplify multistep synthesis and can lead to increased cost-savings, improved carbon economy and enhanced selectivity and yield of a process.

Combining the advantages of Corning's reactors with photochemistry gives users access to a wide variety of photochemical reactions and the ability to precisely control the process conditions and light input.

Available across all scales – from laboratory up to industrial throughputs, Corning Advanced-Flow Reactors offer a controlled and repeatable set of conditions to explore a wide range of photochemical reactions.

Additionally, the fundamentals of photochemistry will be presented. The impact of its combination with flow chemistry will be reviewed and used to show an alternative approach to optimizing reactions using photocatalysts in flow synthesis. Using applications, this different concept is shown to lead to a more cost-efficient competitive process.

624151 Systematic Evaluation of Isoflavone Extraction from Soybean Meal

Abstract:

Resource sustainability concerns continue to increase as natural materials are used continuously without reclamation plans. Research efforts have been made toward finding sustainable chemical feedstocks to produce high-performance materials derived from biomass, with competitive economics and properties comparable to existing petroleum-based polymers [1]–[6]. The United States produces over 125 million metric tons of soybeans annually for animal feed, export, and human consumption [7]. Soybeans are rich in nutrients such as amino acids, proteins, and carbohydrates. There are also small quantities of underutilized chemicals, such as isoflavones, that can prove beneficial in the health and nutraceutical sector due to their anti-inflammatory and cancer inhibition effects in humans [8], [9]. Recent research progress has shown that isoflavones have been used to synthesize high-performance materials with high thermal stabilities [10], [11]. The extraction of isoflavone from soybean meal is a process that has been optimized at a lab-scale with little consideration for costs or the environment as they generally involve the usage of large quantities of hazardous organic solvents [12]–[14]. The commercialization of bio-based products can prove competitive against traditional petroleum-based products if the bio-based chemical extraction is designed to be economically viable and environmentally friendly at a larger scale. The isoflavone extraction methodology from soybean meal was represented as a mixed-integer non-linear programming (MINLP) problem. Alternative pathways were compared simultaneously using a superstructure-based approach and optimized in the General Algebraic Modeling Systems (GAMS) to determine the most economical path [15], [16]. The sustainable process index (SPI) was used to assess the environmental footprint of the recommended process. Each technology option is represented by mathematical models that include mass and energy balances, equipment design, and cost. We considered four essential stages of acquiring purified isoflavones from soybean meal, including pre-processing, extraction, acid hydrolysis, and purification.

Figure 1A displays the optimal pathway to extract and purify isoflavone from soybean meal at the commercial scale. The commercial-scale extraction of isoflavone from soybean meal presents a feasible solution by grinding (GRD), turbo-extraction (TE), filtration (FLT,1), drying (DRY,1), acid hydrolysis (AHY), neutralization (NT), filtration (FLT,2), organic solvent nanofiltration (OSN), and drying (DRY,2). The analysis of each model considered material and energy balances, utilities, design options, industrial constraints, and costs. Figure 1B and 1C presents a cost evaluation of the isoflavone extraction process under two scenarios with and without internal material recovery. The overall process cost can be minimized by reusing material when possible. By analyzing alternative options simultaneously, this study shows that commercial-scale extraction of soy isoflavones can be economically viable without detrimentally impacting the environment.

References

[1] E. D. Hernandez, A. W. Bassett, J. M. Sadler, J. J. La Scala, and J. F. Stanzione, "Synthesis and Characterization of Bio-based Epoxy Resins Derived from Vanillyl Alcohol," *ACS Sustain. Chem. Eng.*, vol. 4, no. 8, pp. 4328–4339, Aug. 2016, doi: 10.1021/acssuschemeng.6b00835.

[2] S. Curia *et al.*, "Betulin-based thermoplastics and thermosets through sustainable and industrially viable approaches: new insights for the valorization of an underutilized resource," *ACS Sustain.*

Chem. Eng., Sep. 2019, doi: 10.1021/acssuschemeng.9b03471.

[3] E. A. Baroncini, "Bio-Based Thiol-ene Polymer Electrolytes," Rowan University, NJ, 2019.

[4] A. W. Bassett *et al.*, "Synthesis and characterization of molecularly hybrid bisphenols derived from lignin and CNSL: Application in thermosetting resins," *Eur. Polym. J.*, vol. 111, pp. 95–103, Feb. 2019, doi: 10.1016/j.eurpolymj.2018.12.015.

[5] J. R. Mauck *et al.*, "Preparation and Characterization of Highly Bio-Based Epoxy Amine Thermosets Derived from Lignocellulosics," *Macromol. Chem. Phys.*, vol. 218, no. 14, p. 1700013, Jul. 2017, doi: 10.1002/macp.201700013.

[6] J. F. Stanzione III, J. M. Sadler, J. J. La Scala, K. H. Reno, and R. P. Wool, "Vanillin-based resin for use in composite applications," *Green Chem.*, vol. 14, no. 8, p. 2346, 2012, doi: 10.1039/c2gc35672d.

[7] National Agricultural Statistics Service (NASS), "Crop Production," United States Department of Agriculture (USDA), 2018. Accessed: Oct. 04, 2019. [Online]. Available: https://www.nass.usda.gov/Publications/Todays_Reports/reports/crop0918.pdf.

[8] S. Joy *et al.*, "The Isoflavone Equol Mediates Rapid Vascular Relaxation: Ca2+ -INDEPENDENT ACTIVATION OF ENDOTHELIAL NITRIC-OXIDE SYNTHASE/Hsp90 INVOLVING ERK1/2 AND Akt PHOSPHORYLATION IN HUMAN ENDOTHELIAL CELL," *J. Biol. Chem.*, vol. 281, no. 37, pp. 27335–27345, Sep. 2006, doi: 10.1074/jbc.M602803200.

[9] R. L. Birru *et al.*, "The impact of equol-producing status in modifying the effect of soya isoflavones on risk factors for CHD: a systematic review of randomised controlled trials," *J. Nutr. Sci.*, vol. 5, 2016, doi: 10.1017/jns.2016.18.

[10] J. Dai *et al.*, "High-Performing and Fire-Resistant Biobased Epoxy Resin from Renewable Sources," *ACS Sustain. Chem. Eng.*, vol. 6, no. 6, pp. 7589–7599, Jun. 2018, doi: 10.1021/acssuschemeng.8b00439.

[11] J. Dai, N. Teng, J. Liu, J. Feng, J. Zhu, and X. Liu, "Synthesis of bio-based fire-resistant epoxy without addition of flame retardant elements," *Compos. Part B Eng.*, vol. 179, p. 107523, Dec. 2019, doi: 10.1016/j.compositesb.2019.107523.

[12] L. Y. Yoshiara, T. B. Madeira, F. Delaroza, J. B. da Silva, and E. I. Ida, "Optimization of soy isoflavone extraction with different solvents using the simplex-centroid mixture design," *Int. J. Food Sci. Nutr.*, vol. 63, no. 8, pp. 978–986, Dec. 2012, doi: 10.3109/09637486.2012.690026.

[13] T. N. T. Tran *et al.*, "Optimization of isoflavones extraction from soybeans using full factorial design," *J. Food Process. Preserv.*, vol. 43, no. 9, p. e14078, 2019, doi: 10.1111/jfpp.14078.

[14] S. H. Yuliani, M. R. Gani, E. P. Istyastono, and F. D. O. Riswanto, "Optimization of genistein and daidzein extraction," *J. Pharm. Pharmacogn. Res.*, vol. 4, pp. 231–241, 2018.

[15] J. Chea, A. Lehr, J. Stengel, M. J. Savelski, C. S. Slater, and K. Yenkie, "Evaluation of Solvent Recovery Options for Economic Feasibility through a Superstructure-Based Optimization Framework," *Ind. Eng. Chem. Res.*, Mar. 2020, doi: 10.1021/acs.iecr.9b06725.

[16] K. M. Yenkie, W. Wu, R. L. Clark, B. F. Pfleger, T. W. Root, and C. T. Maravelias, "A roadmap

for the synthesis of separation networks for the recovery of bio-based chemicals: Matching biological and process feasibility," *Biotechnol. Adv.*, vol. 34, no. 8, pp. 1362–1383, Dec. 2016, doi: 10.1016/j.biotechadv.2016.10.003.

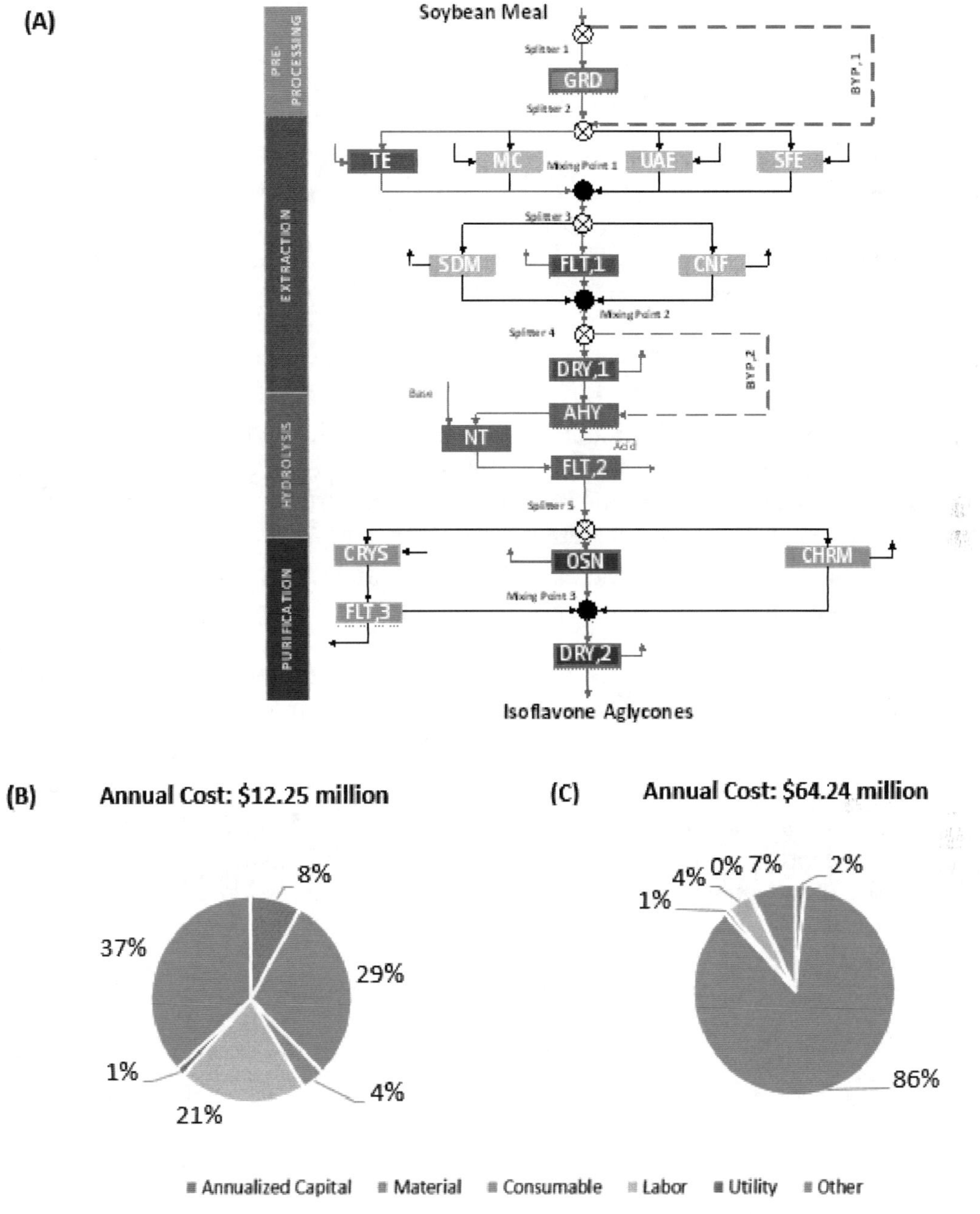

Figure 1. (A) A superstructure of the conventional isoflavone extraction and purification pathways. Bypass streams are used to skip optional separation stages that do not appreciably impact the outcome of the overall process. The unit processes considered include: grinding (GRD), turbo-extraction (TE), maceration (MC), ultrasound-assisted extraction (UAE), supercritical fluid extraction (SFE), sedimentation (SDM), filtration (FLT), centrifugation (CNF), acid hydrolysis (AHY), neutralization (NT), crystallization (CRYS), organic solvent nanofiltration (OSN), chromatography (CHRM), and drying (DRY). The optimal pathway is shown and highlighted as red arrows. (B) Cost distribution of the optimal solution with the material. (C) Cost distribution of the optimal solution with no material recovery.

624737 An Experimental Study of High-Pressure, Multi-Phase Reactions between Sulfur and Hydrocarbons; Strategies to Enhance Reaction Rate for Process Intensification

Abstract:

Sulfurized hydrocarbon materials play a wide role in the chemical process industry. Due to the thermodynamic and kinetic limitations of sulfur in reacting with most hydrocarbon systems, most industrial processes are operated in batch reactors. In order to gain the benefits of continuous operation and production, the reaction rate must be significantly increased through changes in the operating conditions and/or the use of catalysts.

In the current work, the control processes were conducted under the same conditions as in industry, and compared to reactions in 'intensified' process conditions. The control reactions were carried out at 160-190°C at low pressure, controlled by the vapor pressure of the hydrocarbons, with a vapor space comprising approximately 20% of the vessel. Initially, the system shows a liquid sulfur-rich phase and a hydrocarbon-rich phase is formed, with a vapor space rich in hydrocarbon and intermediates such as H_2S above it. Within an hour, the liquid sulfur phase is consumed, leaving a single liquid phase containing sulfur, unreacted hydrocarbons and sulfurized products. High conversion requires a hold time of approximately 5-10 hours.

In order to significantly increase the reaction rate such that a batch-to-continuous transition in manufacturing can occur, three methods were studied using a windowed, variable-volume, Inconel vessel rated to 300°C and 40000psi. First, at 160-190°C, the pressure was increased to prevent the formation of a vapor phase, forcing all the volatile by-products and intermediates to remain dissolved in the liquid phase(s). Second, the temperature was increased up to 250°C to enhance reaction rate at a sufficient pressure to suppress the formation of a vapor phase. Third, attempts were made to identify effective homogenous or heterogeneous catalysts. The reaction results will be discussed in detail in the presentation.

625911 Scheduling for Dynamic Risk and Resource Management for Simultaneous Front-End Refinery Production and Maintenance

Abstract:

The front-end crude transport process (FCTP) is a critical sector of a refinery, which covers both discrete and continuous operations such as crude unloading, storage, pipeline transportation, blending, and feeding to crude distillation units. As the rapid development of the petroleum refinery in recent years, the optimality study of the FCTP system has become a hot topic. In reality, the FCTP system should also be concerned for its operating safety, robustness, and agility under different uncertainties. Thus, unit maintenance plans, which aims at recovering performance-decaying unit performances and prevent potential risks from escalating into catastrophic disasters, should be dynamically prescribed subject to risk management control and resource availability, as well as be seamlessly integrated with ordinary optimal manufacturing schedules. Unfortunately, such important studies are still lacking.

In this study, a new scheduling model for dynamic risk and resource management for simultaneous production and maintenance in the FCTP system has been developed. The developed model will optimize the crude movement in the FCTP system with the minimum operating cost; meanwhile, it can adaptively and timely adjust the unit maintenance plan for the FCTP system, which can minimize the dynamic system risk score with the satisfactory of available dynamic resource constraints. Furthermore, the model is general and can be iteratively employed with rolling time horizon techniques to accomplish dynamic scheduling mechanism internal or external uncertainties. The scheduling model can also handle multiple maintenance tasks at once with identifying the best time to isolate a unit for maintenance and the best time to resume its duty as well. The developed model has been demonstrated through a series of case studies to assure its efficacy.

CFD SIMULATION OF WIND EFFECT ON THE FLAMMABLE GAS CLOUD FORMATION FROM TWO-PHASE FLOW RELEASE

N. A. B. ALMEIDA[1], *P. L. BARROS*[2] *e J. J. N. ALVES*[1]
[1]Federal University of Campina Grande, Chemical Engineering Department, Campina Grande – PB, BR
[2]Ryerson University, Department of Chemical Engineering, Toronto – Ontario, CA
E-mail adresses: natalya.amelia@eq.ufcg.edu.br

ABSTRACT

Flammable liquids are often handled in the industry and it is mandatory to classify the areas where it is likely to form explosive atmospheres according to international or national standards for hazardous area classification. In this regard, the assessment of risk scenarios involved in operations with liquefied gases is necessary. The use of numerical techniques such as Computational Fluid Dynamics (CFD) is recommended by the international standard IEC 60079-10-1 (2015) as an auxiliary tool in hazardous area studies. It allows accurate evaluation of specific risk scenarios including two-phase releases, enabling to verify the influence of different variables on the hazardous extent and zone types. In view of that, this work aims to evaluate the wind effect on the flammable gas cloud formation as a result of the release of liquefied propane from numerical experiments. The CFD model employed a multiphase Eulerian-Lagrangian approach and considered the condition of superheated liquid in the release. The wind velocity was varied regularly up to 5 m.s-1, whereas the wind direction was modified with respect to the jet release direction. As a result, the gas cloud's extent and volume were obtained considering a critical concentration of interest as a boundary, defined as a fraction (0.5 – 1.0) of the Lower Flammability Limit (LFL) of the substance. Regardless of the direction, the increase in wind velocity resulted in smaller flammable cloud volumes due to the dilution effect. A non-monotonic behavior was observed for the hazard extent when the wind and jet are in opposite directions, which occurs due to zero flow velocity. Furthermore, there was a tendency for the cloud to grow behind the release orifice when increasing the wind velocity in the opposite direction to the jet, which is more prominent in lower critical concentrations. The results agree with others reported in the literature for gas emissions.

Keywords: Fluid Mechanics, Fuels and Hazard Identification and Risk Analysis.

Introduction

The formation of flammable gas clouds resulting from the release of chemical substances is a concern for industry due to the transport and storage of chemicals on a large scale by industries. Thus, attention must be paid to environments likely to form an explosive atmosphere. The hazardous area classification appears as a method that aims to analyze and classify environments likely to form an explosive atmosphere. The application of area classification allows determining the zone type in which the area is inserted, and its extent. In this way, it is possible to adapt industrial installation projects, equipment selection, operation, and maintenance, assuring the safety of the process and the industrial environment. The international standard IEC 60079-10-1 (2020) is one of the existing standards that regulate the classification of areas. Its classification criterion is the definition by zone types, according to the occurrence frequency of an explosive atmosphere and its duration. The standard also suggests the application of numerical techniques such as computational fluid dynamics (CFD) as an alternative form of classification.

There are several possible release scenarios capable of forming an explosive atmosphere. Some parameters that characterize the release scenarios correspond to the substance that is released and its properties and state, the storage and release conditions, and the characteristics of the release orifice, such as diameter, shape, and position of the orifice in the equipment (KAISER, 1989; BOSCH and WETERINGS, 2005). A release can be either single-phase or two-phase. Most of the works are aimed at risk analysis either experimentally or/and numerically evaluating such releases (HOURRU, ANGERS and BÉRNARD, 2009; WITLOX et al., 2009; BÉRNARD et al., 2016). However, there are few studies applied to the hazardous area classification, and even less when it comes to two-phase releases evaluated in CFD (OLIVEIRA, NETO, and ALVES, 2019).

A two-phase release can be characterized by the material storage conditions and flow conditions at the leak orifice. When the liquefied gas stored under pressure is released it can be found in the orifice as a two-phase jet in the equilibrium condition or as a superheated liquid jet in the non-equilibrium conditions. These conditions depend on the orifice position in the equipment and the thickness of the wall. If it is a tube, it will also depend on the length (BOSCH and WETERINGS, 2005). At the release orifice, the flow through it can be superheated if the fluid temperature at the orifice (T_o) is greater than the saturation temperature of the fluid at the orifice pressure ($T_o > T_{sat}(P_o)$), or at equilibrium, if these temperatures are equal ($T_o = T_{sat}(P_o)$)) (BRITTER et al., 2011). The implication of having different orifice conditions under the same storage conditions lies in different release parameter values and, consequently, different area classifications.

Another parameter that directly affects the gas cloud is the wind. As pointed out by Barros et al. (2020), who assesses the wind effect on a gas cloud formed from the release of gas, there is a non-monotonic influence of the wind under the cloud. The authors show that the increase in wind velocity can result in either an increase or a decrease in the hazardous extent and volume of the gas cloud, it will depend on the wind velocity magnitude and direction.

Through the described context, there was a need to understand the effect of wind on a gas cloud resulting from a two-phase release in the condition of superheated liquid in the orifice, taking into account a 3D domain and the effect of gravity and ground. Due to the lack of studies available in the literature that address such assessment for two-phase releases, the present work aims to evaluate the influence of wind on the formation of a flammable gas cloud resulting from the liquefied propane release via CFD simulation.

Mathematical Model

Jet release

The model discussed is based on a two-phase release, whose condition at the leak orifice is non-equilibrium (overheated liquid). The pseudo-source approach was considered for the CFD simulations to represent the phenomenon from the post-expansion region, with the release parameters calculated in this region. There are described below the equations that make up the two-phase release jet model for the superheated liquid condition.

Initially, it is necessary to calculate the saturation temperature (T_s) given the storage pressure (P_s), or vice versa. For this purpose, the Antoine equation (Eq. 1) is used, where A, B, and C are the coefficients of the equation referring to the released substance.

$$\log_{10} P_s = A - \frac{B}{T_s + C} \tag{1}$$

Since this is a superheated liquid flow through the release orifice, the vapor fraction in the orifice ($X_{v,o}$) is zero. Given the non-equilibrium jet condition in the orifice, the pressure in the release orifice (P_o) was considered equal to the ambient pressure (P_{amb}) and the temperature equal to the storage temperature. Then:

$$X_{v,o} = 0, \quad P_o = P_{amb} \quad \text{e} \quad T_o = T_s$$

Equation 2 is applied to calculate the mass flow rate through the orifice (m_o), which is a function of the discharge coefficient (C_d), the orifice area (A_o), the liquid density ($\rho_{l,o}$), and the difference between storage pressure and orifice pressure. The flow velocity (v_o) is calculated according to Equation 3, which results from the mass conservation equation.

$$m_o = C_d A_o \sqrt{2\rho_{l,o}(P_s - P_o)} \tag{2}$$

$$v_o = \frac{m_o}{\rho_{l,o} A_o} \tag{3}$$

The post-expansion region was considered for pseudo source calculations. Then, the post-expansion temperature (T_e) corresponds to the saturation temperature at ambient conditions according to Equation 4, the determination of jet velocity (v_e) and vapor fraction ($X_{v,e}$) are given by Equations 5 and 6, and they are obtained from the solution of the conservation equations for mass, momentum, and energy. It is noteworthy that the post-expansion flow is assumed to be homogeneous, i.e., the velocity and temperature of the liquid droplets and the vapor phase are the same. In Equation 6, $C_{p,l}$ corresponds to the heat capacity of the liquid and $\Delta H^{vap,e}$ to the enthalpy of vaporization under post-expansion conditions.

The area formed by the post-expanded jet (A_e) (Eq. 7) is a function of the mixture density in the post-expansion ($\rho_{m,e}$) (Eq. 8), by this area is possible to obtain the expanded diameter (Eq. 9), considering that the orifice is circular. The post-expansion flux is calculated by Equation 10, as a result of the conservation equation.

$$T_e = \frac{B}{A - \log_{10} P_{amb}} - C \tag{4}$$

$$v_e = v_o + \frac{(P_o - P_{amb})}{v_o \rho_{l,o}} \tag{5}$$

$$X_{v,e} = \left(C_{p,l}(T_o - T_e) + X_{v,o}\Delta H^{vap,e} + \frac{v_o^2 - v_e^2}{2} \right) \frac{1}{\Delta \mathrm{H}^{\mathrm{vap,e}}} \tag{6}$$

$$A_e = \frac{\rho_{l,o} A_o v_o}{\rho_{m,e} v_e} \tag{7}$$

$$\rho_{m,e} = \frac{1}{\frac{X_{v,e}}{\rho_{g,e}} + \frac{1 - X_{v,e}}{\rho_{l,e}}} \tag{8}$$

$$d_e = \sqrt{\frac{4A_e}{\pi}} \tag{9}$$

$$m_e = \rho_{m,e} v_e A_e \tag{10}$$

The determination of the liquid drop size after the expansion was calculated based on the correlation developed in Phase III of the JIP project (Joint-Industry Project) (WITLOX et al., 2009). To determine the Sauter Mean Diameter (SMD) in the correlation is necessary to know the degree of superheat. The degree of superheat corresponds to the difference between the temperature at the orifice and the saturation temperature in the orifice pressure, which will dictate what of the three regimes will fit the drop size calculation. The regimes can be a mechanical break-up, transition, or fully flashing. If it is a mechanical break-up regime, the Equation 11 ($SMD = d_p^m$) is used, if it is the transition regime, it is used the Equation 12 ($SMD = d_p^t$), and, if it is fully flashing, the Equation 13 ($SMD = d_p^f$) is used. Where $d_o$is the orifice diameter, We the Weber number, Re the Reynolds number, L the vessel wall thickness, σ the surface tension, μ the dynamic viscosity, and ρ the density. The subscripts lo and $water, stp$ indicate the properties of the substance released in the liquid phase under the orifice conditions and water under the standard conditions of temperature and pressure (1 atm and 0°C) respectively.

$$\frac{d_p^m}{d_o} = 74\, We_{Lo}^{-0.854}\, Re_{Lo}^{0.441} \left(\frac{L}{d_o}\right)^{0.114} \left(\frac{\mu_{Lo}}{\mu_{water,stp}}\right)^{0.971} \left(\frac{\sigma_{Lo}}{\sigma_{water,stp}}\right)^{-0.368} \left(\frac{\rho_{Lo}}{\rho_{water,stp}}\right)^{-0.107} \quad (11)$$

$$d_p^t = d_p^m - \frac{\Delta T_{sh} - \Delta T_{sh}^A}{\Delta T_{sh}^B - \Delta T_{sh}^A} (d_p^m - 80 \times 10^{-6}) \quad (12)$$

$$d_p^f = \max\{10 \times 10^{-6}, 80 \times 10^{-6} - 10^{-7}(\Delta T_{sh} - \Delta T_{sh}^B)\} \quad (13)$$

CFD model

The CFD simulations were performed in the ANSYS® CFX software. The Eulerian-Lagrangian approach was used, which counted as a continuous phase the mixture of propane vapor / ambient air and as a dispersed phase the liquid droplets of propane. The ANSYS® CFX tool is based on solving the conservation equations of mass, species, momentum, and energy (Eq. 14-17). As these equations are solved, the selected mathematical models are also calculated, in this case, to describe the two-phase release phenomenon (ANSYS, 2020).

The SST (Shear Stress Transport) turbulence model was selected for the continuous phase, the Reitz and Diwakar model for breaking up liquid particles, and the Schiller Naumann model to calculate the drag coefficient. The Liquid Evaporation model was taken into account to calculate the mass transfer resulting from the evaporation of droplets due to the temperature gradient resulting from depressurization. Finally, the Buoyant model was used to observe the buoyancy of the gas cloud through the gravitational effect. The convergence criterion used was$1 \times 10^{-5}$for RMS and 1% for imbalance.

$$\frac{\partial \rho}{\partial t} + \nabla \cdot (\rho \boldsymbol{U}) = 0 \quad (14)$$

$$\frac{\partial (\rho Y_i)}{\partial t} + \nabla \cdot (\rho \boldsymbol{U} Y_i) = \nabla \cdot (\Gamma_{iM} \nabla Y_i) \quad (15)$$

$$\frac{\partial (\rho \boldsymbol{U})}{\partial t} + \nabla \cdot (\rho \boldsymbol{U} \otimes \boldsymbol{U}) = -\nabla p + \nabla \cdot \tau + \rho \mathbf{g} \quad (16)$$

$$\frac{\partial (\rho h)}{\partial t} + \nabla \cdot (\rho \boldsymbol{U} h) = \nabla \cdot (\lambda \nabla T) + \tau : \nabla \boldsymbol{U} \quad (17)$$

Boundary conditions

The case under study deals with the two-phase release of propane stored in its liquefied form at a pressure of 14 bar and 312.92 K. It is being released to an open environment, without obstacles, under atmospheric conditions (298.15 K and 1 atm). The superheated liquefied propane flow through an orifice with a diameter equal to 1.3 mm, positioned 2 m above the ground. The wind velocity range considered varies from 0 to 5 m/s (0, 0.5, 1, 1.5, 2, 3, 4, and 5 m/s) in order to assess the wind effect under the formed gas cloud, varying the wind direction with respect to the release jet in the same direction or in the opposite direction to the jet. The values of the parameters applied in all the cases simulations are in Table 1.

To simulate this scenario was assumed half of the study domain, whose boundary types are described in Figure 1. The term h in the figure corresponds to the height of the leak orifice to the ground ($h = 2m$), while the H and the L express the height and width of the domain which are equal to 3 m and 8 m respectively. Based on the pseudo source approach, the orifice diameter considered was the expanded diameter referring to the post-expansion region. The generated geometry counted on the presence of a cone starting from the orifice to further refine the mesh in this region. Another refinement was carried out close to the ground due to its influence on the gas cloud.

Table 1. Post-expansion boundary conditions for the simulations.

Post-expansion condition	Value
Temperature (K)	231.11
Vapor fraction(wt%)	0.4614
Velocity $(m.s^{-1})$	74.652
Particle diameter (μm)	65.3935
Expanded diameter (mm)	13.5
Total mass flow $(kg.s^{-1})$	0.0462

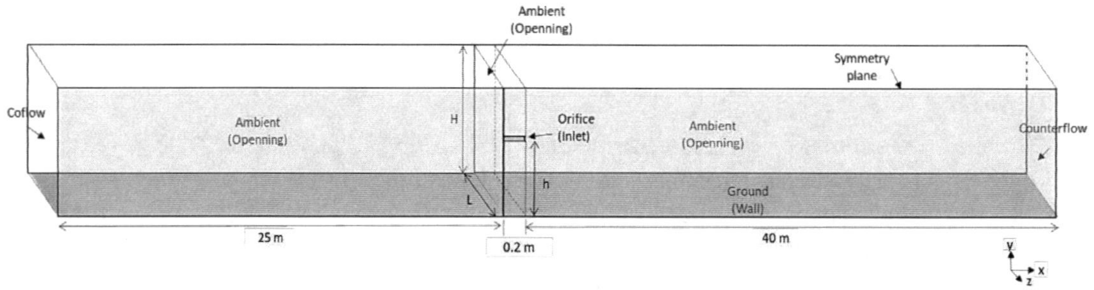

Figure 1. Representation of the computational domain with its dimensions and boundary conditions.

Grid independence study

A mesh dependency test was performed. Three unstructured meshes were tested varying the refinement in the environment and the vicinity of the leak orifice and the ground. The Table 2 shows the characteristics of the tested meshes, their respective results of extent and volume of the formed gas cloud, and convergence.

Table 2. Grid independence test (propane, P_S = 14 bar, d_o = 2.5 mm, T_S = 312.92, h = 0.5 m).

Mesh/Simulation Information	Mesh 1	Mesh 2	Mesh 3
Number of elements	24.690.501	20.908.179	17.789.875
Number of nodes	5.327.212	4.349.374	3.679.018
RMS convergence	< 1E-05	< 1E-05	> 1E-05
Imbalance convergence	< 1%	< 1%	< 1%
Extent to LFL (m)	11.72	11.71	11.69
Extent to ½ LFL (m)	21.22	21.17	21
Extent to ¼ LFL (m)	34.61	34.58	34.27
Volume to LFL (m^3)	4.77	4.76	4.76
Volume to ½ LFL (m^3)	33.82	33.79	33.58
Volume to ¼ LFL (m^3)	200.92	201.18	200.2

Firstly, the coarsest mesh (Mesh 3) was excluded because the RMS did not reach the convergence criterion established for this work. Between the Mesh 1 and 2, despite their presenting similar results except for the first or second decimal place, the Mesh 2 presented results in the interval between the more and less refined meshes. In addition, the Mesh 2 has a smaller number of elements, which indicates a shorter computational effort required and, consequently, shorter simulation time when compared to Mesh 1. Therefore, the Mesh 2 was selected for the development of simulations in this study.

Results

Through the simulations, the extent and volume results of gas cloud were evaluated at the fractions of 0.25, 0.5, and 1 of the Lower Flammability Limit (LFL) of propane (2.1%). The Figure 2 represents the hazardous extent of the gas cloud for the concentrations of interest varying with the wind velocity. It is possible to observe the increase in extent with the wind velocity in the direction of the release jet. Barros et al. (2020) obtained similar behavior and describe the increase in the hazardous extent as a result of air entrainment into the jet, which drives the gas cloud forward and leads to an increase in extent by increasing the wind velocity. When observing the extent to LFL for the counter-flow wind (Fig. 2a), the behavior was different from that observed in the authors' work. A decreasing trend of the extent was observed with the increasing of the wind velocity, while for the gas release shown in the Barros' paper an increasing trend was observed. This difference may be related to the type of release and the characteristics of the simulated component.

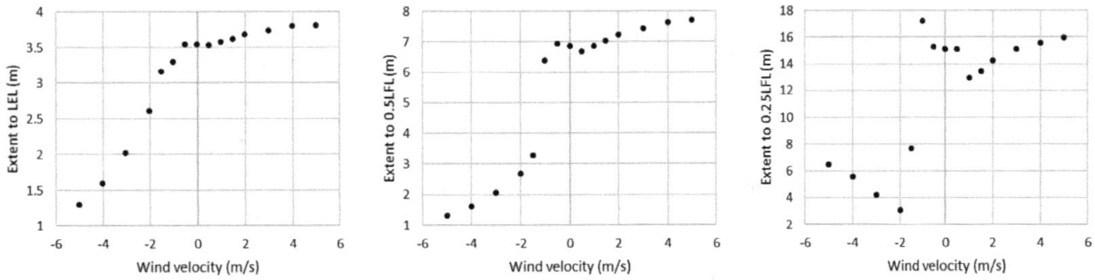

(a) (b) (c)

Figure 2. Extent of the propane gas cloud resulting from a two-phase release under the effect of wind velocity to (a) LFL, (b) 0.5LFL and (c) 0.25LFL.

The Figure 2c represents the extent delimited by 25% of the LFL. There is a discontinuity in the trend of extent points given the wind velocity. The discontinuity occurs in both evaluated wind directions, for the counter-flow wind velocities of 1.5 to 2m/s, and co-flow wind between the velocities of 0.5 and 1.0 m/s. The reached extent at 0.25LFL were 7.68, 3.08, 15.1, and 12.97 m, respectively. Such behavior occurs because the gas cloud hits the ground at 0.25 LFL for cases where the wind velocities ranges from 1.5 m/s (wind counter-flow) to 0.5 m/s (wind co-flow). When the wind velocity increases in both directions with values greater than 0.5 and 1.5 m/s, the gas clouds do not reach the ground anymore. The presence of the ground caused a different trend of the reached extent from that observed in the cases which it does not hit the ground. Similar behavior happened to clouds delimited by 50% of the LFL (Fig. 2b) for 1.0 and 1.5 m/s wind counter-flow, where the first hit the ground and resulted in an extent of 6.38 m, and the second do not reach it and came in an extent of 3.28 m.

The lack of air entrainment in the region of the plume in contact with the ground, coupled with a suction pressure that causes the cloud to stick to the ground, leads to a large extent when compared to the cases where the contact with the ground does not occur, as observed by Hourri, Anger, and Bérnard (2009) and Bérnard et al. (2016). The authors evaluated the effect of the surface on the extent of hydrogen and methane gas jets and verified that the presence of the ground for an orifice height of 0.5m, implied in extents approximately 200% greater than in the case of a free jet for methane, and 42% for hydrogen.

For some cases, considering the counter-flow wind velocity, the propane cloud grew in the region behind the leak orifice. By the velocity profile of the gas cloud along the central axis of release (Fig. 3) are identified points where the cloud velocity is zero, or approximately. These points indicate the non-dispersion of the cloud in the jet direction due to the wind in the opposite direction. As a consequence, the relative cloud/air velocity is zero, and a cloud inversion occurs. The cloud inversion grows around the orifice and in the opposite jet direction, which is seen for the 1.0 m/s wind velocity whose cloud extent behind the orifice is the largest among all.

The cloud volume had the same behavior at all concentrations evaluated when the wind velocity was varied (Fig. 4). When the wind velocity increase, the cloud volume is reduced. It happens due to the intensification of turbulence and, consequently, with the dilution increases. There is an increase in volume when increasing wind velocity in the opposite jet direction until reaches 1.0 m/s, after that the behavior reverses.

The presented scenarios, which analyze the effect of wind velocity in the opposite direction to the release jet, are not foreseen by the IEC 60079-10-1 (2020) standard, i.e., it would not be possible to verify the increase in the extent and volume of the gas cloud for counter-flow winds. The standard states that calm atmospheric conditions can increase the zone extents. Figure 2 shows the predicted effect of wind velocity on the zone extent, which is in accordance with the standard statement. However, the increase in extent is mainly observed for higher co-flow wind velocity. The standard mentions that the increase in wind velocity normally reduces the zone extent for co-flow wind cases. In the present study cases the opposite behavior was verified (Figure 2).

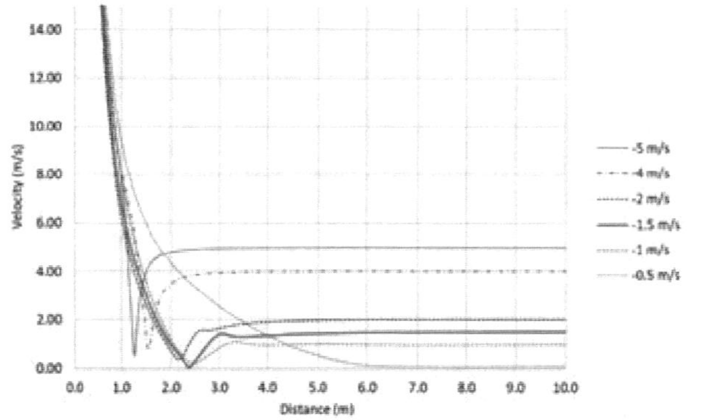

Figure 3. Gas cloud velocity profile along the central release axis for wind in the opposite direction of the jet.

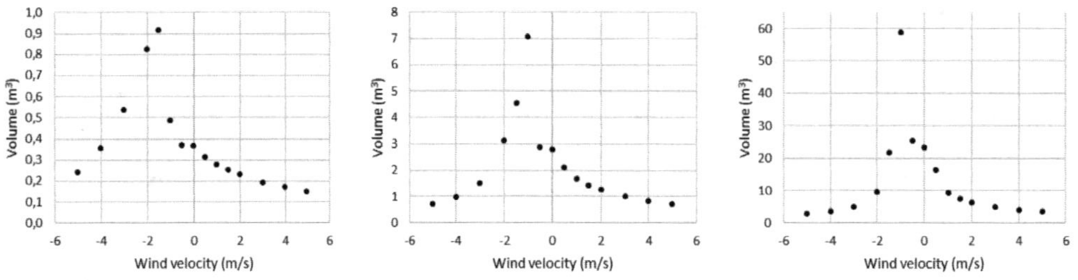

(a) (b) (c)

Figure 4. Volume of the propane gas cloud resulting from a two-phase release under the effect of wind velocity to (a) LFL, (b) 0.5LFL and (c) 0.25LFL.

Some aspects of the dense gas cloud formed by the two-phase propane release under study should be highlighted. According to Hanna and Britter (2002), dense clouds tend to have a larger lateral size than neutral plumes, and when in the presence of a reduced effective velocity of the environment a larger cloud is formed in the y and z directions given the x value of the height of the orifice to the ground. These characteristics of the dense cloud were verified in the present work, and they are highlighted in the cases where the cloud reaches the ground and spreads along with it. Worth noting that in some wind counter-flow cases, the cloud is concentrated around the leak orifice and keeps spreading on the ground in the opposite direction of the release jet. A curious behavior was observed to the wind velocity of 1 m/s counter-flow, in which the gas cloud at 0.25LFL spreads laterally behind the orifice, growing in this region and laterally, not joining behind the orifice. And like the cases where it does not reach the ground, it was noticed that the cloud starts to have an increasingly thinner shape as the wind velocity increases.

Conclusion

Based on the evaluation made on the effect of wind velocity under a gas cloud resulting from a two-phase release under the overheating condition at the leak orifice can be concluded that:

- The increase in velocity downwind implies an increase in the cloud extent, and in the opposite direction, tend to decrease;
- Cloud volume decreases with increasing wind velocity, except in the range of 0 to 1 m/s of counter-flow wind. This last scenario is not foreseen by the international standard IEC 60079-10-1 (2020);
- The ground has a considerable effect in increasing the cloud extent, being demonstrated by the transition points of the wind velocity between the cases that reach and do not reach the ground;
- The wind opposite the release jet can result in gas cloud growth in the region behind the leak orifice, which can spread laterally as it flows through the ground, as well as converge behind the orifice;
- The inversion region of the cloud be identified in cases of counter-flow wind due to the zero velocity of the gas cloud, indicating the absence of dispersion towards the release jet;
- The action of low winds led to increased cloud extents, corroborating the standard IEC 60079-10-1 (2020);
- The characteristics observed in a two-phase release and the formed gas cloud are consistent with the literature.

References

1. ANSYS CFX – Solver Theory Guide (2020), ANSYS Inc., Canonsburg, United States.
2. Barros, Paloma L.; Luiz, Aurélio M.; Nascimento, Claudemi A; Neto, Antônio T. P.; Alves, José J. N. (2020), "On non-monotonic wind influence on flammable gas cloud from CFD simulations for hazardous area classification", Journal of Loss Prevention in the Process Industries 68.
3. Bérnard, Pierre; Hourri, Ahmed, Angers, Benjamin; Tchouvelev, Andrei (2016), "Adjacent surface effect on the flammable cloud of hydrogen and methane jets: numerical investigation and engineering correlations", International Journal of Hydrogen Energy 41, pp.18654-18662.
4. Hanna, Steven R.; Britter, Rex E. (2002), "Wind flow and vapor cloud dispersion at industrial and urban sites", AIChE, 140pp.
5. Hourri, A.; Angers, B.; Bérnard, P. (2009), "Surface effects on flammable extent of hydrogen and methane jets", International Journal of Hydrogen Energy 34, pp.1569-1577.
6. IEC 60079-10-1/Ed2 (2020), "Explosive Atmospheres – Part 10-1: Classification of Areas – Explosive Gas Atmospheres", International Electrotechnical Comission.
7. Oliveira, Caio T. L.; Neto, Antônio T. P.; Alves, José J. N. (2019), "CFD simulation of flashing jet applied to area classification", The Canadian Journal of Chemical Engineering, pp.465-476.
8. Van den Bosch, C. J. H., & Weterings, R. A. P. M. (2005), "Methods for the calculation of physical effects (Yellow Book)", Committee for the Prevention of Disasters, 3ª ed.
9. Witlox, Henk W. M.; Harper, Mike; Oke, Adeyemi; Bowen, Phill J.; Kay, Peter; Jamois, Didier; Proust, Christophe (2009), "Two-phase jet releases and droplet dispersion: scaled and large-scale experiments droplet-sixe correlation development and model validation", Symposium Series Nº 155, Hazards XXI, IChemE, pp.615-623.

628895 Simulation and Analysis of Hydrogen Refueling System

Abstract:

This research aims to study the dynamic simulation of hydrogen refueling system and analyze design of the refueling station. The time a customer waits at a station is considered an essential parameter in its design. Currently, most of the hydrogen source is supplied through a compressed hydrogen tube trailer. The point is that the more hydrogen is supplied to the fuel cell electric vehicles, the lower the pressure at the hydrogen supply stage from the tube trailer, which affects the operational time frame of the station, until depleted tube trailer is replaced into fully charged one.

The developed dynamic model comprises the pressure vessels, compressor, hydrogen cooler, and dispenser. The results of the simulation were compared with the actual refueling station data in Korea. Also, this study examined the effect of the time frame with the variation of pressure vessels and compressor specifications and economic evaluation.

The simulation results matched the actual data well. When the suction pressure of the compressor (tube trailer pressure) decreases, the pressure recovery rate of the tank at the discharge line is delayed increasingly as the next FCEV arrives to refuel.

This study gives the guideline to design a hydrogen station in consideration of the specifications of the pressure tank and the compressor and economic evaluation. It will also help the operational strategy of the station.

620816 Development of Heterogenous Equilibrium Model for Lithium Solvent Extraction Using Organophosphinic Acid

Abstract:

Circular economy is a promising approach to relieve the stresses in the supplychain, by which the end-of-life lithium-ion batteries are recycled to manufacture new batteries. This study applies solvent extraction technology to recycle critical metals (lithium and cobalt) from spent lithium-ion leachate. An organic solvent containing Cyanex 272/TBP was found to extractant to recover both cobalt and lithium together from spent lithium-ion batteries. A new heterogenous equilibrium model for the extraction has been developed and validated experimentally developed in the paper. This model is able to determine the extraction performance of lithium and other metals over a range of operating conditions. The impact from pH, other metals and extractant content is significant, while the impact from temperature, leachate ionic strength and TBP content is minimal.

623144 Corrosion Under Insulation (CUI) – out of Sight Should It be out of Mind?

Abstract:

Every year billions of dollars are wasted by the chemical industry due to corrosion related incidents or mitigations. Process safety is the foremost responsibility of engineers supporting production facilities. While engineering controls are put in place it is assumed pipes and valves will hold reliably. In order to conserve energy of the process or ensure personnel protection pipes need to be insulated. The metal surface of the insulated pipes being out of sight obviously gets out of mind. CUI (Corrosion Under Insulation) is an omnipresent risk in every chemical facility that needs to be evaluated and appropriate actions need to be taken to prevent CUI. Oftentimes, it is already too late, by the time CUI issue is surfaced. In this talk we will identify techniques to prevent CUI from becoming a problem and also ways to remediate CUI in case it is already an issue. Real world industrial examples of solutions available will be discussed to bring awareness on the topic that can seldom go overlooked. Good engineering practices will also be discussed that should be considered while designing chemical manufacturing facilities to avoid CUI.

627745 Process Intensification of CO_2 Capture By Low-Aqueous Solvent

Abstract:

Low-energy solvent-based CO_2 absorption processes have drawn attention as a next-generation post-combustion CO_2 capture technology to reduce CO_2 emissions from fossil fuel– or biomass-fired power generation and industrial emissions streams. A low-aqueous (or water-lean) solvent process may substantially reduce the thermal energy consumption for solvent regeneration. Low-aqueous solvent–based processes are thermally sensitive, requiring thermal management of the solvent because of the fast, exothermic amine-CO_2 reaction and low-heat-capacity organic diluent. These affects may result in heat accumulation in a packed absorption column and undesirable CO_2 desorption occurring as the solvent moves through the column, reducing the solvent's CO_2 capture efficiency if its temperature is not controlled. Using a 3D printed intensified packing device in this work, enhanced heat and mass transfer were demonstrated in an amine-CO_2 scrubbing process with low-aqueous solvent. The multifunctional intensified device facilitates contact of the reactive solvent and gas phases in a single stage and facilitates heat removal by a cooling fluid flowing through channels in the interior of the corrugated plates of the packing device. These functionalities led to effective thermal management along the column via intrastage cooling and significant improvement in CO_2 uptake under a wide range of operating conditions. Intrastage cooling effectively reduced the solvent average temperature along the column by ~10°C and, as a result, the solvent's capture efficiency improved by up to 25%.

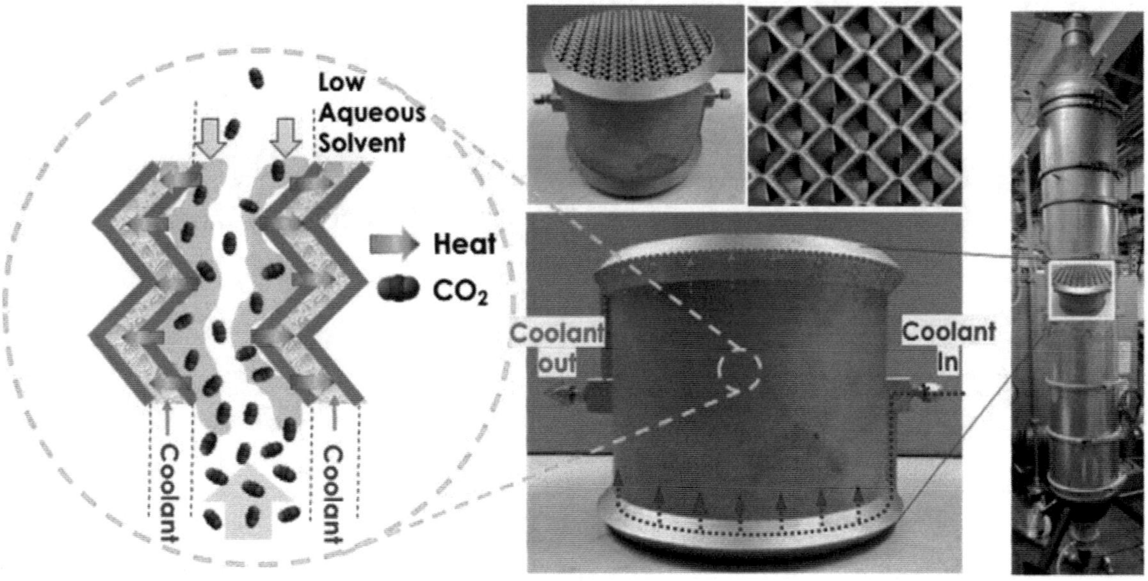

627472 Thermoplastic Polyurethane Gel Quality Improvement Based on Pilot Line Filter Evaluation

Abstract:

Gel quality in thermoplastic polyurethane (TPU) is a critical customer requirement in several applications including films, fibers, easy-to-extrude, and sheet. Gels are visual defects that have several root causes including large agglomerations of TPU hard segments and various foreign contaminates. While the agglomerations of TPU hard segments are significantly impacted by the mixing profile during the initial reaction of the TPU, the amount of agglomerations and size distribution of those agglomerations can be affected by the filter type, media type, and nominal pore size before the TPU is pelletized. However, even with filtration, often the magnitude of impact is not sufficient make a TPU with the desired gel quality outcome for the applications. The aim of this study was to predict the filtration impact on production TPU gel quality based on pilot scale gel quality results and determine the scaling correlation. Not only would this be beneficial to implement new filter types in production for existing products, but having a correlation would also be instrumental in new product commercialization. Early in the development phase, a project could be redirected or stopped if the gel quality could not be met on the production scale based on the available equipment. There is also a safety factor of evaluating new filtration equipment on the pilot scale with lower throughput while generating less waste and eliminating production down time. Factors to consider for the correlation are flow rate per square meter of filter media, pressure through the filter unit, amount of material that can be produced before a filter change, and impact on the TPU properties. An additional challenge to overcome was for a given TPU, the gel quality would change as the equipment size changed, from the 25mm pilot extender, to the 40mm pilot extruder, and to the 92mm production extruder. Through data analysis and trials on production and pilot scale, a correlation was established. From this work, a new filter was successfully implemented in production after extensive testing on the pilot line that reduced the number of gels in the product.

629885 Exploring Flammability Hazards from the Reactive Chemicals Perspective

Abstract:

Over the past half century, Dow's Reactive Chemicals Program has supported a strong safety culture and a scientifically rigorous, risk-based review process for chemical hazards in the company's labs and plants. The foundation of Dow's chemical and process hazard analyses is the data generated by the Reactive Chemicals Group, who leverage measurement science, expertise in chemistry and engineering, and an understanding of Dow's processes to support safe operation at all scales. Despite the critical role the Reactive Chemicals discipline plays in safe chemical research and manufacturing, these approaches and techniques are absent from many chemistry and chemical engineering undergraduate curricula. In this talk, we explore the Reactive Chemicals perspective on assessing flammability hazards, with a focus on common values such as the flash point, upper and lower flammable limits, minimum ignition energy, and auto-ignition temperature. Despite the ubiquity of these values, students are rarely presented with the science behind how they are measured and the corresponding limitations in how they should be applied in the manufacturing or research environment. We include industrially relevant examples to illustrate the importance of understanding flammability data on a fundamental level and to encourage the inclusion of these concepts in students' coursework.

622936 Integration of a Green and Efficient Synthesis Step Enables Enhanced Production of the Antimalarial Precursor Artemisinin from Plant-Based Extraction

Abstract:

Artemisinin is the precursor to several important active pharmaceutical ingredients (API), which are essential in the treatment of malaria. Although an industrial process based on biotechnological and chemical conversion is already available, artemisinin is still mainly produced by extraction of the plant *Artemisia annua*, where it is formed as a secondary metabolite. The low content of the API in the dried matter (max. 1.4 wt-%) and the long cultivation time of the plant caused high and fluctuating prices in recent years. Reducing the cost of the treatments and increasing their availability also for the population of developing countries, is a major issue in the fight against the disease.

Our approach to improve the production pathway is to use the plant not only as source for artemisinin but also for the co-extracted substance dihydroartemisinic acid (DHAA) – a natural precursor to artemisinin. *A. annua* contains up to equal amounts of DHAA as of artemisinin depending on the plant strain and the cultivation conditions. This former waste product of the extraction can be converted chemically to additional artemisinin *via* a photooxidation and subsequent acid-catalyzed reaction sequence [1]. The aim is to design the reaction step in manner so that it can be easily integrated to existing equipment without major adjustments to the upstream extraction and the downstream purification unit.

In previous studies [2] 9,10-dicyanoanthracene (DCA) and trifluoroacetic acid (TFA) were utilized as catalysts - both toxic and expensive substances, which require recycle after the synthesis. In addition, the reaction was performed at a low temperature of -20°C and necessitated highly concentrated conditions, the use of pure oxygen and pure solvents. In this study we demonstrate that these catalysts can be substituted by green alternatives. Co-extracted chlorophyll initiates the photooxidation making the addition of other photocatalysts unnecessary. Immobilized heterogeneous acids are used as catalysts for the second reaction to improve reusability and process safety (Figure 1). In addition, we will show that purified starting material and extensive cooling are not required to convert the DHAA contained in the extract.

The extract used for the reaction experiments was obtained by batch-extraction of dried, ground leaves of *A. annua* in toluene. To perform the partial synthesis, the obtained crude extract was fed into a mini-channel flow system combined with oxygen and irradiated by highly intensive visible light. The small dimensions of the reactor ensure high mass transfer rates between gas and liquid phase and strong illumination of the reactant solution.

The obtained extract is very diluted with respect to artemisinin and DHAA. To compare the catalyst performance with previous published studies performed in purified conditions [2], the partial synthesis was first investigated by adding additional DHAA to the extract (0.5 M DHAA) before testing the concept on the diluted conditions (0.0085 M DHAA).

The performance of co-extracted chlorophyll as photocatalyst was evaluated in comparison to 9,10-dicyanoanthracene at -20°C and 7 bar oxygen. In purified solutions, DHAA is completely converted in 2 min residence time and a yield of 90% of the desired hydroperoxide PO1 is obtained. Using the crude extract as reaction medium, gave the same conversion and yield after a residence time of 5 min, although chlorophyll is 4 times less concentrated than the photocatalyst in the reference case. When trifluoroacetic acid is added to the extract, artemisinin is gained with a yield of 67% equaling the results obtained in purified solutions [2]. The impurities present in the crude extract affected neither the yield nor the selectivity of artemisinin significantly. So prior purification of DHAA is not required before performing the partial synthesis. Additionally, the photooxidation benefits from the strong and broad absorption of chlorophyll enabling the use of different wavelengths in the visible range.

In the next step, trifluoroacetic acid was substituted by heterogenous acids. To identify suitable catalysts, various solid acids ranging from oxides to heteropoly acids were screened in batch experiments at ambient conditions. The best performing candidates, a β-zeolite (Si/Al = 360) and the ion-exchange resin Amberlyst 36, exceeded the performance of TFA in these low-pressure conditions with an artemisinin yield of 56% and 46%, respectively, in comparison to 33% with TFA. Both materials were applied as fixed-bed reactor directly coupled to the photoreactor. The reactor configuration was varied to identify flow conditions with sufficient mass transfer between gas and liquid, as well as liquid and solid phase. The best results were obtained with the β-zeolite (Si/Al = 360) packed in a single-particle reactor achieving a maximum artemisinin yield of 40%.

All results presented so far were performed with a starting concentration of DHAA exceeding the natural concentration in the extract 50-fold. Performing the partial synthesis in diluted solutions at otherwise optimized conditions (-20°C, 7 bar O_2, 0.25 M TFA) resulted in a drop of the final yield to 40%. Experiments with varied initial concentration of DHAA added to the extract showed, that this drop in selectivity to artemisinin was caused by the decreased excess of DHAA in comparison to other co-extracted metabolites.

Adaption of milder reaction conditions did not affect the performance of the partial synthesis at these diluted conditions. DHAA is converted still with 40% yield to artemisinin, when increasing the reaction temperature to ambient conditions, using pressurized air instead of oxygen and reducing the added amount of acid 4-fold to 0.06 M. The application of β -zeolite as acidic catalyst resulted in equal observed yields to artemisinin.

In conclusion, the conversion of the former waste product dihydroartemisinic acid to artemisinin in an additional synthetic step is an attractive extension of the conventional extraction process. Utilizing co-extracted chlorophyll as photocatalyst eliminates the addition and recycle of expensive artificial photosensitizers leading to a significant simplification of the process. The substitution of organic acids by heterogenous acidic catalyst increases the safety and sustainability of the process further. By applying these tools, 40% of DHAA extracted from *A. annua* can be converted to additional artemisinin by treating the extract only with visible light and air at ambient temperatures. Since no components are added for the reaction step, avoiding interference with the downstream purification units, the partial synthesis can be easily integrated to existing plants for extraction-based production of artemisinin.

References:

[1] Levesque, F. and P. H. Seeberger, *Continuous-flow synthesis of the anti-malaria drug artemisinin.* Angewandte Chemie - International Edition, 2012. 51(7): 1706-1709.

[2] Kopetzki, D., F. Levesque, and P. H. Seeberger, *A continuous-flow process for the synthesis of artemisinin.* Chemistry-a European Journal, 2013. 19(17): 5450-6.

[3] Triemer, S., K. Gilmore, G.T. Vu, P.H. Seeberger, and A. Seidel-Morgenstern, *Literally Green Chemical Synthesis of Artemisinin from Plant Extracts.* Angewandte Chemie - International Edition, 2018. 57(19): 5525-8.

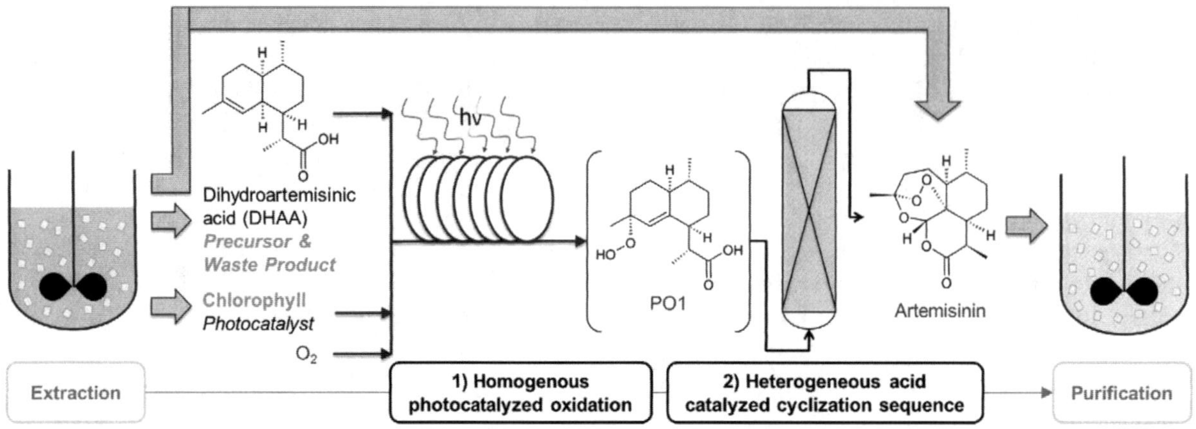

Figure 1: Utilization of co-extracted dihydroartemisinic acid as additional source for the antimalarial artemisinin in a two-step partial synthesis

633045 Designing Green Nanosilicas Using Statistical and Machine Learning Approaches

Abstract:

The *in-vitro* bioinspired synthesis of silica, inspired from *in-vivo* biosilicification, is a sustainable alternative to the conventional production of high value porous silicas [1]. The short reaction time, mild reaction conditions of room temperature and its use of benign precursors makes this an eco-friendly, economical and scalable route with great industrial potential [2-4]. However, a systematic optimisation of critical process parameters and material attributes of bioinspired silica is lacking. Specifically, statistical approaches such as Design of Experiments (DoE) and Global Sensitivity Analysis (GSA) using machine learning could be highly effective but have not been applied to this "green" nanomaterial yet.

This presentation will discuss a novel sequential DoE strategy that was developed with pre-screening experiments to outline the feasible design space [5]. A successive screening using 2^3 full factorial design determined that from the initially investigated three factors (the ratio of the reactant concentrations, pH, and precursor concentration), only the first two were statistically significant for silica yield and surface area. The subsequent concatenated optimisation using central composite design located a maximum yield of 90 mol% and a maximum surface area of 300-400 m^2/g. Since for successful commercialisation, high yields and large specific surface areas are desirable, their simultaneous optimisation was also achieved with high predictability regression models.

For complementation, a variance-based GSA was successfully applied to bioinspired silica for the first time. This method rapidly identified key parameters and interactions that control the physicochemical properties and provided insights in the wide parameter space, which was validated by the extensive DoE campaign. This work is the starting point in holistically modelling the multidimensional factor-response relationship over a large experimental space in order to complement efforts for resource-efficient product and process development and optimisation of bioinspired silica and beyond.

References

1. S. V. Patwardhan and S. S. Staniland, *Green Nanomaterials*, IoP, Bristol, 2019.
2. S. V. Patwardhan, J. R. H. Manning and M. Chiacchia, *Bioinspired synthesis as a potential green method for the preparation of nanomaterials: Opportunities and challenges,* Current Opinion in Green and Sustainable Chemistry, **12**, 110-116, 2018.
3. J. R. H. Manning, T. Yip, A. Centi, M. Jorge, M. and S. V. Patwardhan, *An eco-friendly, tunable and scalable method for producing porous functional nanomaterials designed using molecular interactions.* ChemSusChem, **10**(8), 1683-1691, 2017.
4. 4. J. R. H. Manning, C. Brambilaa and S. V. Patwardhan, *Unified mechanistic interpretation of amine-assisted silica synthesis methods to enable design of more complex materials,* Molecular Systems Design & Engineering, 2021, Advance Article DOI: 10.1039/D0ME00131G.
5. L. Dewulf, M. Chiacchia, A. S. Yeardley, R. A. Milton, S. F. Brown and S. V. Patwardhan, Designing bioinspired green nanosilicas using statistical and machine learning approaches. Mol. Sys. Des. Eng., 2021, **6**, 293.

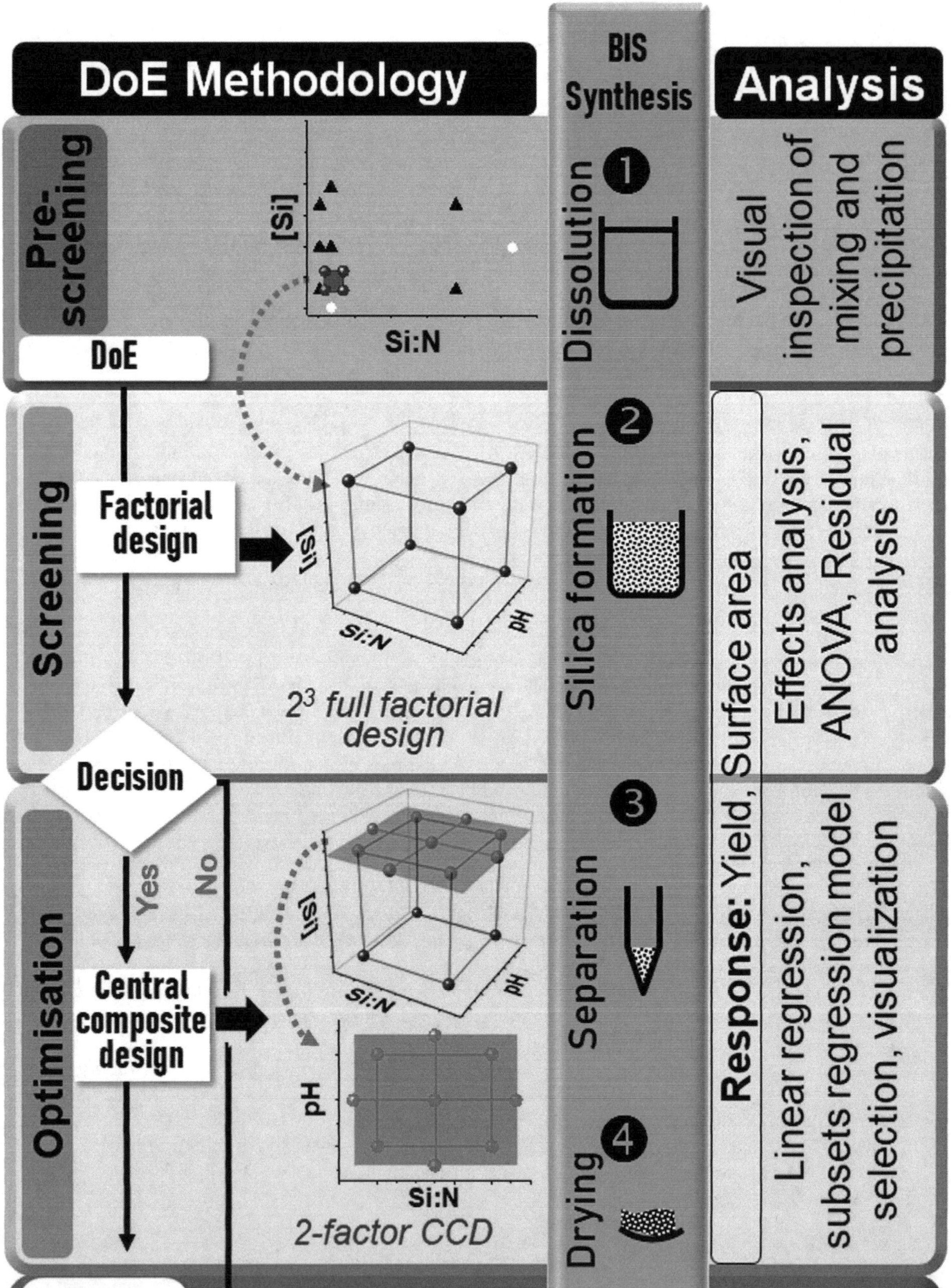
DoE Methodology
BIS Synthesis
Analysis
Pre-screening
DoE
[Si]
Si:N
Dissolution
Visual inspection of mixing and precipitation
Screening
Factorial design
[Si]
Si:N
pH
2^3 full factorial design
Silica formation
Surface area
Effects analysis, ANOVA, Residual analysis
Decision
Yes
No
Optimisation
Central composite design
[Si]
Si:N
pH
pH
Si:N
2-factor CCD
Separation
Drying
Response: Yield, Surface area
Linear regression, subsets regression model selection, visualization
Model

629764 Deposition of Solid Fabric Enhancers in the Domestic Clothes Dryer

Abstract:

Clothes drying is a common household process based on a simple principle; fabrics tumble in a drum through a flow of hot air, aiding the evaporation of water remaining after washing. Existing literature has routinely considered the dryer using black box thermodynamic models. This has led to recent improvements in dryer efficiency and the development of so called 'high-efficiency' appliances. However, consumer perception of drying performance considers more than just final fabric moisture content.

Surface feel, odour, wrinkling and static cling are important factors, difficult to control with changes to the drying process alone. These benefits may be delivered by addition of solid fabric enhancers (SFE) to the dryer using products such as dryer sheets. During drying, the SFE mixture is transferred to items of clothing in surface-surface interactions. These include direct contact between dryer sheets and fabrics, indirect transfer via the drum wall and secondary transfer between fabrics, as shown in Figure 1. Without full understanding of drying, development of these products has been limited to time and labour-intensive trial and error approaches. The behaviour of items within the dryer, including interactions between fabrics, SFE coated articles and the dryer drum, will be significant in both the drying process and delivery of fabric care benefits.

The dynamics of fabric and dryer sheet motion within the dryer have been examined to aid the development of new appliances, fabric enhancers and delivery articles. The objective is to further understanding of behaviour of these items throughout different regions of the dryer and along the length of the drying cycle (i.e. as moisture content decreases) for a range of consumer representative load conditions.

Positron Emission Particle Tracking (PEPT), developed at the University of Birmingham using a modified medical PET scanner, has been employed to track the location of a radioactively labelled tracer particle affixed to items within the dryer drum. Analysis of the timestamped Cartesian coordinates output by the PEPT machine learning tracking algorithm allows the tracer position, velocity and acceleration to be calculated. Individual tracer behaviour and time averaged Eulerian velocity profiles were used to understand motion during drying, as shown in Figure 2.

Six regions within the dryer drum were identified, demonstrating the range of movements experienced by items during tumbling. These showed clear correlation with existing flow regimes describing behaviour of granular media in rotating drums. Fabric motion was primarily cataracting to maximise the surface area of fabric available to interact with drying air in the falling region, with some conditions moving towards centrifuging or cascading flows. Movement in the axial direction was significantly slower than the primary radial flow. Dryer sheets were more prone to centrifuging than fabrics, with significant time spent in contact with the drum wall. Conversely, a wool dryer ball was more likely to cascade, spending time mixed into the top of a fabric bed which forms in the impact and lifting regions. Behaviour in this bed is primarily determined by frictional interactions with the drum wall, which subsequently affects behaviour in the 5 remaining regions. The most significant changes to this behaviour were observed when changing fabric moisture content and volumetric fill ratio, with wet fabrics and smaller load sizes both exhibiting faster falling speeds and spending more time in the fabric bed. The changes were most significant in the lifting, falling and detachment regions, with varying acceleration and shearing likely to influence both fabric wear and SFE delivery.

A deposition device is under development to simulate the range of interactions observed within the dryer at bench scale. This mimics the abrasive motion observed in PEPT experiments at a range of representative relative velocities and normal forces. The device is housed in a temperature and humidity controlled environment to ensure conditions are fully representative of those observed in the dryer. Following deposition an indicator will be used to identify SFE on the fabric surface and images analysed to quantify this across the range of conditions observed. This will enable the rapid testing of new SFE formulations and delivery articles without the need for full scale trials.

<u>References</u>

Ingram, A.; Seville, J. P. K.; Parker, D. J.; Fan, X.; Forster, R. G. Axial and Radial Dispersion in Rolling Mode Rotating Drums. *Powder Technol.*, **2005**, *158* (1–3), 76–91. https://doi.org/10.1016/j.powtec.2005.04.030.

Govender, I. Granular Flows in Rotating Drums: A Rheological Perspective. *Miner. Eng.*, **2016**, *92*, 168–175. https://doi.org/10.1016/j.mineng.2016.03.021.

Windows-Yule, C. R. K.; Seville, J. P. K.; Ingram, A.; Parker, D. J. Positron Emission Particle Tracking of Granular Flows. **2020**. https://doi.org/10.1146/annurev-chembioeng.

Nicuşan, A. L.; Windows-Yule, C. R. K. Positron Emission Particle Tracking Using Machine Learning. *Rev. Sci. Instrum.*, **2020**, *91* (1), 13329. https://doi.org/10.1063/1.5129251.

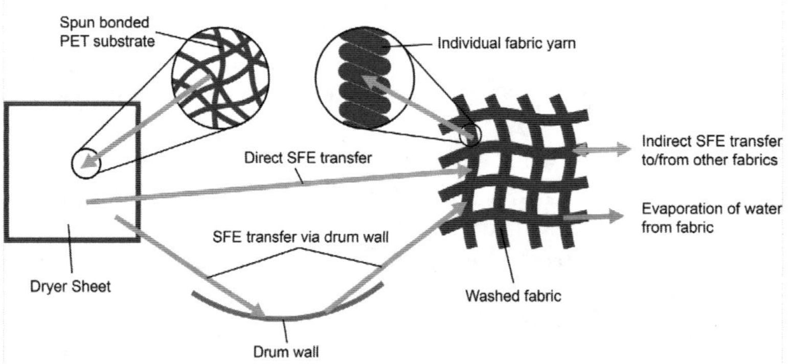

Figure 1. Schematic representation of the interactions involved in SFE delivery in the dryer drum

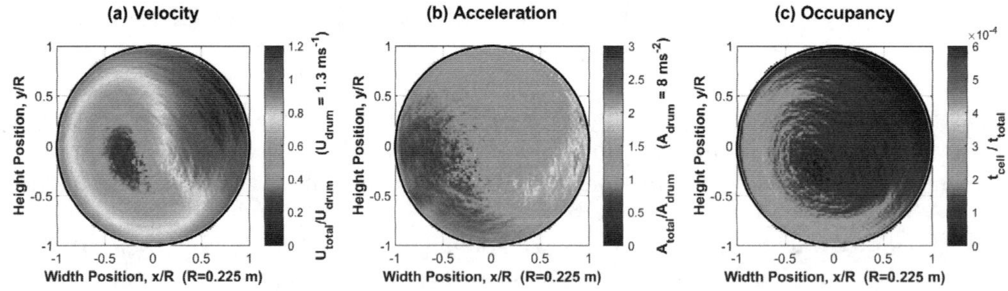

Figure 2. Eulerian velocity, acceleration and occupancy profiles, normalised against the equivalent values for the drum wall, for a tracer particle attached to the centre of a fabric sheet a domestic clothes dryer

629020 Characterization of the Spinnability of Cellulose/Ionic Liquid Spinning Dopes during a Dry-Jet Wet-Spinning Process through in-Situ Rheo-Optic Techniques

Abstract:

Cellulose-based fabrics, such as cotton, make up 37% of all textile fabrics in the global market. The expansion of production is increasingly limited by the demand of arable land for foods. This has motivated efforts to investigate recycling of these textiles by dissolution of previously-produced fabrics and fibers. However, this is hampered by the paucity of solvents that are capable of dissolving cellulose without destroying the constituent polysaccharide chains. In the past few decades, ionic liquids have demonstrated advantageous solubility and chemical stability for the spinning of cellulose in the *dry-jet wet-spinning* process. Previous studies have attempted to optimize the spinning parameters by analyzing the mechanical properties of the spun and coagulated fibers. However, the thermomechanical properties of the ultimate fibers convolve all of the contributions from each individual step in the complex spinning process. Process optimization thus typically requires a large number of empirical tests to optimize spin-line geometries, processing conditions and environmental control. It remains unclear if the resulting process parameters are robust with varying material sources or spinning configurations.

In this work, we address these limitations through the implementation of an *in-situ* rheo-optical technique that monitors the temporal and spatial structural evolution of cellulose fibers during a dry-jet wet-spinning process. The experimental results obtained using a number of ionic solvents (e.g., [EMIM][OAc] and [DBNH][OAc]) demonstrate distinct evolution of the optical retardation in the thinning and solidifying fiber with varying coagulation times and draw ratios. By comparison of the computed birefringence with the stress-optical relationship, we separate the contributions to the flow-induced cellulose reorientation from the dry-jet process and the subsequent coagulation process. Following the evolution of the optical signature at several spatial locations along the spin-line reveals the saturation of birefringence when the fibers are sufficiently drawn, which provides a measure of the maximal alignment of the cellulosic chains that can be achieved at a given draw ratio. We subsequently connect the *in-situ* characterization of birefringence with *ex-situ* measurements of the non-linear shear and extensional rheology of the spinning dopes, as well as the mechanical properties of the ultimate fibers through the application of appropriate constitutive models and structural characterization. This information is used to develop a bottom-up mechanical framework for understanding the complex deformation history experienced by the cellulose/IL spinning dopes during the dry-jet wet spinning process, and to provide further insights in the spin-line optimization.

622570 Optimal Design of Controlled Environment Agricultural Systems Under Market Uncertainty

Abstract:

Global population growth and increased climate variability have posed major challenges to traditional agricultural production [2,6]. As these trends persist, meeting increased consumer demand under heightened uncertainty will further stress the capabilities of our existing food production infrastructure [4,5]. For this reason, alternative methods are being sought to augment conventional food production. Controlled environment agriculture (CEA) is an alternative strategy that decouples crop production from environmental conditions through indoor cultivation. CEA offers a more sustainable alternative to conventional agriculture and presents opportunities for improved food security by increasing public access to nutritious fresh produce. Startup costs for CEA systems, however, are significantly higher than those of traditional growing operations, and therefore present a larger barrier to adoption. To accelerate the adoption of this potentially disruptive technology for next generation food production, a better understanding of the long-term economic viability of CEA technology is desired. To provide this valuable insight, we have developed the first economics-based optimal design and planning model for CEA systems.

To date, the majority of agricultural planning models have focused on traditional outdoor food production. Many of the existing models have been tailored to highly specific operations, few have employed mathematically rigorous optimization techniques, and only a handful have accounted for uncertainty, which is desired to realistically assess the long-term economic viability of the system [1,3,7]. Since existing models are not applicable to CEA systems, we have developed an optimal design and planning model that can be used to assess the long-term economic viability of a broad range of CEA systems. Our proposed model employs robust optimization to account for uncertainty in the design and planning stages of CEA production, which enables risk-averse investment decision making for long-term economic viability.

In this work, we present a novel methodology for the simultaneous robust design and planning of CEA systems under multi-period market price uncertainty/risk. This problem is cast as a bilevel program with a nonconvex outer program pertaining to the design objective and multiple semidefinite inner programs pertaining to mean-variance risk exposure over each period considered in the planning horizon. The bilevel program is reformulated as a nonconvex semi-infinite program with several semi-infinite constraints pertaining to the multi-period risk exposure over the horizon. The problem is solved to global optimality using a custom branch-and-bound strategy constructed with EAGO.jl [10] that exploits a combination of McCormick-based envelopes and Gurobi's nonconvex solver [8], to solve the relaxed (upper-bounding in the maximization sense) subproblems, and IPOPT [9] for local solution of the lower-bounding (in the maximization sense) subproblems. A solution to the model represents a design and operating schedule that is robust to worst-case market uncertainty and therefore provides a conservative basis for CEA engineering and investment decision-making.

Our proposed model has been formulated generally and can be used to optimize large-scale CEA systems for the cultivation of any crop portfolio via single or multi-mode cultivation. The optimization objective employed is to maximize the net present value of the CEA system over its expected lifespan. Constraints limiting risk exposure and production capacity are also included to ensure that the optimal design is both robust and practical. Capital and operating expenses, including those associated with system construction, cultivation equipment, resources, labor, and distribution, are accounted for in the economic objective through sub-models that capture the location, capacity, and crop allocation cost dependencies of the specific CEA system being modeled. These sub-models can be easily adjusted for different CEA configurations without structural changes to the main optimization algorithm. This not only enables efficient determination of the optimal design for a given CEA configuration, but also enables efficient comparison between multiple candidate systems for strategic adoption of CEA technology.

To evaluate the efficacy of our approach, the economic performance of the robust design was compared against that of naïve design and operating strategies through two case studies. In each of the presented case studies, an economically feasible design was identified, even when the naïve design was economically infeasible, and the improvement in system performance attributed to the implementation of a robust design was quantified. This approach represents, to our knowledge, the first robust optimization approach to CEA systems and is demonstrated to effectively increase the robustness of CEA systems to market uncertainty, improve the long-term economics of CEA systems over naïve operating strategies, and validate the economic viability of single and multi-mode CEA production of distinct crop portfolios. This work establishes a means of evaluating the economic viability of a wide array of CEA systems and serves as a baseline for future development of CEA design and planning models to accelerate the adoption of CEA technology.

References:

[1] Ahumada, O., Villalobos, J.R., 2009. Application of planning models in the agri-food supply chain: a review. Eur. J. Oper. Res. 196 (1), 1–20. doi: 10.1016/j.ejor.2008.02.014 .

[2] Deschênes, O., Greenstone, M., 2007. The economic impacts of climate change: evidence from agricultural output and random fluctuations in weather. Am. Econ. Rev. 97 (1), 354–385. doi: 10.1257/aer.97.1.354 .

[3] Dury, J., Schaller, N., Garcia, F., Reynaud, A., Bergez, J.E., 2012. Models to support cropping plan and crop rotation decisions. A review. Agron. Sustain. Dev. 32 (2), 567–580. doi: 10.1007/s13593- 011- 0037- x .

[4] Fedoroff, N.V., 2015. Food in a future of 10 billion. Agric. Food Secur. 4 (1), 11. doi: 10.1186/s40066- 015- 0031- 7 .

[5] Field, C.B., Barros, V., Stocker, T.F., Dahe, Q., 2012. Managing the Risks of Extreme 1083 Events and Disasters to Advance Climate Change adaptation: Special Report of the Intergovernmental Panel on Climate Change (IPCC). Cambridge University Press.

[6] Food and Agriculture Organization of the United Nations (FAO), 2018. The state of agricultural commodity markets 2018: Agricultural Trade, Climate Change and Food Security. Technical Report. FAO. URL http://www.fao.org/3/I9542EN/i9542en.pdf .

[7] Glen, J.J., 1987. Mathematical models in farm planning: a survey. Oper. Res. 35 (5), 641–666. doi: 10.1287/opre.35.5.641 .

[8] Gurobi Optimization, LLC, 2020. Gurobi optimizer reference manual. URL http://www.gurobi.com .

[9] Wächter, A., Biegler, L.T., 2006. On the implementation of an interior-point filter line-search algorithm for large-scale nonlinear programming. Math. Program. 106 (1), 25–57. doi:

10.1007/s10107- 004- 0559- y.

[10] Wilhelm, M.E., Stuber, M.D., 2020. EAGO.jl: Easy advanced global optimization in Julia. Optim. Methods Softw. doi: 10.1080/10556788.2020.1786566 .

632232 Reproducible and Fast Mapping of Material Flows in Industrial Networks Using Piot Hub: A Novel Cloud Based Computational Tool

Abstract:

An important goal of industrial ecology is to ensure a balance is maintained between economic and environmental sustainability. Facilitation of this balance requires an understanding of the structure of the material economy and the material flows/waste generation in industrial networks in order to minimize the environmental consequences of these industrial systems. Physical Input-Output Tables (PIOTs) are uniquely suited for this purpose by mapping physical flows between industrial sectors. In previous work, PIOT-Hub was developed to automate the generation of both Physical Supply Tables (PSTs) and Physical Use Tables (PUTs) as well as their conversion into PIOTs by integrating the strength of mechanistic models and macroeconomic Input-Output framework [1]. Hence, PIOT-Hub, a cloud-based tool can be used to generate the PST and PUT for a region using a bottom-up approach where relevant industrial processes are simulated via mechanistic models to determine steady-state material input and output requirements. Thus, the tool establishes a connection between engineering and economic model in such a way that it is insightful for transitioning to a low carbon and zero waste circular economy. In past work, PIOT-Hub was applied to 11 agro-based economic sectors in Illinois, thus capturing elemental flows across the economy, demonstrating the automation aspect. Here we evaluate the reproducibility and time aspect of PIOT-Hub to map the material flows of equivalent economic sectors in Indiana. We develop simulation models for industries in Indiana and utilize PIOT-Hub for mapping the industrial network, thus evaluating the functionality of this novel computational platform. This work not only demonstrates the reproducibility of the tool's utilization to other geographic regions, but also serves as a case study of how the fast generation of PSTs, PUTs, and PIOTs by PIOT-Hub can be used to compare the material economies of multiple regions. We perform a comparative analysis of material flows in Illinois and Indiana using our work and the PIOT generated for Illinois in previous work. We will also demonstrate the collaborative aspect of this computational tool by live demo of the tool during this talk, thus highlighting the scaling impact of collaboration through cloud-based platform in making sustainable manufacturing decisions.

References

[1] Shweta Singh; Lan Zhao; Jaewoo Shin; Venkata Sai Gargeya Vunnava (2021), "PIOT hub for Automated Generation of Regional PIOTs," *https://mygeohub.org/resources/piotpy.*

625295 Solid Oxide Fuel Cell – Thermal Energy Storage (SOFC-TES) Integrated System Modeling and Energy Efficiency Optimization.

Abstract:

Concerns over greenhouse gas emissions and their associated environmental challenges continue to increase. As a result, cleaner energy generation technologies, such as fuel cells, are receiving attention as the next generation's energy sources. One of these attractive energy technologies is the solid oxide fuel cells (SOFCs) due to its several advantages such as 1) reduced greenhouse gas emissions, 2) higher efficiency, and 3) flexible fuel selection compared to other types of fuel cells. However, the temperature of flue gas from SOFCs is significantly high due to its operating temperature between 600°C and 1000°C. Therefore, it is necessary to recover this waste heat so that we can maximize the process energy efficiency.

Many studies have been conducted to utilize the high-temperature exhaust gases from SOFCs so far. However, they only focused on the direct utilization of the heat through the electricity generation, using processes such as gas turbine[1], Rankine cycle[2], and combined heat and power (CHP)[3] in order to maximize the process energy efficiency. However, the aforementioned utilization methods result in losses where the generated electricity is further converted to other forms of energy such as thermal energy for space heating. The energy losses during this process are often significant.

Motivated by this, we propose the SOFC-integrated thermal energy storage (TES) method as a more efficient utilization technique for recovering waste heat from SOFCs. We employed a 4 kW-SOFC and combined it with the thermal energy storage system for storing the excess energy from the high-temperature exhaust gas. In order to integrate both processes, SOFC and TES mathematical models were developed using MATLAB. Next, the developed models are integrated, simulated, and the overall efficiency of the SOFC-TES model is compared to the general utilization methods. To further improve the proposed model's thermal efficiency, we developed a heat exchanger network using Aspen Energy Analyzer. Through the developed heat exchanger network, the integrated model was optimized by pinch analysis.

The optimization results revealed that the process efficiency of the integrated system, that is, SOFC-TES, showed superiority compared with the traditional methods of utilizing the waste heat from SOFCs. By this, excess heat from the exhaust gases can be reliably stored during low-demand seasons and reutilized during high-demand seasons to supplement residential and industrial energy demand or deficits. We concluded that the SOFC-TES could be widely used for various purposes due to its inherent characteristics, i.e., storing heat, such as a greenhouse in the smart-farm and human life, compared with the traditional method of waste heat utilization from the SOFCs.

References

[1] S.H. Chan, H.K. Ho, Y. Tian, **Modeling of simple hybrid solid oxide fuel cell and gas turbine power plant**, Journal of Power Sources, 109 (1) (2002), pp. 111-120

[2] Masoud Rokni, **Thermodynamic analysis of an integrated solid oxide fuel cell cycle with a Rankine cycle**, Energy Conversion and Management, 51 (2010), pp. 2724-2732

[3] S.H. Chan, C.F. Low, O.L. Ding, **Energy and exergy analysis of simple solid-oxide fuel-cell power systems**, Journal of Power Sources, 103 (2002), pp. 188-200

630767 Upcycling Food Waste for Hydrogen Storage: Technoeconomic Assessment

Abstract:

While annually 1.6 billion tons of food waste is estimated to be discarded globally by 2030, according to U.S. Environmental Protection Agency (EPA) the corresponding figure is projected forecast to reach approximately 66 million tons for U.S. With the monumental challenge of food waste management, researchers are seeking avenues of upcycling FW into higher value materials such as hydrogen storage medium. Therefore, the objective of this research was to evaluate the technoeconomic prospect of activated carbon production from food waste for the ultimate purpose of hydrogen storage through the application of hydrothermal carbonization of food waste followed by chemical activation of the hydrochar using KOH. The process was simulated using Aspen Plus and the process flow diagram (PFD) was constructed such that it mainly constituted of 2 rotary kilns and filter press, 1 rotary cooler, dryer, homogenizer, as well as other necessary equipment - pumps, heat exchangers, storage tanks. The facility processes 96 ton of food waste per day to be converted to activated carbon powder with a carbon yield of 20.4%. Also, it was assumed that the activated carbon manufacturing facility obtains food waste of Brevard County, Florida and the facility operates continuously 330 days a year. Several parameters were incorporated in the study including equipment sizing, capital costs and operating costs, such as labor, utilities, maintenance and equipment depreciation. Break-even cost from the economic analysis was evaluated and sensitivity analysis was carried out to reveal the effect of 7 different process variables on the break-even cost where raw materials' (KOH and food waste) and wastewater treatment cost significantly affected it. The unprecedented analysis shed light on the economic feasibility of producing activated carbon from food waste as a cost-effective route of upcycling it.

625694 Economic and Environmental Impact of Biochar Addition on Anaerobic Digestion Process

Abstract:

Anaerobic digestion (AD) technology is becoming a promising pathway for producing clean and renewable natural gas from waste resources. Despite growing industrial adoption, AD technologies still suffer from inefficiency due to process instability and microbial inhibitions. Scientists have demonstrated that biochar addition has positive impacts on AD. Biochar addition increases biogas yield and methane concentration and reduces gas contaminants. This influence is attributed to several mechanisms: providing support for microbial colonization, adsorbing excess inhibitory products, and improving electron transfer between syntrophic microorganisms. However, due to the cost and energy-intensive conversion process, the additional costs for biochar addition could offset the resulting energy benefits. This study investigates an integrated biorefinery where corn stover biochar is added to the AD of animal manure to produce biogas-derived Renewable Natural Gas (RNG) for pipeline injection.

A kinetic model of AD, biogas upgrading, and pipeline injection is simulated in Aspen Plus ^{TM} for a mid-size farm with 5000 cattle heads. The process simulation model estimates a 28% increase in biogas yield for 3% (wt. % of wet mass) of biochar addition. High pressurized water scrubbing process is used for biogas upgrading into RNG. Techno-economic analysis (TEA) determines the minimum RNG selling price for a 30-year plant lifetime with a 10% Internal Rate of Return (IRR). TEA estimates the minimum RNG selling price as \$42/MMBTU and \$35/MMBTU for the without biochar and with biochar scenario, respectively. Biochar addition lowers the RNG selling price by 16% whereas adding revenue from digestate selling and renewable identification number (RIN) credits result in up to 32% reduction. Biochar addition increases the capital and operating costs by 8% and 13%, respectively. Cradle-to-gate life cycle analysis (LCA) is performed in SimaPro ^{TM} to evaluate the GHG emissions of biochar amended AD for RNG production. GHG abatement for the without biochar and with biochar scenario is 0.038 kg $CO_{2\ eq}$/MJ of RNG and 0.123 $CO_{2\ eq}$ /MJ of RNG, respectively. LCA results show biochar addition leads to 2.18 times GHGs abatement compared to the without biochar scenario. Environmental benefits mostly come from the additional carbon sequestration with biochar land use and fossil fuel displacement by syngas from pyrolysis. Sensitivity analysis shows the influence of different major processes and economic parameters on the RNG selling price. The process is cost-prohibitive for small-scale farms due to the significant economy of scale penalty for the biogas upgrading and RNG injection facility. Increased biogas yield reduces the manure input leading to less digestate production. This would decrease the environmental benefits of emissions avoidance from conventional manure management and carbon sequestration per MJ of RNG while reducing the emissions from digestate transportation and applications. Uncertainty analysis considers the variations of key technical and economic parameters within a reasonable range and gives an estimated range of fuel selling prices and GHG abatement. Minimum RNG selling prices vary between \$8.82/MMBTU and \$16.28/MMBTU and GHG abatement vary between 0.06 kg $CO_{2\ eq}$/MJ and 0.16 kg $CO_{2\ eq}$/MJ within 10-90% confidence intervals.

627985 Optimizing Syngas Production Process By Downdraft Gasifier Using Response Surface Methodology

Abstract:

The gasification process of biomass to produce combustible gases has an attracted interest recently. The downdraft gasifier has an advantage over other types of gasifiers because of its small size, high quality gas production, and low tar yield. One of the challenges in the gasification process is the production of low-quality syngas (low concentration of H2 and CO) and large amount of tar. To overcome this problem, optimization of the gasifier operating conditions was employed to achieve high-quality product gases and hence high gasification process efficiency. In this work, the influence of initial gas temperature, airflow rate, and mole fraction of oxygen in primary air, on the gasification efficiency and product quality was optimized by response surface methodology. The obtained results showed that the reduction in the temperature caused a low concentration of carbon monoxide and a high amount of tar in the product, thereby there was a decline in the efficiency of the gasification process.

623925 Techno-Economic Evaluation of Nano-Porous Silica Production from Rice Husk and Sand

Abstract:

Global rice production in 2019 was estimated at about 500 million tons. Rice husk is a by-product from rice, including about 20 wt% of rice and is needed to be disposed as waste. Rice husk contains approximately 14-25% of ash. The silica content in rice husk ash (RHA) is above 90%. Nanoporous silica (NPS) used in industrial fields such as reinforced rubber additives (tires), concrete mixture (cement binder), nano-fibers, nano-filters, and membranescan be produced from RHA. Rice hisk is combusted at a relatively low temperature (650-800°C) and RHA is produced, whereas the silica sand is operated at 1400-1500°C to produce silicate. In addition, compared to the conventional NPS production from sand, The NPS production from RHA has an advantage in lowering the production cost due to low material cost and energy consumption.

In this study, the techno-economic feasibility for the sodium silicate ($Na_2O \cdot xSiO_2$)-based nano-porous silica (SS-NPS) production from sand and rice husk was performed. The return on investment (ROI) and payback period (PBP) were evaluated according to the plant size for the two cases. The effects of raw material and product prices on ROI and PBP were analyzed and economic conditions to be economically viable were suggested.

AIChE
The Global Home of Chemical Engineers

A CAPITAL AND ENERGY EFFICIENT ALTERNATIVE EMISSIONS CONTROL TECHNOLOGY FOR CONTROLLING EMISSIONS USING BEAD ACTIVATED CARBON

Kim K. Tutin, Captis Aire LLC, East Point, GA United States
Harold R. Cowles, Environmental C&C Inc., Scottsdale, AZ United States
John M. Berger P.E., Environmental C&C Inc., Fishers, IN United States

TABLE OF CONTENTS:

I. INTRODUCTION:

In manufacturing, process exhaust gas (air) streams released may contain volatile organic compounds (VOCs), Hazardous Air Pollutants (HAPs), particulate, etc. These are considered pollutants/emissions and must be removed from the air prior to release to avoid air pollution emissions into our communities. Add-on controls either collect the VOCs for recovery and reuse or destroy the VOCs. A number of control options are available including thermal oxidizers, catalytic oxidizers, and biofiltration which destroy pollutants. Other control systems including condensation, scrubbing, and adsorption systems may recover pollutants.

This paper provides an introduction to an adsorption technology, the Fluidized Bed Concentrator (FBC) with Bead Activated Carbon (BAC) media. It describes 1) the FBC technology, 2) the BAC media used in these systems, 3) the performance characteristics, 3) typical applications of this technology, and 4) a case study review of the FBC technology demonstration for controlling emissions and collecting terpenes from wood drying.

II. DESCRIPTION OF FBC TECHNOLOGY: FOUR PRIMARY PARTS

1. Adsorber:

The adsorber section has trays filled with very adsorbent BAC media. Contaminated process exhaust enters from the bottom, passes upward through the adsorption trays, and fluidizes the Bead Activated Carbon (BAC) adsorbent media. When the process air interacts with the BAC, it adsorbs (captures) the terpenes Volatile Organic Compounds (VOCs) in its pores. VOC saturated BAC flows to the bottom of the adsorber vessel and is transferred using fans to pneumatically transfer it to the desorber. Cleaned (reusable) BAC from the desorber feeds continuously into the top of the adsorber, providing counter-current VOC removal. Cleaned air exits out the top of the adsorber.

2. Desorber:

In the desorber, the BAC media is heated (regenerated), causing it to release the VOC vapors into a low volume, gas stream. This gas stream is then circulated continuously to the next stage in the process. The regenerated (cleaned) BAC is then returned continuously to top of the adsorber for reuse. Returning this regenerated BAC to the top of the adsorber provides a final "polishing" of emissions to remove residual emissions prior to release of the process exhaust air.

"All Electric" Option

Desorber Recovery System: In the recovery system, a low volume inert carrier (nitrogen) gas is introduced into the bottom of the desorber. Gas flows up through the desorber as the BAC flows downward. The system is heated via either hot oil or electric heating. The heat causes the VOCs to volatilize off the BAC. The VOC vapors are liberated resulting in a concentrated waste flue gas stream that moves from the desorber to the condenser. This system may eliminate the on-site use of natural gas entirely. Thus, it offers an "All Electric" control technology.

Desorber Oxidation System: In the oxidation system, clean hot stripping air at 100-700°F is introduced to the bottom of the desorber. It flows upward across the BAC media to both volatilize the VOCs and fluidize the BAC. The VOC vapors are liberated resulting in a concentrated waste flue gas stream that moves from the desorber to the oxidizer.

3. Condenser or Oxidizer:

The gas stream from the desorber is circulated continuously to this system for further treatment to either capture and recover VOCs or oxidize them.

"All Electric" Option

Condenser Recovery System: In the recovery system, nitrogen gas circulates from the desorber to the condenser continuously. The VOC vapors are sufficiently concentrated such that the vapors can be cooled via a tube and shell condenser converting the vapor to liquid. The liquid drains continuously into a storage system. The liquid VOCs can then be reused or sold. The nitrogen recirculates back to the desorber where it entrains the VOCs vapors again and the cycle continues.

Oxidizer Oxidation System: In the oxidation system, the vapors are instead sent to a small thermal oxidizer to treat the VOCs. While this system does not provide recovered (collected) VOCs, it nevertheless retains FBC technology benefits including significantly reducing natural gas usage. The burner component of the thermal oxidizer is selected to provide enough energy to increase the thermal oxidizer to the 1400°F operating temperature. The burner is designed to minimize the "fire rate and thus consume negligible fuel. Optionally, a "low NO_x" design can be selected to minimize production of NO and NO_2 (oxides of nitrogen). Thus, the system minimizes energy usage and GHG emissions. [1]

4. Side Stream Reactivator (SSR) [2]:

A small portion of the BAC media exiting the desorber may be transferred to an optional SSR. The SSR reactivates the BAC by heating it to 1450°F with the introduction of a small amount of water which reactivates the BAC continuously to regenerate it and thus maintain its ability to adsorb VOCs. Reactivated BAC from the SSR is transferred back to the top of the adsorber for reuse.

III. ADSORBENT MEDIA:

The BAC adsorbent media utilized in this system is an essential element of the FBC technology. Specialized Bead Activated Carbon (BAC) efficiently adsorbs VOCs (i.e. terpenes) from the process air. The BAC has high-fill (adsorption) capacity, is a small sphere-shaped particle that is less than 1 mm in diameter with a narrow particle size distribution, has low dust, has high purity, has high flowability, and has high strength and wear resistance as measured by ASTM D 3802 – 79 (2005) Standard Test Method for Ball-Pan Hardness of Activated Carbon.[3] The spherical shape enables the BAC to fluidize and flow.

The BAC media can be reactivated and reused. This is accomplished either by either continuous or batch processes.

---

[1] "VOC Removal and Destruction System and Method". U.S. Patent and Trademark Office. (USPTO) https://appft.uspto.gov/netacgi/nph-Parser?Sect1=PTO2&Sect2=HITOFF&p=1&u=%2Fnetahtml%2FPTO%2Fsearch-bool.html&r=1&f=G&l=50&co1=AND&d=PG01&s1=20190022579.PGNR.&OS=DN/20190022579&RS=DN/20190022579 (Accessed October 4, 2021).

[2] "GASTAK® Solvent Recovery and Odor Removal by Fluidized Bed Process". Kureha Ecology Management Co. , LTD website. https://www.kurekan.co.jp/en/download/gastak.pdf (Accessed September 18, 2021).

[3] "ASTM D3802-79 (2005) Standard Test Method For Ball-Pan Hardness Of Activated Carbon. https://webstore.ansi.org/standards/astm/astmd3802792005 (Accessed September 18, 2021).

Continuous Reactivation: In the continuous process, a small amount of BAC media is diverted from the BAC exiting the desorber and sent to an in-line side stream reactivator (SSR).[4] It is then returned to the top of adsorber. Returning this reactivated BAC to the top of the adsorber provides a final "polishing" of emissions to remove residual emissions prior to release of the process air.

Batch Reactivation: In the batch process, for operators that do not have an SSR, the BAC media is unloaded periodically, shipped offsite for reactivation, and then reloaded once returned. In the interim, a second load of BAC is utilized. Reactivation is typically performed by California Carbon who has been providing reactivation services since 1975.[5] This service includes a reactivation warranty *"Carbon will be reactivated to 95% of Virgin Activity per ASTM D 3467 and Carbon will lose no more than 10% of incoming volume of the specified mesh size in the reactivation process."*[6] BAC lifetime is site and application specific. Some operators have been reusing their BAC for years.

Bead Activated Carbon (BAC) Adsorbent Media	**Bead Activated Carbon (BAC), the Adsorbent Media, Fluidized in Tray**

IV. FBC OPTIONS: RECOVERY AND OXIDATION

The Fluidized Bed Concentrator can be utilized to either recover the VOCs or oxidize them as shown in the figures. Operators typically select the desired option based on a number of factors especially including the anticipated value and reusability of the recovered VOCs. The value of the VOCs depends on a number of factors including the anticipated quantity recovered, purity, salability, and/or reusability.

Recovery: When the value of the recovered VOCS is high and if the cost of recovery is less than the cost of purchasing new material, then recovery makes sense.

---

[4] "GASTAK® Solvent Recovery and Odor Removal by Fluidized Bed Process". Kureha Ecology Management Co. , LTD website. https://www.kurekan.co.jp/en/download/gastak.pdf (Accessed September 18, 2021).

[5] "California Carbon Co. Inc. Welcome". California Carbon website. http://www.californiacarbon.com/index.html (Accessed September 18, 2021).

[6] "Custom Spent Carbon Reactivation". California Carbon website. http://www.californiacarbon.com/pdfs/Custom%20Reactivation.pdf (Accessed September 18, 2021).

Oxidation: If the VOC has low or no recovery value, as, for example, when it is a mixture, or if it poses disposal concerns, such as for toxic compounds, then destruction probably makes the most sense. Excluding value derived from VOC recovery, the oxidation option retains FBC system performance characteristics including reduction in energy usage.

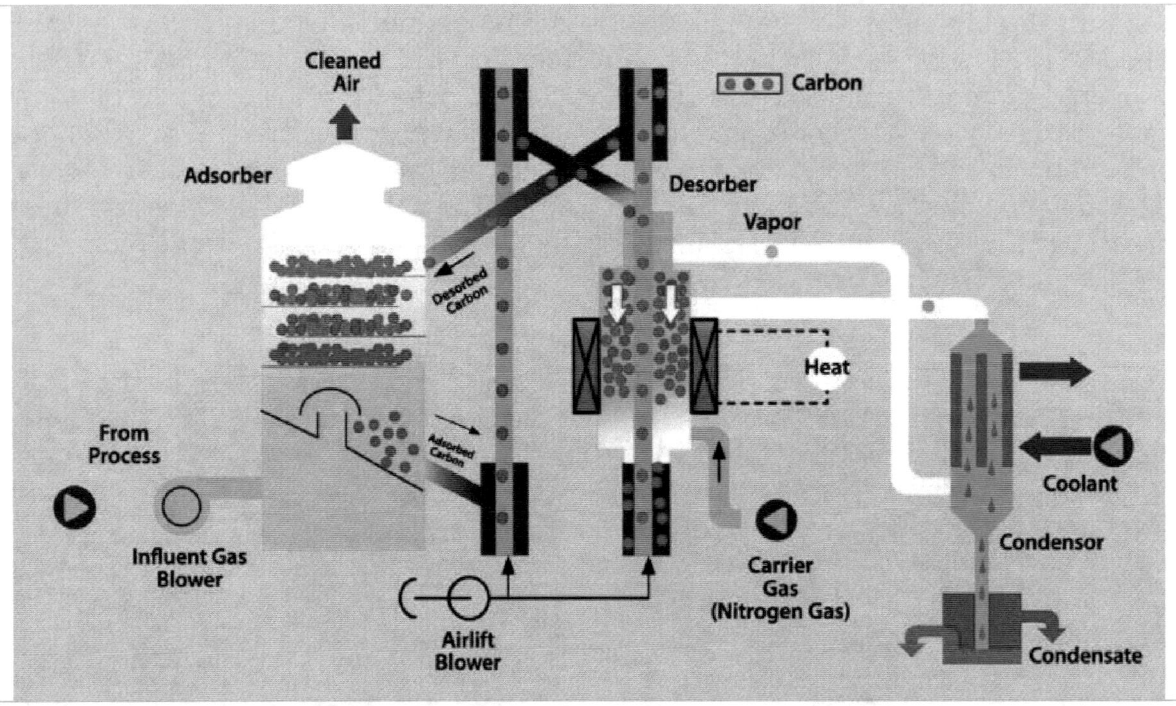

RECOVERY SYSTEM: Adsorbs, Desorbs, and Condenses VOCs

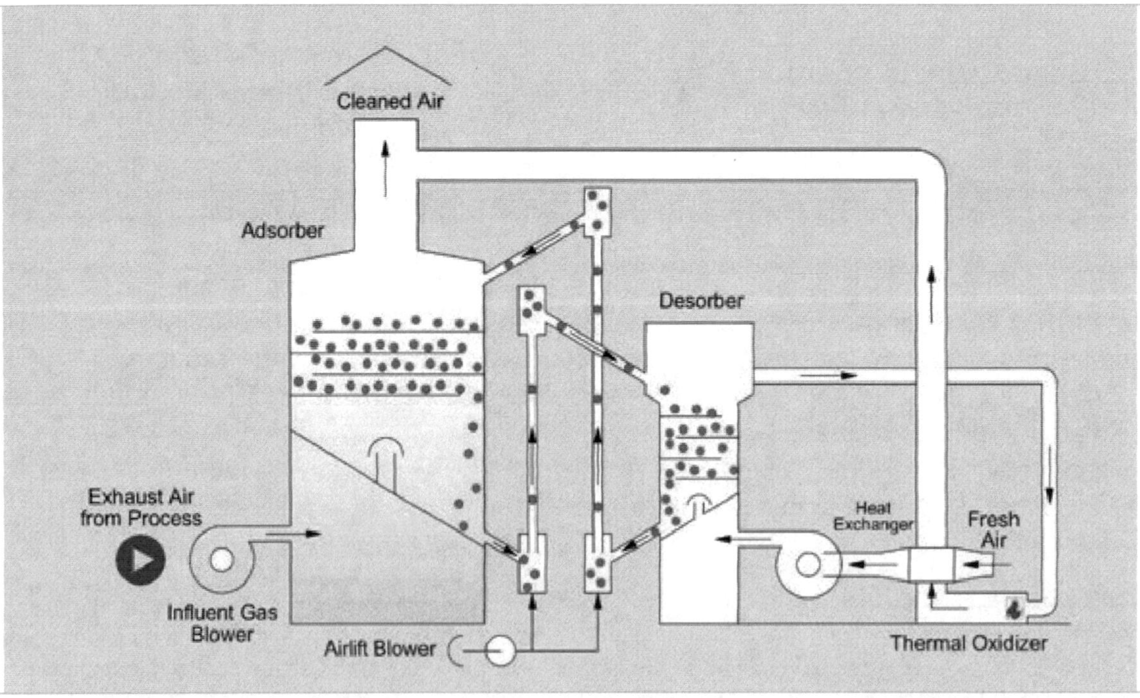

OXIDATION SYSTEM: Adsorbs, Desorbs, and Oxidizes VOCs

V. PROCESS PERFORMANCE CHARACTERISTICS:

1. <u>The FBC reduces energy usage (i.e. natural gas) versus oxidation systems due to reduction or elimination of energy used to heat the air going through thermal oxidizers:</u>

Baseline Oxidation Systems:

Thermal oxidizers burn VOCs by heating them in enclosed combustion chambers in the presence of excess oxygen. The precise temperature and residence time needed for effective reduction of VOC emissions varies with the nature of the VOC. However, temperatures of 1400-2000°F and residence times of 0.5-2 seconds are typical. Since auxiliary fuel is necessary to maintain combustion temperatures, most thermal oxidizers incorporate some form of heat recovery to lower auxiliary fuel costs.

FBC Technology:

The air going through the adsorber is not heated. Thus, the natural gas usage is reduced or eliminated.

Recovery – "All Electric" Option: For a recovery system, the control device can be "All Electric". It could eliminate the use of natural gas entirely.

Oxidation: The FBC oxidation system significantly reduces natural gas usage but does not eliminate it.

Energy Reduction Calculation: The amount of energy that is eliminated, that would have been used by the thermal oxidizer, can be calculated utilizing one of many credible sources. One of the best sources is the Institute of Clean Air Companies' Guidance Methodology[7]. Calculators are also available on-line including those provided by the EPA[8], Cycle Therm[9], and Captis Aire[10]. With these calculators, users can input their specific data and the calculators will calculate the estimated amount of energy required to run the thermal oxidizer. All these sources rely on a single primary factor to calculate energy required to heat the air. This factor is the specific heat of air which is the number of British Thermal Units (BTUs) required to heat one pound of air by one °F. This factor along with the system specific factors including the standard cubic feet per minute (sCFM) of air, the temperature increase, the density of air, and conversion factors provide the number of BTUs required to heat the air going through the device. These factors are used together with the energy efficiency and the heat generated by the oxidation of the emissions to calculate the BTUs of energy required to heat the air going through the RTO. Users can then multiply BTUs times the forecasted long-term cost of natural gas to estimate the long-term cost of natural gas required over the 20+ year life of the thermal oxidizer. This calculation provides the value of the natural gas elimination.

---

7 "Institute of Clean Air Companies (ICAC) Guidance Method for Estimation of Gas Consumption in a RTO". Institute of Clean Air Companies website. https://cdn.ymaws.com/www.icac.com/resource/resmgr/RTO-F1.pdf (Accessed September 18, 2021)

8 "Air Pollution Control Cost Estimation Spreadsheet". Environmental Protection Agency (EPA) website. https://www.epa.gov/sites/production/files/2018-01/oxidizers_calc_sheet_finalversion_1-16-2018.xlsm (Accessed Feb 3, 2021).

9 "Operating Cost Calculator". Cycle Therm website. http://www.cycletherm.com/resources/operating-cost-calculator.aspx (Accessed Feb 3, 2021).

10 "Calculate Engineering Estimate of Natural Gas Usage and Greenhouse Gas Generation". Captis Aire website. https://www.captisaire.com/Value-Calculator/ (Accessed Feb 3, 2021).

2. The FBC generates revenue from sales of recovered organics:

When using a recovery system, including the FBC, operators have the option to collect and sell or reuse the volatile organic compounds recovered. The liquid VOCs are collected from the FBC condenser for reuse or sale. The value of recovered VOCs will depend on the type, amount, and purity of VOCs collected.

3. The FBC reduces Greenhouse Gases (GHG) emissions versus oxidation systems due to the reduction or elimination of burning natural gas:

Due to the reduction or elimination of natural gas usage and reduction in electrical usage by the FBC, greenhouse gas emissions are also reduced. The greenhouse gas emissions reduction can be calculated based on the amount of carbon in natural gas eliminated and its stoichiometric conversion to carbon dioxide (CO_2). The Environmental Protection Agency (EPA) has an on-line calculator[11] that enables the user to calculate greenhouse gas emissions that are eliminated by eliminating natural gas used to heat the air in the thermal oxidizer.

4. The FBC enables creation of carbon offsets versus oxidation systems:

In conversations with American Carbon Registry[12] and ClimeCo[13], it may be possible to measure, register, and verify carbon offsets for a site converting to the FBC technology from an oxidation system due to the reduction in energy usage.[14]

The quantification methodology and value of carbon offsets is evolving quickly. But these carbon offsets can be sold or traded externally. Or they may be utilized for internal credit accounting purposes. Potential values for these carbon offsets vary widely.[15] However, as of early 2021, American Carbon Registry and ClimeCo estimated the value of carbon offsets for an FBC project at $5/ton.[16]

5. The FBC reduces electricity usage versus oxidation systems:

The horsepower required to run the fans to move the process exhaust air through the BAC media is reduced for the FBC due to the fact that it's BAC is fluidized (moving) rather than packed (non-moving). The pressure drop across the system is reduced from a typical 10-20" to 3.5". Thus, the horsepower required to run the fans for the FBC may be ~ ¼ of that required for the RTO.

To estimate the electrical savings, the sum of electrical demands from the FBC is subtracted from the sum of electrical demands of the thermal oxidizer. Net kW·h along with site-specific electricity costs and operating hours per year are used in combination to estimate electrical savings.

---

[11] "Greenhouse Gases Equivalencies Calculator - Calculations and References". EPA website. https://www.epa.gov/energy/greenhouse-gases-equivalencies-calculator-calculations-and-references (Accessed December 28, 2020).

[12] "Harnessing the Power of the Markets to Improve the Environment". American Carbon Registry website. https://americancarbonregistry.org/ (Accessed February 3, 2021).

[13] "Scalable greenhouse gas projects to earn real financial returns for your business." ClimeCo website. https://climeco.com/ (Accessed February 3, 2021).

[14] "AM0044: Energy efficiency improvement projects - boiler rehabilitation or replacement in industrial and district heating sectors --- Version 2.0.0". Clean Development Mechanism (CDM) website. https://cdm.unfccc.int/methodologies/DB/3HZ4USHZ2W449HMAXZN420E5PJB1QF (Accessed Sept 18, 2021)

[15] "Allowance Price Explorer". International Carbon Action Partnership website. https://icapcarbonaction.com/en/ets-prices (Accessed Jan 20, 2021).

[16] Eric Ripley (Director Industrial Programs, American Carbon Registry), in a phone conversation with Kim Tutin November 17, 2020.

6. **The FBC enables superior mass transfer of VOCs through counter current flow fluidized BAC media that is continuously regenerated:**

Baseline Adsorption Systems:

Adsorption: In sorbent treatment systems, process exhaust air moves across the sorbent. VOCs are removed via attraction (adsorption) of the VOCs on carbon or zeolite sorbent media in packed (non-moving) beds. This may be accomplished by concentrators that have wheels with sectors containing activated carbon or zeolite media. In this arrangement, most of the space on the wheel functions in adsorption mode, with one small sector being regenerated at any given time. Or the process exhaust air may be forced through large vessels containing sorbent media. In this arrangement, the VOCs are treated via a once-through process.

Regeneration / Reactivation: Regeneration is accomplished by heating the sorbent media with heaters, heated gases, steam, etc. If using steam, water may be a byproduct. This can be accomplished either via off-site regeneration or by alternating the flow between multiple vessels where one vessel functions in adsorption mode while the other is being regenerated at any given time. In both these arrangements, the sorbent media is being regenerated in batch or semi-continuous fashion. Reactivation may be accomplished by sending the sorbent media off site for reactivation. Or the sorbent media may be discarded when it becomes saturated.

Effect of water vapor: Especially for process exhaust air heavily laden with water vapor, i.e. 10% B_{ws} moisture by volume[17] or more, the sorbent media may become partially or fully saturated with water. At high moisture content, water molecules begin to compete with the hydrocarbon molecules for active adsorption sites. This reduces the capacity and the efficiency of the adsorption system.

FBC Technology:

Adsorption: The BAC media in the fluidized bed concentrator adsorber is fluidized (moving) by the process air moving opposite (counter current) to the BAC as it moves downward through the trays in the adsorber. This provides excellent mixing of the BAC and contact with the air. As this occurs, VOCs or odors are transferred from the process air to the BAC. The cleaned air exits through the top of the adsorber.

Regeneration / Reactivation: The BAC media is regenerated continuously via continuous recirculation to and from the desorber. It is reactivated continuously if system includes the Side Stream Reactivator (SSR). This ensures the BAC retains the capacity to maintain adsorption efficiency. This is done at temperatures well above the boiling point of water, often around 400°F or more. This minimizes or eliminates accumulation of water and/or organics on the BAC.

Effect of water vapor: Continuous regeneration and reactivation at temperatures above the boiling point of water, i.e. 400°F, volatilizes the water off the BAC media and prevents the BAC from becoming saturated with water. Thus, it maintains adsorption efficiency even when treating process exhaust streams with high moisture contents, as are typical for wood drying applications, i.e. ≥20% B_{ws} moisture by volume[18]. This system minimizes or eliminates need for dehumidification systems.

---

[17] "METHOD 4—DETERMINATION OF MOISTURE CONTENT IN STACK GASES". EPA website. https://www.epa.gov/sites/default/files/2016-06/documents/m-04.pdf. (Accessed October 17, 2021).
[18] "METHOD 4—DETERMINATION OF MOISTURE CONTENT IN STACK GASES". EPA website. https://www.epa.gov/sites/default/files/2016-06/documents/m-04.pdf. (Accessed October 17, 2021).

Effect of Fluidized (moving) vs Packed (non-moving) media: BAC media fluidization prevents pockets of high concentration VOCs, provides higher dry particulate tolerance (reduced plugging of interstitial spaces between media), and enables continuous on-site regeneration and reactivation.

TABLE I
PHYSICAL AND OPERATIONAL COMPARISON OF THE FBC VS THE RTO

Parameters:	**RTO**	**FBC**	**Advantages of FBC**
Treatment Temperature, °F	>1400°F	175°F	• Adsorption at process air temperature • No natural gas required for heating
Media Desorption Temperature, °F	>1400°F	400°F	• Much less energy required for desorption • Continuous media regeneration, above the boiling point of water, prevents the media from getting saturated with water. Thus, it maintains its ability to adsorb pollutants.
Media depth, feet (ft)	16 ft	1 ft	• Less Electricity (fan HP) required to move air through less media
Media Configuration	Packed (Not Moving)	Fluidized (Moving)	• Less Electricity (fan HP) required to move process air through fluidized media
Pressure Drop, Inches (")	10-20"	3.5"	• Less Electricity required to move process air against less pressure
Fan size, Horsepower, (HP)	2400 HP	600 HP	• Less HP due to media that is fluidized. • Less susceptible to plugging from particles
Emissions Treatment	Oxidation to CO^2	Adsorption (Capture)	• Emissions are NOT oxidized to CO_2, so reduces GHG emissions • Emissions collected can be sold or reused
Media Reuse	Discarded	Reused	• Media desorption with 'in unit' continuous regeneration. • Flowable bead shaped media can be removed and reused
Safety		Inherently Safer	• Lower operating temperatures • No large heavy moving parts • Easier cleaning and maintenance

VI. AWARDS:

The FBC has been recognized for excellence in environmental performance. Honda Motors of Alabama was recognized as *"Air Conservationist of the Year"* when they implemented the FBC to control emissions when they increased production.[19] Ford's "Fumes to Fuels" work was awarded the EPA's *"Clean Air Excellence"* award.[20] In this work, they utilized the FBC to capture solvent-laden air, strip the solvent from the air, concentrate the solvent up to 2,000 times, and send it to a fuel reformer. The fuel reformer used the concentrate to generate hydrogen-rich fuel, which was directed into a fuel cell to generate electricity.

VII. APPLICATIONS:

Concentration systems, including the FBC, are advantageously used in front of control devices when treating very high volumes of very dilute exhaust gas to reduce energy usage. Recovery systems are beneficial when the value of the recovered VOCs is high. The following are a few examples where the FBC technology has been utilized historically.

TABLE II
FBC APPLICATIONS

Application	Chemistry
Adhesive Tape Mfg	Toluene
Agricultural Products	Organic Acid & Ammonia Odor
Aluminum Casting	Phenol, Formaldehyde, Ammonia
Animal Lab	Odor Control
Atomic Power Plant	Styrene
Auto Parts Degreasing	Trichloroethylene
Bakery	Odor Control
Brewery	Odor Control
Car Wash	Kerosene
Cellophane Coating	Toluene, Ethyl Acetate, Butyl Acetate
Ceramic Condenser	Terpineol
Confectionery	Odor control
Chemical Production	Toluene, Higher Molecular Weight Alcohols, Carbon Tetrachloride

[19] "HMA Honored as Air Conservationist of the Year". Honda website. https://www.hondaalabama.com/article/hma-honored-as-air-conservationist-of-the-year (Accessed September 21, 2021).

[20] "Clean Air Excellence Award Recipients". Environmental Protection Agency (EPA) website. https://www.epa.gov/sites/default/files/2015-06/documents/clean_air_excellence_award_recipients_year_2003.pdf (Accessed September 21, 2021).

TABLE II (CONTINUED)
FBC APPLICATIONS

Application	Chemistry
Dye Production	Perchloroethylene
Electronics Mfg Plant	DMF, MEK, etc.
Electronics Parts Cleaning	Trichloroethylene, Toluene, 1,1,1-Trichloroethane
Fabric Washing	Perchloroethylene
Feed Processing	Odor Control
Film Coating	MEK, Methyl Cellosolve, Toluene, Tetrahydrofuran
Film Laminating	Toluene, Ethyl Acetate, n-Hexane
Gravure Printing	Toluene, Xylene, n-Propanol, n-Propyl Acetate, mixed
Iron Casting	Phenol, Formaldehyde, Ammonia
Lacquering of Film	Toluene, Ethyl Acetate, MEK
Landfill	Odor Control
LSI Mfg Plant	Phenol, Acetone, Methanol, Mixed Solvents
Lubrication Mfg	Mixed Solvent Odor Control Magnetic Tape, Chlorinated Hydrocarbon
Medicine Mfg Plant	Butanol, Methylene Chloride
Paint Booth	Thinner
Printing Ink Mfg	Mixed Solvents
PVC Resin Mfg Plant	Vinyl Chloride
Resin Mfg	N-Hexane, Methylene Chloride
Rubber Coating	Toluene
Rubber Vulcanizing	Odor Control
Sandpaper Mfg	Toluene, Xylene, Ethyl Acetate
Semiconductor Mfg	IPA, HMDS, NMP, Ethanol, Methanol, Ethyl lactate
Silk Screen Printing	Xylene, Mixed Solvents
Teflon Tubing Mfg	Perchloroethylene
Wastewater Aeration	Odor Control

TABLE III
EXAMPLES OF FBC SYSTEMS INSTALLED

FLUIDIZED BED CONCENTRATOR SYSTEMS INSTALLED: [21]	
Industry **Location** **Inlet Stream** **VOCs** **Performance Achieved** **Notes**	Semiconductor Chip Manufacturing Israel 20,000 CFM Semiconductor solvents, 50 ppmv 95% Reduction Efficiency System located in populated area of a major city has eliminated nuisance odor
Industry **Location** **Inlet Stream** **VOCs** **Performance Achieved** **Notes**	Semiconductor Chip Manufacturing Arizona 36,000 CFM Semiconductor solvents, 25-100 ppmv 90%+ Reduction Efficiency System met strict local compliance requirements while achieving Zero Unscheduled Downtime (ZUD)
Industry **Location** **Inlet Stream** **VOCs** **Performance Achieved** **Notes**	Teflon Tubing Manufacturing Massachusetts 1,800 CFM Perchloroethylene, 1500 ppmv 98% Reduction Efficiency Recovered solvent re-used in process. System payback in 1 year.
Industry **Location** **Inlet Stream** **VOCs** **Performance Achieved** **Notes**	Furniture Painting California 50,000 CFM Paint Solvents, 25-50 ppmv 90%+ Reduction Efficiency Recovered solvent and reformulated with paint
Industry **Location** **Inlet Stream** **VOCs** **Performance Achieved** **Notes**	Military Vehicles Michigan 150,000 CFM Paint Solvents 95% Reduction Efficiency Controlled multiple different interchangeable VOC sources.

[21] "Environmental C&C Inc". Environmental C&C website. https://www.environmentalcc.com/ (Accessed September 18, 2021).

Automotive Site, Alabama, 100,000 CFM
Paint Solvents, >90% Reduction Efficiency
Two adsorbers and one desorber

Image shows three adsorbers one desorber	
Image shows ductwork that enabled control of any of the 3 sources to any of the 3 adsorbers. **Each adsorber has its own process fan.** **System provided control of multiple different interchangeable VOC sources**	

Military Vehicles, Michigan, 150,000 CFM
Paint Solvents, 95% Reduction Efficiency

TABLE IV: VALUE CREATION CALCULATION EXAMPLE

Fluidized Bed Concentrator (FBC) vs Regenerative Thermal Oxidizer (RTO)	Value per Year
1) Reduce Energy Usage (Natural Gas) due to Elimination of Energy Used to Heat the Air in the RTO: [22] [23] a. $\frac{mmBTU}{Year}\ to\ heat\ air\ in\ RTO = (\Delta\ T)\ (sCFM)(1.08)(\%Heat\ Loss)(\frac{Operating\ Hours}{Year})/1000000$ b. $(1500^\circ F - 175^\circ F) * (300{,}000sCFM) * \left(\frac{1.08\ BTUs}{sCFM \cdot Hour \cdot Degree^\circ F}\right) * (7.5\%) * (8000\frac{Hrs}{Yr})/1000000$ c. $= \left(257{,}580\frac{mmBTU}{Yr}\right) - [\left(250\frac{lbs\ VOC}{Hour}\right) * \left(15{,}000\frac{BTUs}{lb\ VOCs}\right) * \left(\frac{8000\ Hrs}{Year}\right)] = \mathbf{227{,}580\frac{mmBTUs}{Year}}$ d. $\left(\frac{227{,}580\ mmBTUs}{Year}\right)(\$3.5\frac{Long\ Term\ Cost\ of\ Natural\ Gas}{mmBTU}) = \mathbf{\$796{,}530\ Savings/Year}$	Save by Eliminating Natural Gas **$796,530**
2) Reduce Greenhouse Gases (GHG) Generated from Burning Natural Gas: [24] a. $Tons\ GHG = (mmBTU) * \left(\frac{117\ lbs\ CO2}{mmBTU}\right) * (\frac{1\ Metric\ Ton}{2205\ lbs})$ b. $\left(227{,}580\frac{mmBTUs}{Yr}\right) * \left(117\ lbs\frac{CO2}{mmBTU}\right) * \left(1\frac{Metric\ Ton}{2205\ lbs}\right) = \mathbf{12{,}076\frac{Tons\ GHG\ Reduction}{Year}}$	Sell Carbon Offsets **$60,378**
3) Sell Carbon Offsets (Carbon Credits) Utilizing Carbon Accounting Service Providers, i.e. ClimeCo: [25] a. $\left(\frac{12{,}076\ Tons\ GHG}{Year}\right) * \left(\frac{\$5}{Ton\ GHG}\right) = \mathbf{\$60{,}378/Year\ Selling\ Carbon\ Offsets}$	
4) Reduce Electricity Usage by Reducing Fan Power to Move Process Air through Fluidized Media: $(2400RTO - 600FBC\ HP\ fans - 800\ HP) * \left(\frac{0.75\ kW \cdot h}{HP}\right)\left(\frac{\$0.07}{kW \cdot h}\right)\left(\frac{8000\ Hrs}{Year}\right) = \$420{,}000\frac{\mathbf{Savings}}{\mathbf{Year}}$	Save by Reducing Electrical Usage **$420,000**
5) Sell Terpenes [26] **Yielded from Green Pine Wood** [27]**:** a. $\left(\frac{1\ Gallon\ Terpenes}{Oven\ Dry\ Ton\ Wood}\right) * \left(\frac{300{,}0000\ Dry\ Tons\ Wood}{Year}\right) * (75\%Yield) * \left(\frac{\$3}{Gal}\right) = \mathbf{\$675{,}000/Yr}$ **Selling Terpenes**	Sell Terpenes **$675,000**
6) Reduce Maintenance Costs:[28] Media changeouts, large heavy moving part replacement, corrosion repairs, etc.	Save up to **~$1 Million**
Total Value Creation per Year per Site:	**Up to $3 Million**

---

[22] ICAC Guidance Method for Estimation of Gas Consumption in a RTO. Institute of Clean Air Companies website. https://cdn.ymaws.com/www.icac.com/resource/resmgr/RTO-F1.pdf (Accessed September 18, 2021)

[23] "Operating Cost Calculator". Cycle Therm website. http://www.cycletherm.com/resources/operating-cost-calculator.aspx (Accessed November 21, 2021).

[24] "Greenhouse Gases Equivalencies Calculator - Calculations and References". Environmental Protection Agency (EPA) website. https://www.epa.gov/energy/greenhouse-gases-equivalencies-calculator-calculations-and-references (Accessed December 28, 2020).

[25] "Allowance Price Explorer". International Carbon Action Partnership website. https://icapcarbonaction.com/en/ets-prices (Accessed Jan 20, 2021).

[26] "Price Chart for Gum Turpentine". PineChem website. http://www.pinechem.net/images/gall/ChartGNE.GIF Accessed Jan 3, 2021).

[27] "Sulfate Turpentine Recovery". Pulp Chemicals Association. Pages 9-12, 127. https://www.pinechemicals.org/store/viewproduct.aspx?id=1579188 (Accessed Jan 2, 2021).

[28] "Air Pollution Control Technology Fact Sheet". EPA website. https://www3.epa.gov/ttncatc1/dir1/fregen.pdf (Accessed October 17, 2021).

VIII. CASE STUDY: FBC TECHNOLOGY SPECIFIC TO WOOD DRYING

DEMONSTRATION DESCRIPTION:

Captis Aire LLC founder and Environmental C&C Inc. worked on a project together from 2015-2021 with a commercial oriented strand board (OSB) manufacturer to demonstrate the ability to utilize the FBC to control emissions from wood drying.

In wood products manufacturing, process gas exhaust streams released from drying wood contain VOCs, including turpentine (terpenes). These VOCs are considered pollutants/emissions and must be removed from the air, prior to release, to avoid air pollution emissions into our communities. Wood drying operations utilized in the manufacture of oriented strand board (OSB), lumber, wood to biofuels, i.e. wood pellets, plywood, particleboard, etc. are few examples of these processes that have process exhaust streams containing VOCs.

Work on this patent[29] pending[30] technology was executed in three phases. In the first phase, laboratory screening work was done to explore whether emissions from wood drying, primarily terpenes, could be adsorbed by Bead Activated Carbon (BAC) media at the expected operating temperatures. In the second phase, a small-scale pilot adsorber was installed at a commercial Oriented Strand Board (OSB) manufacturing site to demonstrate ability to adsorb emissions from wood drying under commercial manufacturing conditions. In the third phase, the pilot unit capabilities were expanded to include the adsorber, desorber, and condenser. The SSR performance was simulated by semi-continuous additions of virgin BAC during air emissions compliance testing.

In the third testing phase, the pilot unit ran for 6 months with over 1700 hours on-line. Ability to meet air emissions control compliance, per the Plywood and Composite Wood Products Maximum Achievable Control Technology (PCWP MACT) standard, was determined by two independent certified air emissions[31] stack testing companies[32] during two separate testing events.

---

[29] "METHODS AND SYSTEMS FOR RECOVERING TERPENE COMPOSITIONS FROM WOOD DRYING EXHAUST". World International Patent Office (WIPO) Website. https://patentscope.wipo.int/search/en/detail.jsf?docId=WO2019148208 (Accessed September 18, 2021).

[30] "METHODS AND SYSTEMS FOR CONTROLLING EMISSIONS FROM WOOD DRYING PROCESSES". WIPO Website. https://patentscope.wipo.int/search/en/detail.jsf?docId=WO2019148207 (Accessed September 18, 2021).

[31] "AIR Advanced Industrial Resources". AIR website. https://airtest1.com/ (Accessed April 16, 2020).

[32] "Testing and Modeling". John Zink Hamworthy Combustion website. https://www.johnzinkhamworthy.com/testing-modeling/ (Accessed Feb 3, 2021).

TABLE V
RESULTS OF AIR POLLUTION CONTROL COMPLIANCE TESTING (FBC UNIT):

Emissions Measured:	FBC %Reduction Efficiency[33]
Total Hydrocarbons (THC), as carbon including Methane	92%
Volatile Organic Compounds (VOCs), as carbon. (Total non-methane hydrocarbons – THC minus Methane)	92%
Formaldehyde	83%
Methanol	75%
GHC - Nitrogen Oxides (NOx)	55%
Particulate Matter (PM), total filterable	50%
Condensable Particulate Matter	59%
Total Particulate Matter, Filterable + Condensable	51%
Total PM2.5, Particle Size Analyses (PSA) Filterable + Condensable	60%
Total PM10, Particle Size Analyses (PSA) Filterable + Condensable	64%
Carbon Monoxide (CO)	7%
Sulfur Dioxide (SO2)	99%

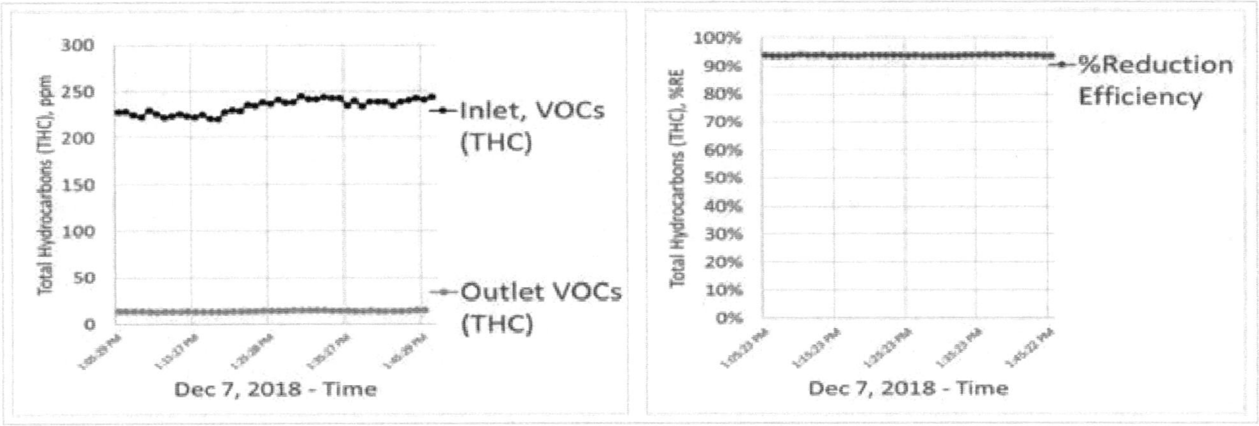

RESULTS OF VOC REDUCTION EFFICIENCY PERFORMANCE MEASURED BY FLAME IONIZATION DETECTOR (FBC DEMONSTRATION UNIT):
Achieved >90% Reduction Efficiency meeting PCWP MACT Standard[34]

---

33 Emissions results FBC 2018 at commercial wood drying site done by Advanced Industrial Resources, Inc. Results were demonstrated during three 1-hour testing intervals. https://airtest1.com/ . (Accessed September 18, 2021)

34 "Electronic Code of Federal Regulations (eCFR)". eCFR website. See https://www.ecfr.gov/cgi-bin/text-idx?node=sp40.13.63.dddd#ap40.14.63_12292.1 (Accessed April 16, 2020).

RESULTS OF VOLATILE ORGANIC COMPOUNDS CAPTURE:

Terpenes were collected from the FBC demonstration unit. These terpenes are a mixture of terpene hydrocarbons, primarily Alpha and Beta pinenes. These terpenes are responsibly-recovered, biobased materials. Potential applications for these terpenes include use in fuels, fragrances, solvents, adjuvants, cleaners, and polyterpene resins.

TERPENES COLLECTED FROM FBC DEMONSTRATION UNIT

VALUE OF TERPENES:

The Pine Chemicals Association has documented the gallons of terpenes per oven dry ton of wood in a variety of tree species. Typically, there is ~1 gallon of turpentine (terpenes)[35] per oven dry ton of wood product produced from green pine wood from the SE part of the United States, i.e. Loblolly Pine, Jack Pine, Slash Pine, Shortleaf Pine, Ponderosa Pine, Pitch Pine, Pond Pine, Red Pine, and Spruce Pine. If 75% of this is evaporated and captured from the drying process, the value of the terpenes per year can be calculated by multiplying this yield by the oven dry tons (ODT) of wood produced and a typical price per gallon[36] for terpenes. For example:

$$\left(\frac{1\ Gallon\ Terpenes}{ODT\ Wood}\right) * \left(\frac{ODT\ Wood}{Year}\right) * (\%Yield) * \left(\frac{\$\ Dollars}{Gallon\ Terpenes}\right) = \$\ /\ Year$$

TABLE VI
TERPENES TYPICAL PROPERTIES:

Property	Result
Appearance	Clear Liquid
Moisture	<0.1%
Weight per Gallon	7.2 pounds per gallon
Specific Gravity	0.86
Sulfur Content	< 10 ppm

---

35 "Sulfate Turpentine Recovery". Pulp Chemicals Association. Pages 9-12, 127. https://www.pinechemicals.org/store/viewproduct.aspx?id=1579188 (Accessed Jan 2, 2021).

36 "Price Chart for Gum Turpentine". PineChem website. http://www.pinechem.net/images/gall/ChartGNE.GIF Accessed Jan 3, 2021).

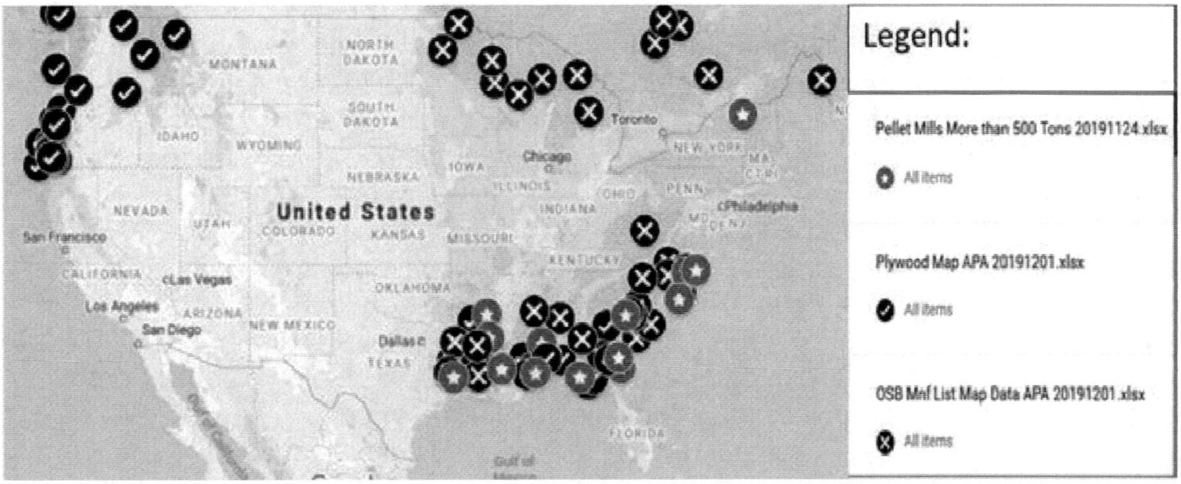

MAP OF POTENTIAL WOOD DRYING SITES FOR FBC USE:

This map shows potential sites where various wood products are produced where the FBC may be used to control emissions. Potential sites include but are not limited to Oriented Strand Board (OSB), lumber (sawmills)[37], plywood[38], wood to biofuels, i.e. wood pellets[39], particleboard, Medium Density Fiberboard (MDF), etc.

RESULTS OF DISCUSSIONS WITH ENVIRONMENTAL AGENCIES:

Official stack testing results were discussed with two state environmental agencies and two federal environmental protection agency (EPA) personnel. In all discussions, environmental agency personnel were supportive of providing a permit for full size commercial FBC unit.

FINAL RESULTS FROM WOOD DRYING DEMONSTRATION. THE UNIT:

- Provided salable terpenes, composed primarily of Alpha and Beta Pinenes.
- Reduced greenhouse gas (GHG) emissions.
- Reduced energy usage. It eliminated natural gas usage and reduced electrical usage.
- Met the PCWP MACT requirement of >90% Reduction Efficiency[40] in both tests. In the second test, the system met the compliance requirements three times in a row, over each of the three one-hour testing intervals.
- Demonstrated good operational reliability.
- Enabled State and Federal environmental personnel that reviewed these results to support permitting for a full commercial FBC system.

---

[37] "The mill map allows you to visually locate and drill down to find the mill you're interested in." Primary Forest Products Network website. https://primary.forestproductslocator.org/mill-map (Accessed September 18, 2021).
[38] "Manufacturer Directory". Engineered Wood Association website. https://www.apawood.org/manufacturer-directory (Accessed September 18, 2021).
[39] "Biomass Magazine". Biomass Magazine website. See http://biomassmagazine.com/plants/map/pellet/ (Accessed September 18, 2021).
[40] "Electronic Code of Federal Regulations (eCFR)". eCFR website. See https://www.ecfr.gov/cgi-bin/text-idx?node=sp40.13.63.dddd#ap40.14.63_12292.1 (Accessed April 16, 2020).

IX. CONCLUSIONS:

The FBC provides an affordable, reliable, and accessible clean technology for controlling VOCS, HAPS, and even particulate in some applications. This technology is well established in many applications including automotive and electronics manufacturing where it has been used successfully for many years. It is an emerging technology for use in biomass (i.e. wood drying) applications.

Especially for process exhaust streams with low VOC concentrations, the FBC technology is a patent[41] pending[42], cost competitive, industrial air pollution control technology that enables operators to advantageously:

1. Generate revenue via sales of recovered organic products (including biobased products such as terpenes in biomass applications) and sales of carbon offsets
2. Improve sustainability by reducing energy usage and thus greenhouse gas (GHG) emissions
3. Reduce operational costs by reducing energy usage (natural gas and electricity) and maintenance costs.
4. Enhance VOC mass transfer to the BAC media
5. Provide continuous in-situ BAC media regeneration and reactivation. This enables the media to continue to be reused.
6. Minimize water adsorbed on the BAC media thus maintaining its ability to adsorb VOCs, even in high moisture content process exhaust streams

Like any air pollution control technology installation, careful consideration of many design parameters, the unit's specific operating profile, and the operator's goals need to be taken into account during the design, installation, and startup/optimization processes. There are excellent resources, including Environmental C&C Inc.[43], Captis Aire LLC[44], and the Institute of Clean Air Companies (ICAC)[45] personnel, to assist with evaluating, implementing, and optimizing FBC applications.

---

[41] "METHODS AND SYSTEMS FOR RECOVERING TERPENE COMPOSITIONS FROM WOOD DRYING EXHAUST". World International Patent Office (WIPO) Website. https://patentscope.wipo.int/search/en/detail.jsf?docId=WO2019148208 (Accessed September 18, 2021).

[42] "METHODS AND SYSTEMS FOR CONTROLLING EMISSIONS FROM WOOD DRYING PROCESSES". WIPO Website. https://patentscope.wipo.int/search/en/detail.jsf?docId=WO2019148207 (Accessed September 18, 2021).

[43] "Environmental C&C Inc". Environmental C&C website. https://www.environmentalcc.com/ (Accessed September 18, 2021).

[44] "Emissions Control Technology Captures Emissions to Clean the Air". Captis Aire website. https://www.captisaire.com/ (Accessed November 21, 2021).

[45] "The Trusted Voice of the Clean Air Technology Industry". Institute of Clean Air Companies (ICAC) website. https://www.icac.com/ (Accessed September 18, 2021)

X. ACKNOWLEDGEMENTS:

The author recognizes inputs from a broad spectrum of personnel at the commercial OSB manufacturer and at Environmental C&C[46] including Harold Cowles, John Berger P.E., and Jim Starek.

CAPTIS AIRE CAPTURE-CONCENTRATE-CLEAN	Environmental C&C, Inc. Solvent Recovery & VOC Abatement Systems
Kim TUTIN Captis Aire LLC [47] 1612 W. Cleveland Ave, Suite 001 East Point, GA 30344-3213 Tel: +01 404 5802795, E-mail: kim@captisaire.com Website: www.captisaire.com	Harold COWLES Environmental C&C Inc. 16705 E. Calle de las Estrellas Road Scottsdale, AZ 85262-5982 Tel: +01 518 2250315, E-mail: halc@ecnc.com Website: https://www.environmentalcc.com/

[46] "Environmental C&C Inc". Environmental C&C website. https://www.environmentalcc.com/ (Accessed September 18, 2021).

[47] "Emissions Control Technology Captures Emissions to Clean the Air". Captis Aire website. https://www.captisaire.com/ (Accessed November 21, 2021).

624380 Crystallization Design to Avoid Liquid-Liquid Phase Separation and Control Drug Substance Physical Properties

Abstract:

Crystallization is a critical purification and separation process used in the pharmaceutical industry. A well-developed crystallization process should not only deliver exceptional chemical purity, high yield and throughput, but also the ability to control physical properties and improve downstream processes (*e.g.,* drug substance filtration and drying and formulation of the drug substance into the drug product). For this case study involving a small molecule drug substance, the challenge of liquid-liquid phase separation (or oiling out) was encountered during process development. Crystallization design changes were required to avoid liquid-liquid phase separation and control residual solvent levels in the final drug substance. The regressed-UNIFAC group contribution theory methodology was used to model the phase behavior and predict liquid-liquid phase boundaries in binary solvent systems including the effects of the drug substance; this data was then used to design a crystallization process that avoided the two-phase liquid region. Process analytical technology (PAT) tools were used to collect *in-situ* data for modeling the crystallization and to guide the selection of optimal process parameters. Crystal morphology and particle size were controlled through seed surface area, control of the crystallization kinetics (nucleation and growth), and application of slurry milling. The combined *in-silico* and *in-situ* techniques from this work provide the basis for understanding and predicting liquid-liquid phase separation during crystallization and can be used to design future processes in complex solvent systems.

627323 Material Sparing Approaches for Predicting Powder Flow Using Machine Learning Methods

Abstract:

The performance of pharmaceutical powders is often determined by the distributions of both shape and size of the constituent particles, but typically they are described by simplistic, and often inappropriate descriptors such as D_{50}. Consequently, most of the information about the materials being characterized is lost and/or misrepresented. Image analysis enables the imaging of all particles within a sample and measurement of multiple particle characteristics resulting in an information rich data set. The complex interdependency between particle size and shape and its combined effect on powder performance metrics such as flowability is a prime problem to be tackled using machine learning techniques.

In this work, we explore the application of unsupervised and supervised machine learning techniques to predict the effect of particle morphology of pharmaceutical powders on their flow function coefficient (FFC). To achieve this, we develop novel feature extraction methods using principal component analysis and convolutional neural networks that capture the particle size and shape distribution of powders and lay foundations for a machine learning framework for aiding material selection in pharmaceutical development. Our results indicate that it is possible to distinguish non-flowing powders (FFC <= 2) from other flow classes (FFC > 2) with high confidence, thereby providing an early indication of poor flow and possible recourse using this material sparing approach. The broad applicability of this approach for predicting other powder performance metrics as well as applicability in other areas of drug development will also be discussed.

631685 In-Depth Mechanistic Understanding of Pibrentasvir Agglomeration to Enable Successful Scale-up

Abstract:

Physical properties of active pharmaceutical ingredients (APIs), such as the particle size distribution (PSD) and extent of agglomeration, often impact downstream drug product processes. Strategies utilized to isolate the API, namely the crystallization, filtration, and drying unit operations, must be designed to ensure consistency in the resultant API physical properties. Development and deployment of non-traditional isolation strategies, for an API prone to agglomerate formation, is often necessary to ensure that the API meets the acceptance criteria for PSD.

In this presentation, we will describe the strategies utilized to develop the API isolation for pibrentasvir, a non-nucleoside inhibitor of the hepatitis C virus NS5A replicon. While the pibrentasvir crystallization consistently yielded non-agglomerated material, drying in the agitated filter dryer resulted in agglomerates on the order of 10 to several 100 μm, with varying degree of agglomerate strength as a result of agitation in the presence of residual solvent. In this presentation, we will discuss a mechanistic understanding of the parameters that impact the extent of agglomeration of pibrentasvir and methods for quantifying the strength of the resulting agglomerates. To address agglomeration, the use of non-traditional isolation strategies were employed to ensure significant removal of the residual solvent and prevent the formation of agglomerates through crystalline bridges. This work demonstrates a systematic approach for understanding the formation and characterization of API agglomerates and provides a novel approach to address agglomeration in an agitated filter dryer. Successful deployment of the isolation strategy enabled scale-up of the drying process to yield API that met the acceptance criteria for PSD.

All authors are employees of AbbVie and may own AbbVie stock. AbbVie and Enanta sponsored and funded the study; contributed to the design; participated in the collection, analysis, and interpretation of data, and in writing, reviewing, and approval of the final publication.

623454 Influence of the Rheology of Lactose Excipients on the Feeding Consistency in Tabletting Processes

Abstract:

Minimizing the variability in the feeding process is a crucial challenge in oral delivery systems (e.g. tabletting, capsule filling) to ensure good quality and consistency of the end product. To control and reduce the variability in the feeding rate, it is important to characterize the material properties and bulk behaviour of the powder. It is known that good flowing properties of pharmaceutical powders are required to achieve uniform flow into the die cavities. Insufficient flow can lead to uneven filling of the dies, resulting in large weight and dosage variations of the final tablets. Moreover, at high tableting speeds, the time to fill dies is reduced, making flow properties of the blend even more important.

Lactose is one of the most widely used excipients in the pharmaceutical industry. There are many reasons for its popularity, such as the fact that lactose is largely inert, relatively inexpensive, safe, available in many different grades, and has a long history of usage in successful formulations world-wide. For direct compression processes like tableting, lactose excipients can be used as a filler-binder to provide bulk density, compaction, and flow to the formulation.

In this study, the rheological properties of lactose powders have been investigated with the rotating drum method (GranuDrum) and correlated with the actual consistency of flow achieved in a tableting system. The investigation of powder flowability at different drum rotating speeds gives useful information on the evolution of powder processability at higher shear rate. Indeed, with the historical repose angle characterization method, the stress state at which the powder is submitted is far from the conditions encountered in a tableting machine. In this work, we observed that some excipient grades exhibit shear-thinning behaviour (i.e. an increase of flowability with increasing applied stress) while others show the opposite (i.e. a shear-thickening behaviour). The classification of powder cohesiveness is therefore inverted at high shear. These observations have been correlated with the mass variation of tablets produced with a RoTab tableting press at different feeder speeds. We present the correlation between the flowability assessments obtained with the rotating drum and the weight consistency of the produced tablets. We thus highlight the importance of powder characterization at a stress state equivalent to those at which the powder is processed.

626104 Characterizing the Effect of Particle Properties and Surface Topography Towards Pharmaceutical Powder and Surface Adhesion Using the Enhanced Centrifuge Method

Abstract:

The behavior of pharmaceutical powders during processing is impacted by the adhesion of the individual particles which comprise the powder. Problems such as poor flowability, dust hazards, and equipment wear arise due to uncontrolled particle adhesion and can lead to production challenges. This adhesion behavior, in turn, is critically influenced by the complementarity between the topography of a surface and the shape and topography of the particles that adhere to that surface. Computational models have been developed to predict the behavior of highly idealized powders (i.e. powders comprised of simple geometries such as spheres) under various processes but are limited in their ability to model and optimize the manufacturing and handling of powders comprised of many complex particles. This work focuses on quantifying the real particle property effects of particle size, shape, and topography of three pharmaceutical powders towards the adhesion of three pharmaceutical powders to stainless steel surfaces through a novel experimental and modeling framework called the enhanced centrifuge method that maps particle and surface scale properties onto experimentally-validated adhesion parameters. The result of this work is a validated semi-empirical model for the powder adhesion which describes the behavior of the full powder, across the full range of particle sizes and shapes, against the steel. This model is presented in terms of van der Waals forces, although the approach can be applied to describe capillary-, electrostatic-, or deformation-driven adhesion.

Disclosures:

Andrew Vogt, Maxx Capece and William Ketterhagen are employees of AbbVie and may own AbbVie stock. AbbVie and Purdue contributed to the design; participated in the collection, analysis and interpretation of data, and in writing, reviewing and approval of the final presentation.

627255 Surfactant/Preservative Instability in Injectable Pharmaceutical Formulation

Abstract:

Recent study of incompatibility between surfactant molecules and antimicrobial agents has caused concerns for future protein formulation of multi-dose therapeutic drugs in the pharmaceutical industry. In particular, polysorbate 80 (PS80), a nonionic surfactant widely used in formulation, has been found to be incompatible with m-cresol, a commonly used antimicrobial agent. This incompatibility results in increased turbidity caused by unsteady micelle aggregation or coalescence that can progress over weeks. Here, small-angle neutron scattering (SANS) is used to fundamentally understand the structural origins of this incompatibility. We observed that PS80/m-cresol solution instability is influenced by storage temperature, ionic strength, and component concentration. SANS analysis of PS80/m-cresol solutions over a pharmaceutically relevant concentration range of each component reveals the cause of aggregation, the coalescence mechanism and aggregate structure. Our results identify two concentration-dependent stability regimes at low (below ≈ 2 mg/mL) and high (5 mg/mL) m-cresol concentrations. In these stable formulations, micelle morphology is highly sensitive to m-cresol concentration. PS80/m-cresol solutions aggregate at m-cresol concentration between the two stable regimes. SANS results reveal the aggregation mechanism, and kinetic modelling of the results provides a simplified empirical equation capable of describing and predicting the multiple aggregation phases observed in these solutions. This work aims to develop insight into surfactant/preservative stability for pharmaceutical formulation, which can be leveraged to better understand the effect of excipients on surfactant self-assembly.

AUTHOR INDEX